A Study of Mammalia and Geology Across the Cretaceous-Tertiary Boundary in Garfield County, Montana

by J. David Archibald

UNIVERSITY OF CALIFORNIA PRESS
Berkeley • Los Angeles • London

UNIVERSITY OF CALIFORNIA PUBLICATIONS IN GEOLOGICAL SCIENCES

Volume 122

Issue Date: June 1982

UNIVERSITY OF CALIFORNIA PRESS
BERKELEY AND LOS ANGELES
CALIFORNIA

UNIVERSITY OF CALIFORNIA PRESS, LTD.
LONDON ENGLAND

ISBN: 0-520-09639-8
LIBRARY OF CONGRESS CATALOG CARD NUMBER: 81-2606

Library of Congress Cataloging in Publication Data

Archibald, J. David.
Fossil mammalia from the Hell Creek formation and lowermost Tullock formation and the geology of the Hell Creek and Tullock formations, Garfield County, Montana.

(University of California publications in geological sciences ; v. 122)
Bibliography: p.
1. Mammals, Fossil. 2. Paleontology--Cretaceous. 3. Paleontology--Paleocene. 4. Geology, Stratigraphic--Cretaceous. 5. Geology, Stratigraphic--Paleocene. 6. Geology--Montana--Garfield County. I. Title. II. Series.
QE881.A8 569'.09786'27 81-2606
ISBN 0-520-09639-8

Contents

Acknowledgments, ix
Abstract, xi
List of Abbreviations, xv

INTRODUCTION 1

TECHNIQUES OF STUDY 3

STRATIGRAPHY, PHYSICAL GEOLOGY, AND LOCALITIES 9
Introduction, 9
Regional Setting, 9
Formations, 11
Fossil Localities, Sedimentary Environments, and Stratigraphic Correlation, 18

SYSTEMATICS 37
Class MAMMALIA, Subclass ALLOTHERIA,
Order MULTITUBERCULATA, 37
Suborder PTILODONTOIDEA (Gregory and Simpson, 1926)
Sloan and Van Valen, 1965, 39
Family NEOPLAGIAULACIDAE Ameghino, 1890, 39
Mesodma Jepsen, 1940, 39
Mesodma hensleighi Lillegraven, 1969, 41
Mesodma cf. *M. formosa* (Marsh, 1889b) Clemens, 1964, 43
Mesodma thompsoni Clemens, 1964, 45
Mesodma garfieldensis, n. sp., 46
Neoplagiaulax Lemoine, 1880, 54
?*Neoplagiaulax burgessi*, n. sp., 54
Family ?NEOPLAGIAULACIDAE, Gen. et sp. indet., 57
Family CIMOLODONTIDAE Marsh, 1889a, 58
Cimolodon Marsh, 1889a, 58
Cimolodon nitidus Marsh, 1889a, 58
Suborder TAENIOLABIDOIDEA (Granger and Simpson, 1929)
Sloan and Van Valen, 1965, 63
Family EUCOSMODONTIDAE (Jepsen, 1940) Sloan and Van Valen, 1965, 63

Subfamily EUCOSMODONTINAE Jepsen, 1940, 63
Stygimys Sloan and Van Valen, 1965, 63
Stygimys aff. *S. kuszmauli* Sloan and Van Valen, 1965, 63
Subfamily MICROCOSMODONTINAE Holtzman and Wolberg, 1977, 68
Acheronodon, n. gen., 68
Acheronodon garbani, n. sp., 68
Family TAENIOLABIDIDAE Granger and Simpson, 1929, 71
Catopsalis Cope, 1882b, 71
Catopsalis cf. *C. foliatus* Cope, 1882b, 71
Suborder, *incertae sedis,*
Family CIMOLOMYIDAE (Marsh, 1889b) Sloan and Van Valen, 1965, 72
Cimolomys Marsh, 1889a, 73
Cimolomys gracilis Marsh, 1889a, 73
Meniscoessus Cope, 1882c, 75
Meniscoessus robustus (Marsh, 1889a) Osborn, 1891, 75
Family ?CIMOLOMYIDAE,
Essonodon Simpson, 1927a, 89
Essonodon browni Simpson, 1927a, 89
Suborder and Family, *incertae sedis,*
Cimexomys Sloan and Van Valen, 1965, 102
Cimexomys minor Sloan and Van Valen, 1965, 102
Cimexomys hausoi, n. sp., 105
Suborder and Family, *incertae sedis,*
Paracimexomys, n. gen., 111
Paracimexomys priscus (Lillegraven, 1969), 112
Comments on the Phylogentic Relationship of Multituberculates, 114
Subclass THERIA, Infraclass METATHERIA
Order MARSUPIALIA, 123
Superfamily DIDELPHOIDEA (Gray, 1821) Osborn, 1910,
Family DIDELPHIDAE Gray, 1821,
SubfamilyDIDELPHINAE (Gray, 1821) Simpson, 1927b, 124
Alphadon Simpson, 1927b, 124
Alphadon marshi Simpson, 1927b, 124
Alphadon wilsoni Lillegraven, 1969, 126
Alphadon lulli Clemens, 1966, 128
Alphadon rhaister Clemens, 1966, 129
Peradectes Matthew and Granger, 1921, 130
Peradectes cf. *P. pusillus* (Matthew and Granger, 1921) Clemens, ms., 130
Subfamily GLASBIINAE Clemens, 1966, 137
Glasbius Clemens, 1966, 137
Glasbius twitchelli, n. sp., 137
Family PEDIOMYIDAE (Simpson, 1927a) Clemens, 1966, 146
Pediomys Marsh, 1889a, 146
Pediomys elegans Marsh, 1889a, 149
Pediomys cf. *P. cooki* Clemens, 1966, 150
Pediomys krejcii Clemens, 1966, 151
Pediomys hatcheri (Osborn, 1898) Simpson, 1927b, 152

Pediomys florencae Clemens, 1966, 153
?*Pediomys* cf. *P. florencae* Clemens, 1966, 154
Family STAGODONTIDAE Marsh, 1889b, 157
Didelphodon Marsh, 1889a, 157
Didelphodon vorax Marsh, 1889a, 157
Infraclass EUTHERIA, 158
Infraclass EUTHERIA, *incertae sedis,*
Superfamily LEPTICTOIDEA (Gill, 1872) Butler, 1956, 160
Family, *incertae sedis,*
Gypsonictops Simpson, 1927a, 160
Gypsonictops cf. *G. illuminatus* Lillegraven, 1969, 160
Infraclass EUTHERIA, *incertae sedis,*
Family PALAEORYCTIDAE (Winge, 1917) Simpson, 1931, 162
Cimolestes Marsh, 1889a, 162
Cimolestes incisus Marsh, 1889a, 162
Cimolestes cf. *C. cerberoides* Lillegraven, 1969, 163
Cimolestes stirtoni Clemens, 1973, 163
Cimolestes propalaeoryctes Lillegraven, 1969, 164
Cimolestes magnus Clemens and Russell, 1965, 165
Batodon Marsh, 1892, 166
Batodon tenuis Marsh, 1892, 166
Procerberus Sloan and Van Valen, 1965, 167
Procerberus formicarum Sloan and Van Valen, 1965, 167
Order CONDYLARTHRA, 171
Family ARCTOCYONIDAE (Giebel, 1855) Murray, 1866, 172
Subfamily OXYCLAENINAE (Scott, 1892) Matthew, 1937, 172
Protungulatum Sloan and Van Valen, 1965, 172
Protungulatum cf. *P. donnae* Sloan and Van Valen, 1965, 172
Protungulatum mckeeveri, n. sp., 173
Subfamily LOXOLOPHINAE Van Valen, 1978, 176
Ragnarok Van Valen, 1978, 176
Ragnarok harbichti Van Valen, 1978, 176
Ragnarok nordicum (Jepsen, 1930b) Van Valen, 1978, 180
Ragnarok engdahli, n. sp., 183
Family ?HYOPSODONTIDAE (Trouessart, 1879) Lydekker, 1889, 188
Subfamily ?HYOPSODONTINAE (Trouessart, 1879) Matthew, 1937, 188
Oxyprimus Van Valen, 1978, 189
Oxyprimus erikseni Van Valen, 1978, 189
Family PERIPTYCHIDAE Cope, 1882d, 197
Subfamily ANISONCHINAE Osborn and Earle, 1895, 197
Mimatuta Van Valen, 1978, 197
Mimatuta morgoth Van Valen, 1978, 197
Mimatuta cf. *M. morgoth* Van Valen, 1978, 205
Mimatuta minuial Van Valen, 1978, 207
Oxyacodon Osborn and Earle, 1895, 212
Oxyacodon agapetillus (Cope, 1884a) Matthew, 1897, 215
Oxyacodon priscilla Matthew, 1937, 219

Family ?PERIPTYCHIDAE,
Gen. et sp. indet, 223
Trends in Early Anisonchine Evolution, 225

BIOSTRATIGRAPHY 237
Introduction, 237
Cretaceous-Tertiary Boundary in Eastern Montana, 238
Hell Creek Faunal-Facies Mammals, 241
Bug Creek Faunal-Facies, Pre-mantuan, and Mantuan Mammals, 248
Mammal Evolution near the Cretaceous-Tertiary Boundary and Comments on Recent Models of Dinosaur Extinction, 255

Summary, 263
Appendices, 273
Literature Cited, 279

Acknowledgments

This study encompasses all but the chelonian portion of my dissertation completed at the University of California, Berkeley, in December 1977 and includes further research through 1979. The dissertation research was part of an ongoing project in Garfield County, Montana, under the direction of Dr. William A. Clemens. I am most grateful to Dr. Clemens, first, for suggesting this project, and, second, for his support, encouragement, and constructive criticism.

The field research in Garfield County has been a cooperative effort that has benefited from the help of innumerable people. A large amount of our success in field operations is due to the interest shown by the residents of Garfield County. In particular I would like to thank the Engdahl family: Mr. Lester and Mrs. Cora Engdahl, Mr. Robert and Mrs. Jane Engdahl, and their children, Duane and Cathy, for their friendship and support while we worked and camped on their ranch. I am also grateful to the Twitchell family for allowing our field crews to use their ranch as a second base of operations. Thanks should also go to the following individuals and families who allowed us to work on their land: the Binyons, the Blisses, the Burgesses, Mrs. Leah Cole, the Edwards family, the Flints, the Harbaughs, the Hausos, the Kerrs, the Larsons, the MacDonalds, the McKeevers, the Trumbos, and other families I may have failed to mention. I am also indebted to many of the residents of Jordan, who helped our field crews in many ways. Special thanks must go to the members of the field parties preceding, during, and after my work in Garfield County: Mr. James Clark, Mr. John Chiment, Dr. William Clemens, Mr. Lowell Dingus, Mr. Jacques Gauthier, Mr. Michael Greenwald, Mr. William Hass, Mr. David Lawler, Mr. Fred McEnroe, Dr. Larry Marshall, and Mr. Hugh Wagner. Without seeming to disparage the work of these individuals, I would especially like to acknowledge the collecting expertise of Mr. Harley Garbani and Mr. Richard Farrand, who have discovered many of our most productive localities over the past several years. Our fieldwork was also augmented by many visitors. Several individuals provided field data critical to this work: Mr. William Rohrer, U.S. Geological Survey; Mr. Kenneth Padgett, U.S. Soil Conservation Service; Dr. Robert Sloan, University of Minnesota; Mr. Don Beckman, Army Corps of Engineers; Drs. Robert Butler and Everett Lindsay, University of Arizona; and many individuals from the State and Miles City District Offices of the Bureau of Land Management.

I am indebted to the many people who spent tedious hours sorting the mammalian specimens used in this study. I would like to thank my dissertation committee, Drs. William A. Clemens, Joseph T. Gregory, and Richard L. Hay, for reading my thesis, and Drs. Richard C. Fox and Jason A. Lillegraven for critically reviewing the manuscript. I would also like to thank the following individuals who provided helpful comments, discussions, and data, not only on the results presented here, but also during the preparation of my dissertation: Dr. D. Bramble, Dr. R. Estes, Dr. R. C. Fox, Dr. E. S. Gaffney, Mr. H. J. Garbani, Mr. M. T. Greenwald, Dr. L. D. Hewitt, Dr. J. H. Hutchison, Dr. Z. Kielan-Jaworowska, Mr. M. Lambert, Dr. J.A. Lillegraven, Mr. S. G. Lucas, Mr. M. Middleton, Dr. M. J. Novacek, Mr. K. Padian, Dr. J. K. Rigby, Jr., Dr. D. E. Savage, Dr. D. Schindel, Dr. R. E. Sloan, Dr. L. Van Valen, Mr. K. Whetstone, Dr. R. W. Wilson, Dr. K. Waage, many of my fellow graduate students at Berkeley, and colleagues at Yale. Thanks also to Booper (a.k.a. Barnum Brown), Cicero, and Mousetrap.

The following people kindly allowed me to study specimens in their respective institutions or in their care: Drs. E. S. Gaffney, M. C. McKenna, and R. Tedford, and Messrs. H. Galiano and E. Manning, American Museum of Natural History; Mr. M. Lambert, Carter County Museum, Ekalaka, Montana; Dr. M. Dawson, Carnegie Museum, Pittsburgh; Drs. J. Bolt and W. Turnbull, Field Museum of Natural History, Chicago; Drs. L. G. Barnes and D. Whistler, Los Angeles County Museum of Natural History; Dr. F. Jenkins and Messrs. C. Schaff and C. Wood, Museum of Comparative Zoology, Harvard University; Dr. D. Baird, Princeton University; Dr. G. Edmund, Royal Ontario Museum; Dr. B. Erickson, St. Paul Science Museum; Dr. L. Van Valen, University of Chicago; Dr. P. Robinson and Mr. M. Middleton, University of Colorado; Dr. D. Webb and Mr. G. McDonald, University of Florida; Dr. D. Bramble,University of Illinois at Chicago Circle; Dr. P. Gingerich, University of Michigan Museum of Paleontology; Dr. R. Sloan, University of Minnesota Museum of Paleontology; Drs. R. Emry and R. Purdy, United States National Museum; and Drs. J. Ostrom and E. Simons and Ms. M. A. Turner, Yale Peabody Museum.

Financial support from the following sources is gratefully acknowledged: Department and Museum (Annie M. Alexander Endowment) of Paleontology, University of California, Berkeley; the University of California (Regents and Wheeler Fellowships); National Science Foundation Grant nos. GB 39789 and BMS 75-21017, to Dr. W.A. Clemens; Grant-in-Aid of Research from Sigma Xi; and the Los Angeles County Museum of Natural History. Portions of this work were completed during my tenure as a J. Willard Gibbs Instructor in the Department of Geology and Geophysics, Yale University.

Most of the line drawings of specimens were done by Ms. Pat Lufkin and Ms. Augusta Lucas-Andreae, whose work is identified by the artist's initials in parentheses following the figure caption. All other illustrative material was prepared by the author.

Finally, I would like to thank my parents, Jim and Donna, for their encouragement before and throughout my academic career.

Abstract

The Hell Creek (Upper Cretaceous and lowermost Paleocene) and overlying Tullock (Paleocene) formations were mapped in detail near the head of Hell and Snow creeks, Garfield County, Montana. Both formations are predominately fluviatile. The contact between the formations is usually placed at the base of the lowest lignite (Z coal) above the last dinosaur. This contact also usually marks the Cretaceous-Tertiary boundary. These two definitions are separated, with the lowest lignite marking the formational contact and the extinction of dinosaurs marking the Cretaceous-Tertiary boundary. The extinction of dinosaurs occurred 6 to 10 ft. (1.8-3.0 m) below the lignite and thus this interval of the Hell Creek Formation is Paleocene in age.

Two distinct latest Cretaceous (Lancian) faunal-facies are present in much of the upper Hell Creek Formation. In the Paleocene portion near the formational contact, two distinct mammalian levels are present, the pre-mantuan and mantuan. Both levels are older than the presently recognized Puercan. Although not considered in this study, Puercan and Torrejonian mammals are known from the Tullock Formation.

The Hell Creek faunal-facies is dominated by multituberculates and marsupials, with rare eutherians (all typical latest Cretaceous mammals). Mammals of this faunal-facies usually occur in small, poorly sorted, siltstone lenses throughout the stratigraphic extent of the formation, except probably the upper 6 to 10 ft. (1.8-3.0m). Dinosaurs are common in this faunal-facies. The richest Hell Creek faunal-facies locality (V73087) is included with four other localities in the Flat Creek local fauna and three of the localities are only 17 ft. (5.2 m) below the base of the Tullock Formation. This local fauna is geographically intermediate between the Lance (Wyoming) and the Trochu (Alberta) local faunas and is an admixture of these two. Of the thirty-one species-level taxa occurring in the Hell Creek faunal-facies, only four are (or may be) new: a new species of ?*Neoplagiaulax*; neoplagiaulacid gen. et. sp. indet.; a new species of *Glasbius*; and ?*Pediomys* cf. *P. florencae*. Based on close faunal resemblance, the Flat Creek, Trochu, and Lance local faunas are considered to be of similar age.

The Bug Creek faunal-facies is only questionably known in Garfield County but is well known in McCone County from the research of others. The Bug Creek faunal-facies is dominated by Paleocene-aspect mammals, particularly condylarths and taeniolabidoid multituberculates. Based on stratigraphic correlations, the Bug Creek faunal-facies localities are approximately coeval with more typical latest Cretaceous mammal localities from Montana, Alberta, and Wyoming.

Pre-mantuan Paleocene localities on either side of the Hell Creek–Tullock contact are more similar to the Bug Creek faunal-facies localities in their mammalian components.

The two richest pre-mantuan localities which occur low in the Tullock Formation are included in the Hell's Hollow local fauna. Biostratigraphically, this local fauna is younger than the latest Cretaceous (Lancian) Harbicht Hill local fauna, McCone County, and older than earliest Paleocene (mantuan) localities from Garfield County and Wyoming (Mantua lentil). New taxa from the Hell's Hollow local fauna include new species of *Mesodma, Cimexomys,* and *Ragnarok* and a new genus and species of microcosmodontine.

Mammals from two poorly known localities in the lowermost portion of the Tullock Formation are also described. Two of the condylarths are referred to two species known from Mantua lentil, Wyoming. Based largely on this, these two localities are considered to be earliest Paleocene (mantuan) in age. A new species of *Protungulatum* occurs at one of these localities.

Systematic considerations (in addition to those mentioned above) include: *Meniscoessus borealis* is synonymized under *M. robustus*; a new genus, typified by *"Cimexomys" priscus* is recognized; this new genus, *Cimexomys*, and the Cimolomyidae are shown to be primitive and not closely related to each other, the Ptilodontoidea, or the Taeniolabidoidea, based on a cladistic analysis; upper dentitions of two species of *Oxyacodon* (from the San Juan Basin) are tentatively recognized and this genus is transferred to the Anisonchinae; and a cladistic analysis of early anisonchines shows that the typical periptychid molar evolved before the premolars were greatly modified.

Determination of the Cretaceous-Tertiary boundary in eastern Montana will only be able to be done accurately by correlation to the type marine sections in Europe through further development of such methods as magnetostratigraphy. Although it is fraught with uncertainties, dinosaur extinction, supplemented by palynofloral studies, remains the most utilitarian method of estimating the Cretaceous-Tertiary boundary in terrestrial sequences.

Mammalian evolution near the Cretaceous-Tertiary boundary (at least in eastern Montana) included a gradual decrease and eventual loss of most typical latest Cretaceous mammals and all dinosaurs *within* the Bug Creek faunal-facies sequence, while Paleocene-aspect mammals were radiating within this faunal-facies. In contrast, typical latest Cretaceous mammals (but no Paleocene-aspect mammals) and dinosaurs persisted under relative stasis in the Hell Creek faunal-facies to within 6 to 10 ft. (1.8-3.0 m) of the Hell Creek–Tullock formational contact. At this level all the dinosaurs and some of the latest Cretaceous mammals (particularly marsupials) became extinct. Some, but scant, evidence suggests that Paleocene-aspect mammals may have expanded beyond the margins of the large channels of the Bug Creek faunal-facies at this time. Dinosaurs and most typical latest Cretaceous mammals may have become locally extinct within the Bug Creek faunal-facies before their total extinction within the Hell Creek faunal-facies.

Possible factors for the extinction of dinosaurs and some of the typical latest Cretaceous mammals are difficult to demonstrate. However, the type of replacement *within* the Bug Creek faunal-facies suggests some sort of competition, at least between typical latest Cretaceous and Paleocene-aspect mammals. A similar sort of competition between Paleocene-aspect mammals and dinosaurs is more difficult to visualize. It still remains possible that such competition took place between younger dinosaurs and Paleocene-aspect mammals. These sorts of causal factors cannot easily be extended to the Hell Creek faunal-facies. It is still possible that the Paleocene-aspect mammals (as well as changes in such factors as climate and flora) eventually had an impact on the Hell Creek faunal-facies. Although theories invoking catastrophic events to explain terminal Cretaceous extinctions cannot be ruled out, the evidence from eastern Montana does not support such theories.

The change-over of mammals at the Cretaceous-Tertiary boundary was probably no more drastic than that which took place between many other Tertiary land mammal ages.

Whatever the cause of dinosaur extinction, the inception of the large channels of the Bug Creek faunal-facies that brought the invasion of the Paleocene-aspect mammals had an impact on both dinosaurs and typical latest Cretaceous mammals. This demonstrates that such factors as ecology can affect large and small vertebrates equally. Thus, models that largely invoke changes in physical barriers such as a decrease in topographic-geographic diversity in explaining tetrapod extinctions are oversimplifications.

List of Abbreviations

(other than those defined in the text)

TOOTH MEASUREMENTS AND RATIOS (SEE FIG. 1):

General:	L	Greatest anteroposterior length parallel to: a line drawn through the apices of the paracone and metacone in upper molars and the metaconid and entoconid in lower molars of therians; the longest cusp row of multituberculate molars and P^4; and the baseline of the P_4 of multituberculates (see Fig. 1a & b and Novacek and Clemens, 1977, for further explanation).
	W	Greatest labiolingual width normal to length.

Multituberculate P_4's only (Novacek and Clemens, 1977, p. 704):

H	". . . measured from the highest point of the crest to the baseline."
L1	". . . parallel to B (baseline), is measured from the anterior edge of the blade to a point directly below its highest point."
H/L	". . . an approximation of the degree of arc of the crest."
L1/L	". . . a measure of its [the blade's] symmetry."

Therians only:	W-A	Greatest labiolingual width from anterolabial corner of crown to lingualmost point of protocone on therian upper molars normal to length.
	W-P	Same as W-A, but measured from posterolabial corner of crown to lingualmost point of protocone.
	W-TA	Greatest labiolingual width of the talonid normal to length.
	W-TR	Greatest labiolingual width of the trigonid normal to length.

Condylarths upper molars only (ratios assume the paracone position as the constant):

A	Distance between lingualmost point of protocone base to its apex normal to length.
B	Distance between protocone and paraconule apices normal to length.
C	Distance between paraconule and paracone apices normal to length.

D	Distance between paracone apex and labialmost point of crown opposite paracone normal to length.
A/W–A	An approximation of the size of protocone's lingual surface and the labiolingual position of its apex.
B/W–A	A measure of the proximity of paraconule and protocone.
C/W–A	A measure of the proximity of paraconule and paracone.
D/W–A	An estimate of the width of the stylar shelf labial to the paracone.
L/W–A	An estimate of lingual expansion of the crown or antero-posterior lengthening or shortening utilizing the anterior width.
L/W–P	Same as L/W–A, but utilizing the posterior width.

Condylarth lower molars only:	L–TA	Distance from point of intersection of cristid obliqua and back of trigonid to posteriormost point of talonid parallel to length.
	L–TR	Distance from anteriormost point of paraconid to point of intersection of cristid obliqua and back of trigonid parallel to length.
	P–M	Distance between paraconid and metaconid apices normal to length.
	L–TR/L–TA	Length of trigonid relative to length of talonid.
	W–TR/W–TA	Width of trigonid relative to width of talonid.

STATISTICS:	CV	Coefficient of variation
	N	Sample size
	OR	Observed range of variation
	s	Standard deviation
	$\bar{x}$	Arithmetic mean

INSTITUTIONS:	AMNH	American Museum of Natural History
	CCM	Carter County Museum, Ekalaka, Montana
	CM	Carnegie Museum, Pittsburgh
	LACM	Los Angeles County Museum of Natural History
	MCZ	Museum of Comparative Zoology, Harvard University
	PU	Princeton University
	SPSM	St. Paul Science Museum
	UA	University of Alberta
	UCMP	University of California Museum of Paleontology, Berkeley
	UF	University of Florida
	UMVP	University of Minnesota Museum of Paleontology
	USNM	United States National Museum
	YPM	Yale Peabody Museum

INTRODUCTION

This study is divided into three parts. The first is the description of the geology of the uppermost Cretaceous and lowermost Paleocene Hell Creek Formation and the lower and middle Paleocene Tullock Formation, particularly near the head of Hell Creek, Garfield County, Montana. Brief descriptions of these formations are also given for other selected areas where microvertebrate fossils have been recovered. The second part deals with the systematic descriptions of mammals recovered from the Hell Creek Formation and the lowermost Tullock Formation, Garfield County. The third part is a comparison of these mammals with those previously described from other latest Cretaceous and earliest Paleocene local faunas of western North America. (A fourth section dealing with the chelonians recovered from the Hell Creek Formation will be presented elsewhere: Archibald and Hutchison, 1979; Archibald, unpublished).

A number of other workers have examined various aspects of the geology and paleontology of the Cretaceous-Tertiary "transition" beds in Garfield County and surrounding regions. The Hell Creek Formation first received notice in paleontological circles through the efforts of Barnum Brown of the American Museum of Natural History. Brown worked sporadically in the Hell Creek Formation from 1902 through 1910. Much of the time he concentrated on the recovery of dinosaurs, most notably *Tyrannosaurus, Triceratops,* and *Anatosaurus.* He also recovered a few isolated mammalian specimens. One sample collected in the valley of Crooked Creek was described by Simpson (1927a). Working from copies of photographs of the locality on file at the AMNH, the probable site was relocated in 1976.

In 1907 Brown described and named the "Hell Creek beds." For the next fifty years no major attempts were made to recover mammals or other microvertebrates from the Hell Creek and Tullock formations. In the 1960s, R. E. Sloan and L. Van Valen recovered a large sample of mammals of Paleocene aspect from the Hell Creek Formation at the now world-famous Bug Creek Anthills, McCone County, Montana. They also collected smaller samples of typical Late Cretaceous mammals at several sites in both McCone and Garfield counties and of Puercan mammals at Purgatory Hill, McCone County. Two preliminary faunal studies appeared in 1965 (Sloan and Van Valen, 1965; Van Valen and Sloan, 1965). Estes and others (1969) and Estes and Berberian (1970) described the lower vertebrates and the paleoecology of the vertebrates from the Hell Creek Formation of McCone County.

The present research in Garfield County, just west of McCone County, was initiated by the Los Angeles County Museum in 1965. In that year Harley Garbani arrived in Garfield County to commence a search for dinosaurs for exhibit. In the following years a number of dinosaurs were recovered; they are now housed in the L.A. County Museum. During the excavation of a skeleton of *Tyrannosaurus rex* in the headwaters of Hell Creek, Mr. Garbani discovered a rich microvertebrate site in the overlying Tullock Formation. In 1972, a joint UCMP-LACM party began extensive excavation of the Garbani Locality and investigation of other microvertebrate sites in the Hell Creek and Tullock formations, Garfield County.

My input into the field aspects of this project began in the summer of 1973 and continued through the summer of 1976, when I was a graduate student at Berkeley; further research was done during the summer of 1979, when I was at Yale. The field research conducted since 1976 has resulted in a number of additions and modifications to my doctoral work. Some of these modifications were made possible by the continued research of W. A. Clemens, and these are cited as personal or written communications from Clemens.

The results presented here are part of a continuing project under the direction of Dr. Clemens. Preliminary papers on the mammalian aspects of this work and other papers based on material from the area have been published elsewhere (Archibald, 1977, 1979a, 1979b; Archibald and Clemens, 1977, 1980; Clemens, 1968, 1974; Clemens and Archibald, 1980; Lillegraven and others, eds., 1979; Novacek, 1977; Novacek and Clemens, 1977).

Author's present address: Department of Biology and Peabody Museum of Natural History, Yale University, New Haven CT 06520.

TECHNIQUES OF STUDY

The following is a summary of the methodology and terminology employed in this study.

Geology. Most geologic and paleontologic data were plotted on U.S.G.S. 7.5-minute quadrangles. When these maps were unavailable, plotting was instead done on aerial photographs or 7.5-minute orthophoto quadrangles. In all instances, stratigraphic sections were measured using a brunton compass as a hand-held level.

Measurements of geologic features, distances, stratigraphic intervals, and miscellaneous non-fossil specimens are given in the English system, with metric equivalents in parentheses.

Collecting. In addition to surface collecting, extensive underwater screening similar to that described by McKenna (1965), Clemens (1964), and Lillegraven (1969) was carried out at a number of microvertebrate localities. The concentrations of bone were moderately low at the Hell Creek Formation sites, and dentulous jaws were uncommon. Therefore, hand-quarrying was not considered to be a feasible method of collecting. One Tullock Formation site, the Garbani Quarry, preserved a sufficient number of dentulous jaws to warrant hand-quarrying. At this site, hand-quarrying was followed by underwater screen-washing of the discarded matrix.

The washing boxes were divided into two groups of approximately twenty-five boxes each. An attempt was made to use one group of boxes solely for matrix recovered from the Tullock Formation and the other for matrix from the Hell Creek Formation. This procedure was usually successful; however, the matrix from some localities that have since been determined to be from the Tullock Formation were washed in the Hell Creek Formation boxes. Boxes were thoroughly cleaned before matrix from a new locality was introduced, so the possibility of contamination is remote. The two groups of boxes differed slightly in size to reduce the likelihood of mixing samples. The screened bottom measured approximately 12 in. × 18 in. (30.5 cm × 45.7 cm) on the Tullock boxes and 12 in × 24 in. (30.5 cm × 70.0 cm) on the Hell Creek boxes. Brass screen with 24 openings per in. (= approx. 9.5 openings per cm) was used for both groups. Since running water was not easily accessible, washing was carried out at small cattle ponds. Processing of the two groups of boxes was done at separate sites at one pond or at separate ponds. The matrix was transported to the washing site in burlap bags. Up to ten pounds (4.5 kg) of matrix were put in each washing box and were allowed to stand in water for an average of 24 hours. Before removal, the boxes were gently agitated to remove the fine silt and sand. The remaining concentrate was allowed to dry in the boxes and was then packed in small paper bags lined with plastic bags for shipment to the UCMP for sorting under low-powered binocular microscopes.

Faunal terminology. In synthesizing fossil data it is difficult, if not impossible, to use terminology which has been universally defined or agreed upon. To avoid misunderstandings, the faunal terms used in this paper are briefly discussed below.

An attempt has been made to use the term *local fauna* in the sense of Tedford (1970). As used here, it refers to one or more fossil samples that have a similar faunal content and are limited in their stratigraphic and geographic extent. Although it is not possible to avoid paleoecological and biostratigraphic inferences, *local fauna* is intended here to be a descriptive term used for convenient grouping of fossil localities, based on the above criteria. Two local faunas are proposed, the Hell's Hollow and Flat Creek local faunas, which include, respectively, two and five vertebrate localities. These local faunas include the richest mammalian microvertebrate localities found to date in the lowermost Tullock and Hell Creek formations of Garfield County. A description of these local faunas is given in the section dealing with stratigraphy and physical geology. Numerous other localities bearing mammals were discovered in the lowermost Tullock and Hell Creek formations, Garfield County. None is very rich and it was deemed unnecessary to add local fauna names to locality numbers and names that are adequate for occasional references. However, data collected during 1977–1979 may warrant the recognition of additional local faunas. Most other local fauna names mentioned in this study are self-explanatory. Only one requires additional comment: the Trochu local fauna is equivalent to the localities from the "upper part of the Edmonton Formation" of Lillegraven (1969). In reading Lillegraven (1969), I could not find any reference to an "upper Edmonton local fauna." Although it is convenient to have a collective name for his localities, it is fortunate that this name was not proposed. Subsequent to Lillegraven's work, Irish (1970) and Gibson (1977) proposed name changes for the lithologic unit that included Lillegraven's localities. The "upper part of the Edmonton Formation," used by Lillegraven (1969), was renamed the "Scollard Member of the Paskapoo Formation" by Irish (1970). Subsequently, Gibson (1977) raised the Scollard Member to the Scollard Formation. In addition, L. S. Russell (1964, 1975) and Clemens and others (ms) have tentatively proposed a sequence of terrestrial vertebrate "stage/ages" for the Cretaceous of western North America, based mostly on mammal assemblages. One of these stage/ages is the Edmontonian. L. S. Russell (1975) used *Edmonton* in the sense of Irish (1970), excluding the "upper Edmonton" above the Kneehills tuff. Thus the mammals described by Lillegraven are not included in the Edmontonian but are referred to the following Lancian "stage/age." To avoid conflicts with Irish's and Gibson's revised lithologic nomenclatures and Russell's "stage/age," it is necessary to propose a different name for Lillegraven's fossil samples and localities. W. A. Clemens (pers. comm.) has suggested "Trochu local fauna" for this aggregation, since the town of Trochu is located southwest of Lillegraven's study area.

The second group of terms in this study that warrants comment is that of the "stage/ages" of L. S. Russell (1975). Material recovered from all but the upper 6 to 10 ft. (1.8–3.0 m) of the Hell Creek Formation of Garfield County is referable to Russell's Lancian "stage/age." Occasional reference is also made to the "stage/ages" that precede the Lancian. Although these "stage/ages" are still not widely used or accepted, I find them to be preferable to a first- or second-order correlation to the standard European marine sequence. However, as their usage continues, it is becoming clear that they suffer from some of the same conceptual and operational problems that have plagued the North American Land Mammal "Ages."

Recently, Van Valen (1978) proposed the "Mantuan" land mammal age. The Mantuan is pre-Puercan in its traditional usage but earliest Paleocene in age, as far as can be determined. Van Valen noted that "only two Mantuan (local) faunas are yet reported, Mantua and Leidy Quarry, both in the Bighorn Basin, Wyoming." Two as yet poorly known localities from Garfield County, Montana (McKeever Ranch 1, UCMP loc. V72210 = LACM loc. 3103, and McKeever Ranch-Harley's Point, UCMP loc. V77087) are probably also Mantuan in age. I support Van Valen's view that the differences between Puerco ("type" Puercan) and Mantua seem to be due to evolution rather than environment or geography, but I do not think it is necessary to recognize a new, separate Mantuan mammal age. The naming of new Cenozoic land mammal ages will not add to our understanding of mammalian biostratigraphy or evolution. Rather, existing ages should be more rigorously redefined and informally subdivided on a regional basis. Mantua lentil and other Mantuan localities should be included in a redefined Puercan land mammal age (eventually maybe even a stage). However, without a more formal redefinition of the Puercan (and that is beyond the scope and intent of this study), the probable biostratigraphic relationship of Mantua lentil (and other similar sites) to "type" Puercan could well be obscured if Mantua was simply referred to the Puercan. Therefore, the term *mantuan* is used in this study, with its informal status indicated by the use of a lower-case initial letter. Similarly, several earliest Paleocene localities (including those of the Hell's Hollow local fauna) that are older than mantuan but younger than Lancian are informally referred to as *pre-mantuan*.

The third and final term to be mentioned is *faunal-facies*. During this study it became apparent that two distinct types of mammal-bearing localities are present in much of the upper Hell Creek Formation. Sloan and Van Valen (1965) noted the presence of two types of mammal-bearing localities in the Hell Creek Formation of McCone County but, particularly from later papers, it appears that they were unaware of the extensive overlap of the two types. One type of locality is dominated by mammals of Paleocene aspect and the other by more typical latest Cretaceous forms such as those found in the Lance Formation, Wyoming. Sloan and Van Valen have assigned these mammals to distinctive "communities" (Sloan, 1970, 1976; Van Valen and Sloan, 1977a). The word *community* connotes a paleoecological interpretation to me. I hasten to add that I applaud Sloan's and Van Valen's (and Estes and Berberian's, 1970) attempts at "community" reconstruction. However, for use in general discussion of the two types of mammal-bearing localities I prefer a less interpretive, more descriptive term. Possibly the best word would be *assemblage*, but unfortunately this term has suffered from overuse. A second choice is *biofacies*, but because only animals are discussed in this presentation, the term *faunal-facies* seems most appropriate. *Facies* can have either an interpretative or descriptive connotation. In the present study it is used descriptively, somewhat analogously to the descriptive phrase *sandstone lithofacies*, but it is named after a geographic feature. The two recognized here are the Hell Creek and Bug Creek faunal-facies, after the creeks of the same names in Garfield and McCone counties, respectively. Both types of faunal-facies appear to be restricted to the Lancian land mammal age.

The Hell Creek faunal-facies includes localities that are dominated both in numbers and diversity by didelphid and pediomyid marsupials and ptilodontoid multituberculates. Eutherians are represented only by rare specimens of the leptictoid, *Gypsonictops*, and palaeoryctids. These localities are most similar in faunal composition to local faunas described from the Lance Formation, Wyoming (Clemens, 1964, 1966, 1973) and the

Scollard Formation (Lillegraven, 1969). These are the local faunas that have been referred to as typical or standard Cretaceous (Sloan and Van Valen, 1965) or have been included as part of the "*Triceratops* community" (Sloan, 1976; Van Valen and Sloan, 1977a). Dinosaurs (mostly teeth) are common members of Hell Creek faunal-facies localities. Hell Creek faunal-facies localities appear to be restricted to the smaller, siltier channel-fill deposits, and paludal and overbank deposits. Localities of this faunal-facies are best known in the upper one-half of the Hell Creek Formation but appear to extend to its base and to within 6 to 10 ft. (1.8–3.0 m) of the overlying Hell Creek–Tullock formational contact.

The Bug Creek faunal-facies localities also have typical latest Cretaceous mammals and dinosaurs. However, at stratigraphically higher localities within this faunal-facies, these taxa become taxonomically less diverse and numerically less abundant. These localities also include Paleocene-aspect mammals. At the stratigraphically lowest locality (Bug Creek Anthills) four species of this type of mammal (*Protungulatum donnae, Procerberus formicarum, Catopsalis joyneri*, and *Stygimys kuszmauli*) are very common taxa. At higher localities Paleocene-aspect mammals increase in taxonomic diversity and numerical abundance. Sloan and Van Valen have considered these mammals to be part of a proximal terrestrial vertebrate community (apparently in the sense used by Shotwell, 1955). These authors have referred to this community by various names: "condylarth—multituberculate community" (Sloan, 1970); "*Protungulatum—Stygimys* community" (Sloan, 1976); and "*Protungulatum* community" (Van Valen and Sloan, 1977a). Bug Creek faunal-facies localities are found in the larger, sandier, channel-fill deposits. Stratigraphically, these localities are known to range from 80 ft. (24.4 m) to within about 10 ft. (3.0 m) (a channel capping) below the Hell Creek–Tullock formational contact. The upper limit is somewhat arbitrary because the overlying pre-mantuan localities appear to be a continuation of the Bug Creek faunal-facies localities.

Systematics. The paleontological aspects of this presentation are concerned foremost with the description of the mammals from the lowermost Tullock and Hell Creek formations of Garfield County and comparisons with taxa from similar faunas. A number of taxa have been adequately described and figured elsewhere by Clemens (1964, 1966, 1973), Lillegraven (1969), and Sloan and Van Valen (1965). In these instances only additional or differing comments are given. Only in two groups (multituberculates and anisonchine condylarths) was there enough additional material to warrant extensive comments on phylogenetic relationships. Since this is not intended as a phylogenetic study, the classificatory schemes employed for the various taxa vary considerably depending on how recently and by whom they have been reviewed. An attempt has been made to follow recent modifications in classification proposed by various specialists. In the few instances in which phylogenetic relationships have been explored, the emphasis has been on phylogenetic reconstruction rather than on classificatory schemes. The procedure for phylogenetic reconstruction basically follows the methodology of Hennig (1966). Gaffney (1975) gives a concise explanation of the Hennigian cladistic method. In the few cases in this study where changes in classification are warranted, such changes do not necessarily follow a strictly cladistic scheme. By *strictly cladistic* I mean the practice of nesting taxa on the basis of recency of dichotomous splitting and then erecting categories for each bifurcation. I have attempted to erect only monophyletic groups, where a *monophyletic group* is defined as one whose members have an ancestor in common only to themselves (after Hennig, 1975) and their similarity has been determined by shared derived characters (synapomorphies).

Analysis of fossil material. The cusp terminology used for multituberculates follows Clemens (1964) unless otherwise noted. Standard length and width measurements of molars and P^4 are also those employed by Clemens (1964). The measurements of P_4 are those proposed by Novacek and Clemens (1977) and repeated in Fig. 1a & b.

The cusp terminology used for the therians is after Van Valen (1966, pp. 7–9, fig. 1) with the following exceptions: the stylar cusps of the marsupials are named following Clemens (1966, p.3; i.e., cusps A, B, C, D, and E going anteriorly to posteriorly), and the cusp developed on the precingulum of upper molars of anisonchines is called the *protostyle*, following Matthew (1937) and Simpson (1937a). The measurements of therian molars were done after Lillegraven (1969, fig. 5) and are repeated in Fig. 1c & d. Additional measurements made of the condylarth molars are also shown in Fig. 1c & d. The numbering sequence for therian post-canine teeth is as follows: marsupials—premolars, 1–3, molars, 1–4; eutherians—premolars, 1–4, molars, 1–3.

Measurements of mammalian specimens are given in millimeters, unless noted otherwise. Most specimens were measured on an Ehrenreich Photo Optical Shopscope. These measurements were rounded to the nearest one-hundredth of a millimeter. The following specimens were measured with a microscope eyepiece grid, and those measurements were rounded to the nearest five-hundredths of a millimeter: all specimens referred to *Pro-*

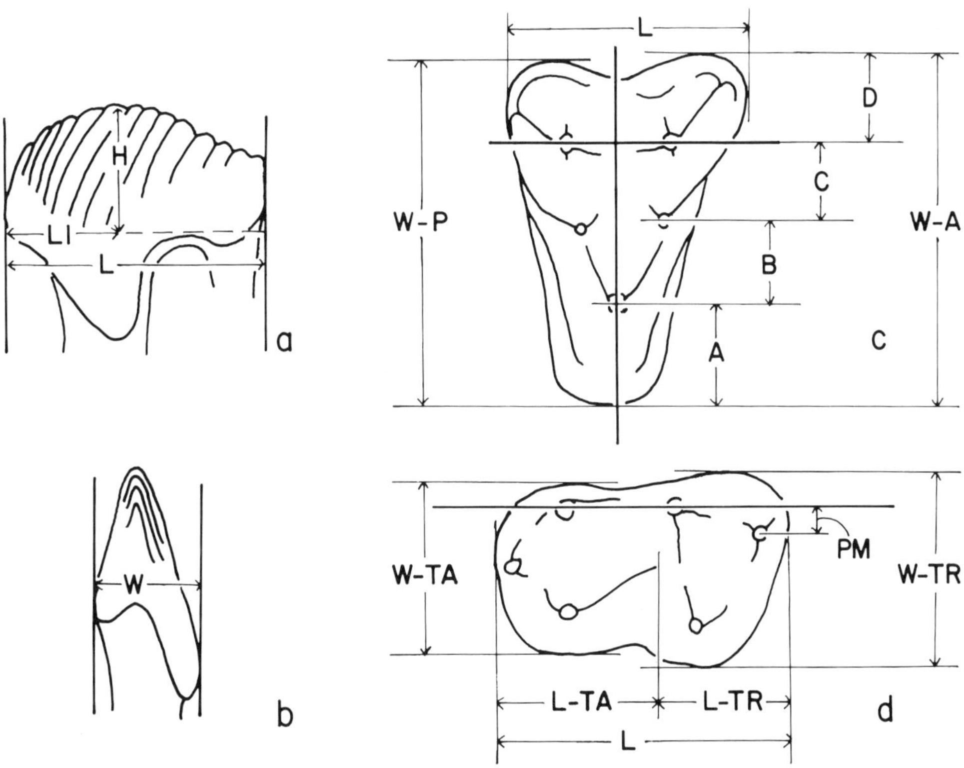

FIG. 1. Labial (a) and anterior view (b) of a multituberculate P_4, and occlusal view of a therian upper (c) and lower (d) molar showing orientations for measurements. Thick lines are the axes of orientation. See List of Abbreviations. Measurements of the P_4 are after Novacek and Clemens (1977) and those of the therian are modified after Lillegraven (1969).

cerberus formicarum; four P^4's of *Meniscoessus robustus* (UCMP 120389, 120390, 120394, and 120395); three isolated molars (UCMP 120391, 120393, 120396) and a maxilla (UCMP 120397) of *Essonodon browni*; two lower molars (UCMP 120408 and 120409) and an upper molar (UCMP 120410) of *Mimatuta minuial*; two lower molars (UCMP 121791) of *Ragnarok harbichti*; and the type dentary (UCMP 121782) of *Protungulatum mckeeveri* n. sp.

STRATIGRAPHY, PHYSICAL GEOLOGY, AND LOCALITIES

INTRODUCTION

In Garfield County, Montana, the Hell Creek (Upper Cretaceous and lower Paleocene) and Tullock (lower and middle Paleocene) formations are exposed in a number of areas. The most extensive exposures occur in a broad belt south of Fort Peck Reservoir (Missouri River). In the major region of study in northern Garfield County, the Hell Creek and Tullock formations are relatively flat-lying and crop out as badlands formed by myriad creeks flowing north to the Missouri River drainage.

The major stratigraphic concerns of this study have been: (1) to map the upper part of the Hell Creek and Tullock formations in several selected areas; (2) to investigate the boundary between these formations; and (3) to place the vertebrate fossil localities in their stratigraphic context. The biostratigraphic considerations of the fossil localities in the Hell Creek and lowermost Tullock formations are given following the systematic descriptions of the mammals from these localities.

Before discussing the above stratigraphic problems, a brief synopsis is given of the regional setting and of the formations cropping out in Garfield County.

REGIONAL SETTING

The Hell Creek Formation overlies the Fox Hills Sandstone and Bearpaw Shale. The latter two formations are exposed along the southern shore of the Fort Peck Reservoir. They represent, respectively, the strandline and marine deposits of the regressive phases of the last marine incursion that covered eastern Montana during the latest Cretaceous. The Hell Creek Formation is a largely fluviatile deposit that encroached eastward in a low, broad plain as the sea regressed. The superjacent Fort Union Group, particularly the Tullock Formation and lower parts of the Lebo Formation, represent, with slight sedimentological changes, a continuation of fluviatile deposition. To the east in North and South Dakota, the Cannonball Formation represents the continued, but probably much reduced, presence of a seaway that received sediments approximately equivalent in age to the Paleocene Tullock and parts of the Lebo Formation (see Fox and Olsson, 1969).

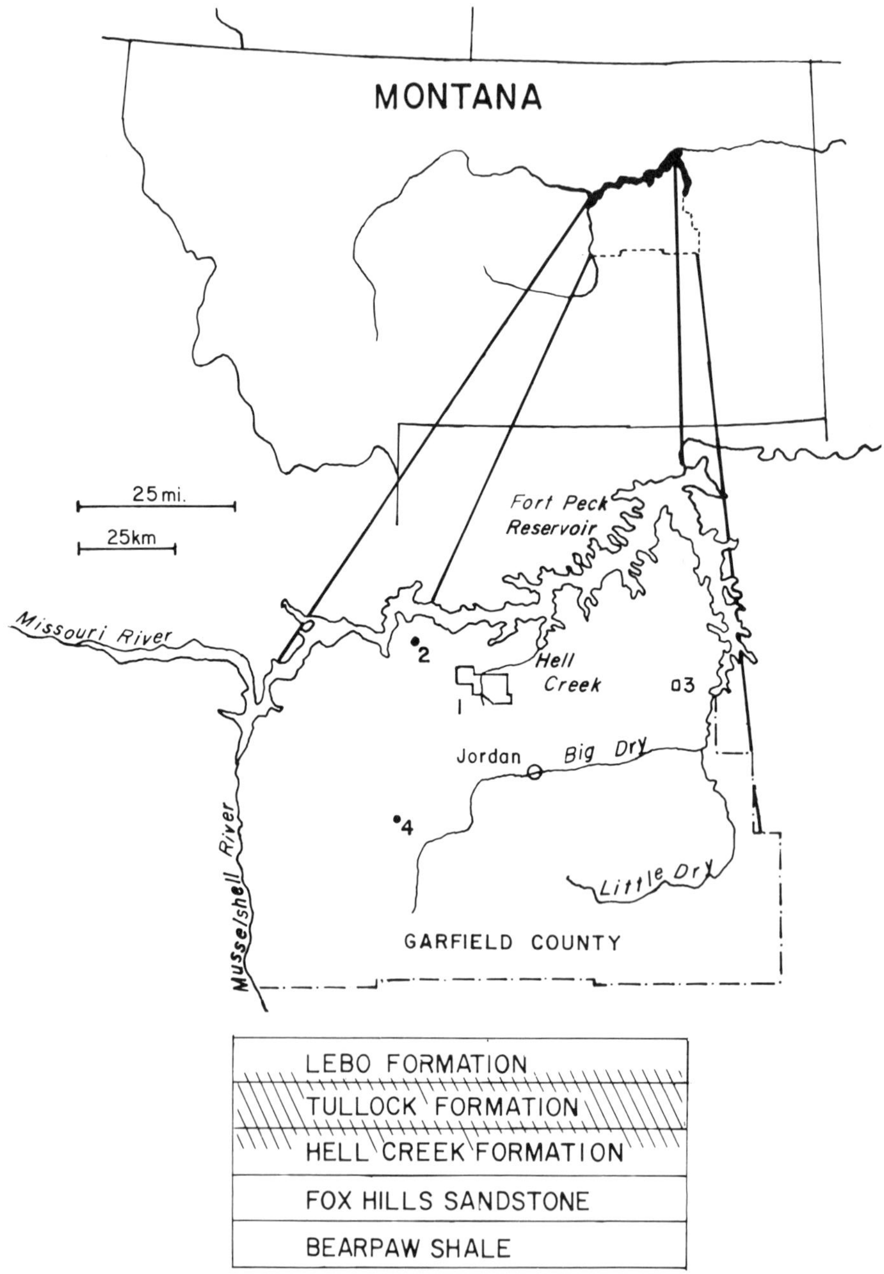

FIG. 2. Index map showing the location of Garfield County, Montana. The numbers refer to the major areas of study considered in the section dealing with stratigraphy and localities, as follows: (1) Hell Creek–Snow Creek area; (2) Billy Creek–McKeever area; (3) Flat Creek area; (4) Horsethief Canyon area. The generalized stratigraphic section shows Cretaceous and Paleocene formations exposed in Garfield County. The diagonal lines show the stratigraphic interval emphasized in the present study.

FORMATIONS

Bearpaw Shale. The formation is exposed in southwestern, western, and northern Garfield County along Fort Peck Reservoir. According to Jensen and Varnes (1964), the bulk of the formation is composed of "dark-gray clayey shale. Bentonite occurs either as distinct beds or is disseminated in the shale." Just northeast of Garfield County in the Fort Peck area, Jensen and Varnes (1964) report a thickness of 1,140 ft. (347 m) for the Bearpaw Shale. They note that, based on electric logs recorded in test wells southeast of the Fort Peck area, the Bearpaw Shale appears to thin somewhat, and they suggest that this may be due either to actual thickening of the sequence as one goes northwest or to errors in measured sections or inconsistencies in drawing formational boundaries.

Utilizing the European standard sequence, there is some controversy as to where to draw the Campanian-Maestrichtian boundary in the Western Interior (see Obradovich and Cobban, 1975). However, fossils recovered from the Bearpaw Shale by Jensen and Varnes in the Fort Peck area and studied by Cobban (reported in Jensen and Varnes, 1964) suggest a late Campanian through Maestrichtian age for the formation in this area. More recent work (Gill and Cobban, 1973; Obradovich and Cobban, 1975) indicates a late Campanian through early Maestrichtian age in this sequence.

Fox Hills Sandstone. The Fox Hills Sandstone conformably overlies the Bearpaw Shale and, according to Jensen and Varnes (1964), disconformably underlies the Hell Creek Formation in the Fort Peck area. Jensen and Varnes compiled measurements of a number of thicknesses of the Fox Hills Sandstone in the Fort Peck and surrounding areas, including four measurements in northern Garfield County (1964, fig. 9). They found that the formation varies in thickness from 35 ft. to 146 ft. (10.7 m to 44.5 m). The variations in thickness were attributed to the erosional nature of the Fox Hills–Hell Creek contact. The lower portion of the Fox Hills Sandstone is composed of marine to brackish-water clay-, silt-, and sandstones. The upper portion can reach a thickness of 85 ft. (29.9 m), according to Jensen and Varnes (1964). This upper sandstone unit was observed in the bluffs above the mouth of Hell Creek. In this area it is an easily recognizable, cliff-forming unit. Some portions of this upper unit may be non-marine, as R. Brown (1939) reported fossil plants from the formation in southeastern Montana. The formation has not been studied in detail in eastern Montana as it has been in South Dakota by Waage (1968) and in North Dakota by Feldman (1972), but by analogy it appears to represent the shoreline environments of the regressing seaway (see Gill and Cobban, 1973).

The Colgate Member of the Fox Hills Sandstone, according to Jensen and Varnes (1964), is not present in the Fort Peck–McCone County area. However, it seemed to them that the upper sandstone unit they observed was its equivalent. They traced the Colgate Member from the Musselshell River eastward to Hell Creek. Although they report it as "snowy white" sandstone in the western area, it "gradationally changes its character eastward and is indistinguishable as a lithologic unit before Hell Creek is reached." However, further to the east in the drainage of Crooked Creek, a white sandstone unit is present below the Hell Creek Formation which corresponds well to Thom and Dobbin's (1924) description of the Colgate Member as a "conspicuous, white, upper sandstone of the Fox Hills." This unit was observed in 1976 while attempting to find B. Brown's Crooked Creek mammal locality (Simpson, 1927a). Although it is odd that B. Brown did not specifically mention this unit (if it was observed by him), it is possible that his "capping of sandstone" (1907, p. 827) is the Colgate Member, at least in the valley of Crooked Creek.

Hell Creek Formation. The Hell Creek Formation was named by B. Brown (1907, p. 829) for deposits "typically exposed on Hell Creek and nearby tributaries of the Missouri River." He noted (p. 844) that: "Lithologically the Hell Creek beds of Montana are similar in almost every respect to the Ceratops beds (Lance Formation) of Converse County, Wyoming." Thom and Dobbin (1924) included the Hell Creek Formation as a member of the Lance Formation, along with the overlying Tullock Formation. Although the Hell Creek Member was returned to the rank of formation in 1935 (Wilmarth, 1938), some later reports continued to recognize it as a member of the Lance Formation. Collier and Knechtel (1939) considered the Hell Creek and Tullock to be members of the Lance Formation, although they indicate (p. 10, footnote 11) that both Hell Creek and Tullock were raised to the rank of formation by the U.S.G.S. only after their report had been written.

The present report is concerned with the upper portion of the Hell Creek Formation, particularly in relation to its contact with the overlying Tullock Formation. No stratigraphic sections were measured that include the total thickness of the Hell Creek Formation. B. Brown (1907, p. 825) reports the thickness of the formation along Hell Creek to be 310 ft. (94.5 m). Thom and Dobbin (1924) report a thickness of 550 ft. (167.6 m) in an unidentified location in Garfield County.

B. Brown (1907) reported that the lower Hell Creek was composed of a sandstone that varied in thickness from 40 ft. (12.2 m) to 180 ft. (54.9 m). This portion of the formation was not studied in detail, but north of the headwaters of Hell Creek and in other locations this lower portion often stands out as bluffs which support growths of pine trees.

The upper 150 ft. (45.7 m) of the Hell Creek Formation in all areas examined were usually dominated by light to medium gray siltstones. This coloration probably explains why some early authors (e.g., Collier and Knechtel, 1939) referred to the Hell Creek Formation as "somber." Locally, however, beds can be reddish purple, tan to medium brown, or greenish gray. No persistent lignites were found in the upper portion of the formation in Garfield County, but localized carbonaceous shales and lignites up to 4 in. (10.2 cm) do occur infrequently. Beds of dark gray mudstone can also be found occasionally. In places these mudstones reach up to 10 ft. (3.0 m) in thickness and contain numerous fresh-water molluscs.

Most beds within the upper portion of the formation do not persist laterally and can seldom be traced more than one-quarter mile (.40 km) in any direction. Sandstone bodies do not predominate as in the lower portions of the Hell Creek Formation; in the upper Hell Creek Formation they are generally lenticular in shape and usually do not exceed 20 ft. (6.1 m) in thickness nor extend laterally for more than 300 ft. (91.4 m). When present, they tend to be tan or buff to medium brown in color because of the presence of more oxidized iron.

An interesting weathering phenomenon that occurs most often in association with the sandstone lenses is the development of loglike "concretionary" bodies. These bodies also occur in the overlying Tullock Formation. B. Brown (1907, p. 832) observed concerning these bodies that: "Crossbedding in the surrounding soft sandstone is frequently carried through the concretions line for line." Observations made during the current study corroborate those made by Brown. In some instances, the cross-section of one of these bodies will show intricate ripple-marks or other forms of cross-stratification, while another within a few feet will be devoid of primary structures. In the latter case, the cross-sections usually show uneven concentric rings, suggesting that if primary sedimentary structures were present they have been obliterated. Jacob (1973) found that "elongate concretions" in the Tongue River Formation, North Dakota, parallel the trend of large elongate sand-bodies and paleocurrent indicators. He suggested that they indicate low-sinuosity fluvial deposits. He observed similar bodies in other formations, including the Hell Creek Formation, and suggested that they parallel the trends of high- and low-sinuosity streams. Jacob noted that the cementing agent is mostly calcite. This is in agreement with the finding of B. Brown (1907) that the "concretions" are composed of 60 percent calcium carbonate. Although I regard the calcite cement as a secondary feature, possibly resulting from the action of ground water, the elongate shape of the "concretions" is probably a primary sedimentary structure. Observations by W. Rohrer (written comm.) of similar structures in the Hell Creek and Tullock formations generally agree with these conclusions.

Hell Creek–Tullock formational contact. By tradition, the Hell Creek–Tullock formational contact has been drawn at the lowest lignite above the last dinosaur. This contact has also traditionally served as the Cretaceous-Tertiary boundary.

Early forerunners of those definitions can be found in the paper by B. Brown (1907) in which the "Hell Creek beds" were named. He clearly distinguished the "Lignite beds," which are now included as part of the Tullock Formation, from the dinosaur-bearing "Hell Creek beds." In his diagrammatic section (1907, fig. 3), he places the contact between these two units at the lowest lignite. At one point he notes in italics (p. 835): "It is a most remarkable and significant fact that in no instance has a fragment of dinosaur bones been found in or above the lignite series by any of our party during five years' work in this region."

In the ensuing decades there was considerable discussion as to the utility of using the lowest lignite as a stratigraphic marker. There was also considerable debate as to whether what are now known as the Hell Creek and Tullock formations were Cretaceous or Tertiary in age. For the purposes of the present study, it is not necessary to discuss at length the historical development of the stratigraphic and time-stratigraphic controversies that were then being considered. Some idea of these debates can be traced in the coal reports of the U.S. Geological Survey beginning in the early part of this century: one of the earliest discussions can be found in an introduction to four coal surveys done along the Montana and South Dakota border (Calvert, 1912, pp. 187–201).

The stratigraphic boundary between the Hell Creek and Tullock formations would probably not have received such undue attention if it had not been for the discovery of dinosaurs by B. Brown and the subsequent determination that the Cretaceous-Tertiary boundary falls somewhere near this lithologic boundary.

The use of the lowest lignite as a time-stratigraphic boundary was popularized by R. Brown in various publications. In one paper (1952, p. 92), he suggested the following "formula" to define this boundary:

> Search for remains of dinosaurs as high as they can be found. Then look for the first coal zone, no matter how thin. As no authentic, indigenous, dinosaurian bones have ever been found above the base of this zone, the contact is considered to be at that level. It marks not only the inauguration of new environmental conditions but also the beginning of Tertiary time, if the primary assumption that the passing of the last dinosaur signalled the close of the Cretaceous, be accepted.

In large-scale field studies, R. Brown's "formula" may still remain a useful criterion, but it is unfortunate that such a mingling of rock-stratigraphic and time-stratigraphic boundaries has occurred. The placement of the Hell Creek–Tullock formational boundary and the recognition of the Cretaceous-Tertiary boundary can and should be decoupled from one another, for a variety of reasons.

To place the formational contact at the lowest coal above the last dinosaur violates the spirit, if not the letter, of the Code of Stratigraphic Nomenclature (1969). To paraphrase Article 4, Sections c and d of the code: concepts of time-spans or concepts based on biological sequence play no part in the definition or differentiation of a rock-stratigraphic unit. The presence of dinosaurs can still be used to help characterize the Hell Creek Formation, but their extinction ("the last dinosaur") is a biological event that should play no part in the definition of a rock-stratigraphic unit. Coal beds are also biological in origin; however, the code (Article 4) allows the use of fossils as lithologic units when they form a physical constituent of the sediment. Therefore, the lowest lignite can be used in defining the Hell Creek–Tullock formational contact because it is used as a lithologic body, not because of its fossil content or any implications of age.

The coupling of rock-stratigraphic and time-stratigraphic concepts in this instance has had a further unfortunate result. At least by implication, the lowest lignite has also become widely used as a time or geochronologic boundary between the Cretaceous and Tertiary. This rests on the unwarranted assumption that the deposition of the lowest lignite and the extinction of dinosaurs were isochronous (although possibly separate) events throughout the lateral extent of the Hell Creek and Tullock formations.

It is not yet known with certainty whether R. Brown's "formula" has been violated, but there may be instances of dinosaurs occurring above the lowest lignite. To my knowledge, Calvert's (1912) assertions that dinosaurs occur above the lower lignite have not been reinvestigated. Rogers and Lee (1923) noted the occurrence of a fragmentary supraorbital horn core of a ceratopsian, probably *Triceratops* (identified by C.W. Gilmore), as a surface find on a "steep slope about 50 ft. (15.2 m) above the base" of the type Tullock Formation. Since these authors were careful to note that the specimen was not found *in situ,* the report is only of passing interest.

During the 1979 summer field season, reports of a lignite below Bug Creek Anthills were investigated. A 9 in. (0.2 m) lignite was found approximately 30 ft. (9.1 m) below Bug Creek Anthills (the exact section was not measured) and 36 ft. (11.0 m) below Ken's Saddle. Both localities were originally described by Sloan and Van Valen (1965) as being within the Hell Creek Formation. Both these localities are in channel deposits that contain dinosaur fragments, particularly teeth, and such material may have been substantially reworked. However, *in situ* dinosaur remains were also found in non-channel deposits above this lignite. The lignite was traced over about 1 sq. mi. (2.6 sq. km) and except for one place where it was cut by a channel deposit, the lignite was consistently present. Based on the definition of the Hell Creek–Tullock contact suggested below, such localities as Bug Creek Anthills and Ken's Saddle, plus *in situ* dinosaurs, could be considered to be within the Tullock Formation. The examination of this lignite was of such cursory and limited scope that such a step is not taken here. However, whatever future investigations of this question may show, I fully agree with Sloan and Van Valen that these localities are latest Cretaceous (assuming that the presence of dinosaurs is used to define Cretaceous in this region).

For several reasons, a formal redefinition (although there may never have been a formal definition) of the Hell Creek–Tullock formational contact is not attempted here. Although much of the mapping and stratigraphic work presented here was done in the type area of the Hell Creek Formation, no stratotype has ever been designated. Further, a stratotype cannot be designated in the present study because a complete section through the Hell Creek

Formation was never measured. Therefore, the following definition is to be considered only as a suggestion utilized here.

As noted previously, the upper Hell Creek Formation is dominated by light to medium gray siltstones in its type area along Hell Creek, as in all other areas examined in Garfield County. As will be described in the following section, the Tullock Formation is distinctly more brown to yellow in appearance; it also includes numerous lignite beds. Particularly when the two formations can be observed together, it is a simple matter to distinguish between them. Within the valley of Hell Creek is a zone of 20 to 40 ft. (6.1–12.2 m) of variegated beds, alternating between brown or tan and gray sands and silts. This zone is above the gray siltstones of the Hell Creek Formation and below the more brown to yellow Tullock Formation. A 4 ft. to 8 ft. (1.2–2.4 m) laterally persistent lignite caps this zone. Two to four lignites, the thickest measuring 3 ft. (0.9 m), are present within the zone of variegated beds, and some have also been found lower within the upper portion of the dominantly gray siltstones. The base of the lowest laterally persistent lignite is taken to be the Hell Creek–Tullock formational contact.

From its type area along Hell Creek, the name "Hell Creek Formation" has been extended to similar units throughout eastern Montana, western North and South Dakota, and northernmost Wyoming. Although coal is used to mark the upper contact of this formation in these various regions, it would be absurd to assume that it is the same coal bed throughout. There may well have been an interval in which the regional inception of coal deposition occurred, possibly due to fluctuations of the intracratonic seaway (the Bearpaw Shale and the Cannonball Formation). However, there is no reason to believe that this began over the whole region at the same time. It may be speculated that the inception of coal deposition began in a broad north-south band in western North and South Dakota in part as a response to a standstill or small transgressive pulse of the intracratonic seaway (probably the base of the Cannonball Formation). With time, this band of coal deposition may have spread westward. Whatever the pattern of deposition may have been, however, the inception of coal deposition should be used to mark the upper limit of the Hell Creek Formation. As argued here, this was a diachronous event. Also, the coal used to mark this contact is not the same coal from place to place, as may be the case within the valley of Hell Creek. This will be discussed in the section dealing with fossil localities.

With the removal of dinosaur extinction as the criterion for determining the lowest coal of the Hell Creek–Tullock contact, it can be argued that this lithologic contact should no longer be regarded as the Cretaceous-Tertiary boundary. This topic is discussed more fully in the final section dealing with biostratigraphy, but dinosaur extinction and the lowest coal have so long been linked that a brief comment on the Cretaceous-Tertiary boundary is appropriate here.

Although the Cretaceous-Tertiary time or geochronologic boundary is by definition isochronous everywhere, it can only imperfectly be translated into a time-stratigraphic boundary in the type area in Europe. Even there, considerable controversy still remains. Thus, correlation of this time-stratigraphic boundary to continental sediments in North America remains imprecise and possibly inaccurate.

Techniques such as magnetostratigraphic studies may afford a means of correlation, and E. Lindsay and B. Butler have begun such a study in northeastern Montana. However, their preliminary results already show that paleontological events are on a much finer scale than paleomagnetic events (pers. comm.). Unless an arbitrary decision is made to place the boundary between two magnetozones, location of the boundary must still rely on paleontological evidence. All paleontological evidence should be used to determine this boundary, but dinosaur extinction remains the most easily recognized feature in field studies. Thus, this is the definition used here—with the full realization that dinosaur extinction may have been diachronous in Garfield County. One result is that approximately the upper 6 to 20 ft. (1.8–6.1 m) of the Hell Creek Formation lacking dinosaurs is now Paleocene by definition.

Tullock Formation. Rogers and Lee (1923) named this unit the "Tullock member" and included it as the upper member of the Lance Formation. It is typified by exposures in the valley of Tullock Creek, Treasure County, Montana, approximately 70 mi. (112.7 km) south of Hell Creek. In the type area the reported thickness is 300 ft. (91.4 m). Rogers and Lee (1923) placed the lower boundary of this unit at the base of the lowest coal. Collier and Knechtel (1939) recognized the Tullock Member of the Lance Formation in McCone County (the county immediately east of Garfield County). They reported about 165 ft. (50.3 m) for the thickness of the unit. Unlike Rogers and Lee, they placed the lower boundary above the lowest coal.

The "Lexicon of Geologic Names" (Wilmarth, 1938) elevated the Tullock member to the rank of formation. Current U.S.G.S. usage considers it to be a member of the Fort Union Formation. The present report follows Wilmarth (1938), and Sloan and Van Valen (1965) in recognizing it as formation of the Fort Union Group.

Although B. Brown (1907) did not apply names to the lignites, geologists preparing coal reports for the U.S.G.S. have given informal names to these beds. In some reports, such as that of Rogers and Lee (1923), the lowest coal was termed "bed A," with successively higher coals receiving the designations "bed B," "bed C," etc. In the coal report of Collier and Knechtel (1939), the lowest coal was termed "bed Z," with higher coals receiving letter designations in reverse alphabetical order. They placed the upper boundary of their Tullock Member at the base of bed U, and bed U was the lowest unit in their overlying Lebo Shale member.

Using informal names is a very practical convention for discussing the large number of lignites within the Tullock Formation. However, as discussed previously, specific coals or even groups of coals probably do not correspond between isolated outcrops, particularly over such a large area as northeastern Montana. The informal naming of coal beds in the present study follows the reverse alphabetic designations utilized by Collier and Knechtel (1939), except that the lower boundary of the Tullock Formation is placed at the base of the lowest lignite (Z coal) and not, as was done by these authors, at its top. It is realized that true equivalency of the various lignites in Garfield and McCone counties can only be established by physically tracing these beds.

In addition to the Z coal, the Tullock Formation includes a number of other relatively persistent coals. Like the Z, these coals are usually lignitic. (In this study the terms *lignite* and *coal* will be used interchangeably.) W. Rohrer of the U.S.G.S. has mapped portions of the coals in Garfield County (oral and written comm.) and has kindly provided his criteria for recognition of these units. The V, W, X, Y, and Z coals are included in the Tullock Formation. In the Hell Creek drainage, the W, Y, and Z are usually well developed. The V and X are much thinner and in places appear to be absent. One procedure used by W. Rohrer, but not followed here, is to give a combined alphabetic designation to coals that fall between those with single alphabetic designations: thus, a coal occurring between the X and the Y would be called the XY coal. The procedure used in the present study has been to name a series of coals by a single alphabetic designation. This means that in many instances a coal such as the Z is actually composed of several lignites that may be separated by as much as 45 ft. (13.7 m) of intervening sediment. In such cases the lignites included under the Z coal are referred to as the upper or lower Z coal (i.e., $\bar{Z}$ and $\underline{Z}$ coal), with intervening lignites receiving no special designation.

In addition to the presence of lignites, the Tullock Formation can be distinguished from the subjacent Hell Creek Formation by the occurrence of more beds of a brown to yellow color. Detailed studies comparing the lithology of the upper Hell Creek and Tullock formations have not been made. Based on field observations, it appears that the clastic component of the Tullock Formation does not differ greatly from the Hell Creek Formation. However, locally the Tullock Formation does include some large sandbodies of fluvial origin; these will be discussed extensively in the following section. The brown to yellow coloration prevailing in the Tullock Formation is probably due to more iron oxide and carbonaceous material than is present in the Hell Creek Formation.

Although the Tullock is clearly distinguishable from the underlying Hell Creek Formation, there is no suggestion of an unconformity at their boundary except for possibly very localized events. These local phenomena appear to be related more to the lenticular nature of the beds in the Hell Creek Formation than to any major hiatus. In the Tullock Formation, beds tend to be traceable for much greater distances than is the case in the Hell Creek Formation.

In the valley of Hell Creek, the Tullock Formation has been measured in several places and its maximum thickness there reaches 305 ft. (93.0 m), with a minimum thickness of about 290 ft. (88.4 m). This is close to the thicknesses reported by Rogers and Lee (1923) for the type area but contrasts markedly with the 165 ft. (50.3 m) given by Collier and Knechtel (1939) for the formation in McCone County. Unless drastic errors in measurement or in the correlation of equivalent coal units have been made, this suggests a thinning of the formation to the east.

Lebo Formation. This formation is typified along Lebo Creek northeast of the Crazy Mountains in southcentral Montana (Stone and Calvert, 1910). Thom and Dobbin (1924) report that it varies from a "tongue of andestic rocks" in its western exposures to "dark clays and fluviatile sandstones" in eastern Montana. According to these authors, "yellow strata and coals are interbedded with the dark shales" east of Miles City, Montana, which is approximately 100 mi. (161 km) southeast of the present area of study. To the east of Garfield County, in McCone County, Collier and Knechtel (1939) described the Lebo as consisting "of thick beds of white sandy clay and brownish sandstone, cross-bedded in places, alternating with nearly black clay shale." They report that the only persistent coal in the unit is the U, the base of which defines the boundary with the subjacent Tullock Formation.

The formation is exposed in various places in Garfield County, particularly in the eastern half. In the headwaters of Hell Creek, only the basal U coal and 10 to 20 ft. (3–6.1 m) of overlying yellow silt and sandstones are present. Vertebrate fossils have yet to be found in the Lebo Formation of Garfield County.

The coal reports of the U.S.G.S. refer to the Lebo as a member of the Fort Union Formation. The present report follows Simpson (1937a) and Sloan and Van Valen (1965) in referring to the Lebo as a formation that is included in the Fort Union Group.

FOSSIL LOCALITIES, SEDIMENTARY ENVIRONMENTS, AND STRATIGRAPHIC CORRELATION

Over 200 fossil vertebrate localities have been discovered in Garfield County. By far the greatest percentage of the localities are from the Hell Creek and Tullock formations. Most have produced very little vertebrate material. Table 1 lists all localities from the Hell Creek and Tullock formations discovered (or catalogued) from 1972 through 1976 that at least questionably have produced mammals. In addition, W. A. Clemens has called my attention to four localities discovered in 1977. These are also given in Table 1. Coordinates for the localities are not given here but are on file at the University of California Museum of Paleontology. All localities preceded by the designation "V" are UCMP vertebrate localities. Locality numbers preceded by "LACM" are Los Angeles County Museum of Natural History localities.

The paleontological sampling conducted in this study spans at the greatest only about 800 ft. (244 m) of strata, and over most areas only about 400 ft. (122 m). Nevertheless, within this rather short stratigraphic interval mammals were found that were referable to two long-established North American Land-Mammal "Ages," Puercan and Torrejonian; to two more recently proposed "Ages," Lancian and mantuan; and to an interval of time (pre-mantuan) that is faunally transitional between Lancian (i.e., Bug Creek faunal-facies) and mantuan, based on the current definitions of these two "Ages." In addition, the localities yielding Lancian mammals were subdivided into the Bug Creek faunal-facies and Hell Creek faunal-facies, based on lithological and faunal criteria.

Definitions of the faunal-facies and comments on the usages of the "Ages" are given in the chapter on Techniques of Study. The "Ages" and faunal-facies are used in the present section only to clarify from which stratigraphic levels and lithologies the mammals have come. Faunal analyses and biostratigraphic correlations are given in the chapter on Biostratigraphy. Although the stratigraphic relationships of all the mammal-bearing localities will be discussed, only the systematics of Lancian, pre-mantuan, and mantuan

TABLE 1. UCMP mammal localities from the Hell Creek and Tullock formations, Garfield County, Montana, discovered (or catalogued) from 1972 through 1976, plus four localities discovered in 1977. Numbers under "Area" are the areas of study discussed in the text. A key to these numbers is to be found in Fig. 2. Double entries under stratigraphic position indicate level of locality and level of channel capping, respectively, above or below the Z coal. The Garbani Quarry includes nineteen UCMP localities, which were the only sites of major hand-quarrying operations in Garfield County.

Abbreviations:

Lithologic occurrence:
- m mudstone or carbonaceous shale
- si siltstone
- sis silty sandstone or sandy siltstone
- ss sandstone
- ch channels over 10 ft. (3 m) thick

Type of sampling:
- s surface
- w screen-washing
- q hand-quarrying

Series/mammal age or level:
- UC Upper Cretaceous
- lP lower Paleocene
- mP middle Paleocene
- Lh Lancian Hell Creek faunal-facies
- Lb Lancian Bug Creek faunal-facies
- pm pre-mantuan
- m mantuan
- P Puercan
- T Torrejonian

Locality number	Locality name	Area	Feet (m) above (+) below (−) base of Z coal	Lithologic occurrence	Type of sampling	Series/ mammal age or level
—	Garbani Quarry	1	+90 to 95 (27.4 to 29.0)	si, ss, sis	s, w, q	lP/P
V72200	Fred's Place	1	−?	ss	s	UC/Lh
V72201	MacDonald Locality	1	+95 (29.0)	sis, ch	s, w	lP/P
V72204	Flint South		−?	sis	s	UC/Lh
V72206	*Meniscoessus* Skull		−?	si	s	UC/Lh
V72207	Windy Hill		−40 (12.2)	ss, ch	s	UC/Lh
V72208	Lead Slug Hill 1		low in Hell Creek	sis	s, w	UC/Lh
V72210	McKeever Ranch 1	2	+1 (0.3)	ss, ch	s, w	lP/m
V72211	Billy Creek 1	2	−24 (7.3)/ +17 (5.2)	ss, ch	s,w	lP/pm?
V72212	Engdahl Anthills 1		> −150 (45.7)	ss	s,w	UC/Lh
V72213	Trumbo's Anthill	1	−54 (16.5)	ss, ?ch	s, w	UC/Lh
V72215	Baldy Butte	1	−98 (29.9)	?	s	UC/Lh
V73051	Fred Fitch Ranch		−?	?ss	s	UC/Lh
V73076	Engdahl Anthills 2		> −150 (45.7)	?	s, w	UC/Lh
V73077	Engdahl Anthills 3		> −150 (45.7)	ss	s, w	UC/Lh
V73078	Engdahl Anthills 4		> −150 (45.7)	ss	s, w	UC/Lh
V73079	Harley's Lizard Anthill		−?	?	s	UC/Lh
V73083	Flat Creek 1	3	−34 (10.4)	?sis	s, w	UC/Lh
V73084	Flat Creek 2	3	−34 (10.4)	sis	s, w	UC/Lh
V73085	Flat Creek 3	3	−17 (5.2)	sis	s, w	UC/Lh
V73086	Flat Creek 4 (no mammals)	3	−17 (5.2)	si	s, w	UC/Lh
V73087	Flat Creek 5	3	−17 (5.2)	sis	s, w	UC/Lh
V73088	Fig Locality		−34 (10.4)	ss	s	UC/Lh
V73089	Gynnoskus Locality		−?	?ss	s	UC/Lh
V73090	Ried Coulee 1		−?	ss	s	UC/Lh
V73091	Ried Coulee 2		−?	?ss	s	UC/Lh

Locality number	*Locality name*	*Area*	*Feet (m) above (+) below (−) base of Z coal*	*Lithologic occurrence*	*Type of sampling*	*Series/ mammal age or level*
V73093	Jordan Coulee		− ?	ss	s	UC/Lh
V73094	Horsethief Canyon 1	4	> + 200 (70.0)	ss, ch	s, w	mP/T
V73095	Horsethief Canyon 2	4	> + 200 (70.0)	ss, ch	s, w	mP/T
V73096	Biscuit Butte	1	+ 120 (36.5)/ + 170 (51.8)	sis, ch	s, w	lP/P
V73097	Brownie Butte	1	− ?65 (19.8)	sis	s	UC/Lh
V73098	Rick's Place	1	− 40 (12.2)/ − 14 (4.3)	?sis, ch	s	UC/Lb?
V73099	Rick's Place 2	1	− ?	?	s	UC/Lb?
V73102	*Didelphodon* Locality		low in Hell Creek	sis	s	UC/Lh
V73103	Dry Fork Coulee		− ?	ss, m	s	UC/Lh
V74106	Harmonica Hollow		− 34 (10.4)	ss	s	UC/Lh
V74107	Mexico Place		− 20 (6.1)	ss	s, w	UC/Lh
V74110	Hell's Hollow Channel	1	+ 7 (2.1)/ + 12 (3.7)	ss, ch	s, w	lP/pm
V74111	Worm Coulee 1	1	+ 7 (2.1)/ + 12 (3.7)	ss, ch	s, w	lP/pm
V74116	Baldy Butte SW–Blazing Saddle	1	− 92 (28.0)	sis	s, w	UC/Lh
V74117	Baldy Butte SW–Top Level	1	− 82 (25.0)	si	s, w	UC/Lh
V74121	Supra Z (?mammals)	1	+ 3 (0.9)	si	s	lP/?
V74122	Biscuit Springs	1	+ 115 to 120 (35.1 to 36.6)	sis, ch	s, w	lP/P
V74123–28	Yellow Sand Hill 1–6	1	+ 115 to 120 (35.1 to 36.6)	si, ss, ch	s, w	lP/P
V75157	Haley Locality		− ?	m	s, w	UC/Lh
V75159	Haley Anthill		− ?	?	s, w	UC/Lh
V75160	Dry Fork, Coulee–North		− ?	?sis	s	UC/Lh
V75161	Rex Hill Locality		− ?	si	s	UC/Lh
V75162	Worm Coulee 5	1	− 6 to 10 (1.8 to 3.0)	sis	s, w	UC/Lh
V75168	Windmill 1 (?mammals)	1	> − 90 (27.4)	sis	s	UC/Lh
V75169	Windmill 2	1	> − 90 (27.4)	?	s	UC/Lh
V75170	Windmill 3	1	> − 90 (27.4)	sis	s	UC/Lh
V75171	Windmill Anthill	1	> − 90 (27.4)	?	s	UC/Lh

Locality number	Locality name	Area	Feet (m) above (+) below (−) base of Z coal	Lithologic occurrence	Type of sampling	Series/ mammal age or level
V75173	Billy Creek 2	2	> −?80 (24.4)	ss	s	UC/Lh
V75177	Pipeline Anthill		−?	?	s	UC/Lh
V75178	Wild Horse Basin		−25 (7.6)	ss	s, w	UC/Lh
V75179	Brownie Butte Anthill	1	−?	?	s	UC/Lh
V75184	Harley's East Anthill 1	1	+?	?	s	lP/P?
V75185	Harley's East Anthill 2	1	+?	?	s	lP/P?
V75190	Biscuit Butte–Dave's Ext. 1	1	+120 (36.5)/ +170 (51.8)	ss, ch	s	lP/P
V75191	Biscuit Butte–Dave's Ext. 2	1	+120 (36.5)/ +170 (51.8)	?ss, ch	s	lP/P
V75192	Horsethief Canyon (South Ext.)	4	> +200 (70.0)	ss, ch	s, w	mP/T
V75193–95	Yellow Sand Hill Localities 7–9	1	+115 to 120 (35.1 to 36.6)	si, ss, ch	s, w	lP/P
V75196–98	Kerr Butte 1–3	3	+210 (64.0)	sis	s, w	lP?/P?
V76136	Custer's Memorial Anthill	1	+?	?	s	lP?/pm?
V76137	Hell's Hollow–East End	1	−?	?si, m	s	UC/Lh
V76138	Clam Hollow	1	−115 (35.0)	m	s	UC/Lh
V76144	Crazy Bone		−12 (3.7)	sis	s	UC/Lh
V76145–46	Flint's Stones and Ext.		−?	ss, sis	s	UC/Lh
V76149	Barney's Rubble–Mammal		−?	si	s	UC/Lh
V76150	Flint Basin–Headwater		−50 (15.2)	si	s	UC/Lh
V76155–58	Penick Coulee 1–4		−?	ss	s	UC/Lh
V76161	Penick Coulee 7		−?	?	s	UC/Lh
V76166–67	Cole Anthill 1 and 2		−?	?	s	UC/Lh
V76170	Mosquito Gulch	1	> +200 (60.1)	ss, ch	s, w	mP/T
V76171	Two Mosquitos	1	> +200 (60.1)	ss, ch	s, w	mP/T
V77087	McKeever Ranch, Harley's Point	2	+1 (0.3)	sis, ch	s, w	lP/m
V77128	Morales 1		−2.5 (0.8)	si	s, w	lP/pm?
V77129	Herpijunk Promontory		−4.5 (1.4)	si	s, w	lP/pm?
V77130	Hauso 1		−40 to 50 (12.2 to 15.2)	sis	s, w	UC/Lh

mammals are given in the present study. The Puercan and Torrejonian mammals will be described elsewhere by W. A. Clemens.

Field studies were conducted in various portions of Garfield County. However, as is generally the case, the infrequent stratigraphic and geographic pattern of the fossils largely dictated the choice of areas of study within the county. Stratigraphic correlations among the various study areas or isolated localities required a much greater degree of extrapolation than those within the study areas, where stratigraphic relationships could often be traced. The following discussion is subdivided into the four geographic areas of study. These areas are shown in Fig. 2. Only the localities listed in Table 1 that have produced moderately large surface or underwater screening collections (or are of interest for some other reason) are discussed. A schematic summary of the stratigraphic relationships of the localities is given in Fig. 72.

Hell Creek–Snow Creek area. This is the largest and most extensively examined study area. The area covers approximately 32 sq. mi. (80 sq. km) including, in the eastern and central portion, part of B. Brown's type area of the Hell Creek Formation along the head of Hell Creek, and in the western portion, the head of Snow Creek (Hell's Hollow). As can be seen in the map of this area (Fig. 4), lignites can be traced throughout, the major exception being the floodplain of Hell Creek. Small outcroppings of the lower Z coal ($\underline{Z}$) can be found at various places within the floodplain. Within or near the Hell Creek–Snow Creek area, localities from the Hell Creek Formation are referred to both the Bug Creek and the Hell Creek faunal-facies Lancian and to the pre-mantuan, while localities from the Tullock Formation are referred to the pre-mantuan, Puercan, and Torrejonian.

Over 50 Lancian UCMP localities in Garfield County have produced at least one mammalian specimen and probably belong to the Hell Creek faunal-facies. Unfortunately, most of these localities are not very rich and their inclusion in the Hell Creek faunal-facies is based more on the absence of Bug Creek faunal-facies mammals than on the presence of an abundant sample of mammals typical of the former facies.

Within or near the Hell Creek–Snow Creek area, eighteen Hell Creek faunal-facies localities have been discovered, most producing only a few specimens. Like other localities of this faunal-facies in Garfield County, these localities are usually restricted to small, silty channel deposits and, on occasion, occur in dark mudstones that suggest quiet, stagnant water conditions.

In the Hell Creek–Snow Creek area, Hell Creek faunal-facies localities and dinosaurs occur throughout the stratigraphic extent of the Hell Creek Formation except for the upper 6 to 10 ft. (1.8–3.0 m). Although other portions of the county have not as yet been as thoroughly investigated as the Hell Creek–Snow Creek area, this 6 to 10 ft. (1.8–3.0 m) limit for Hell Creek faunal-facies localities and dinosaurs appears to hold true for other areas as well. Most of the stratigraphically very high dinosaur remains are skeletal fragments or teeth, some of which are found at Hell Creek faunal-facies localities. The highest *in situ* partial skeleton of a dinosaur, located in the northeastern quarter of Garfield County, is 11 ft. (3.4 m) below the base of the Z coal. This partial skeleton consists of a fragmentary skull and possibly some vertebrae of *Triceratops* sp.

Within the headwaters of Hell Creek it is difficult to measure the stratigraphic position of most localities lower than about 50 ft. (15.2 m) below the Z coal because of lateral distances of up to a mile (1.6 km) between such localities and the nearest exposure of the formational contact. Fortunately, three of the stratigraphically lower Hell Creek faunal-facies localities in this area are situated below two isolated buttes which preserve beds that

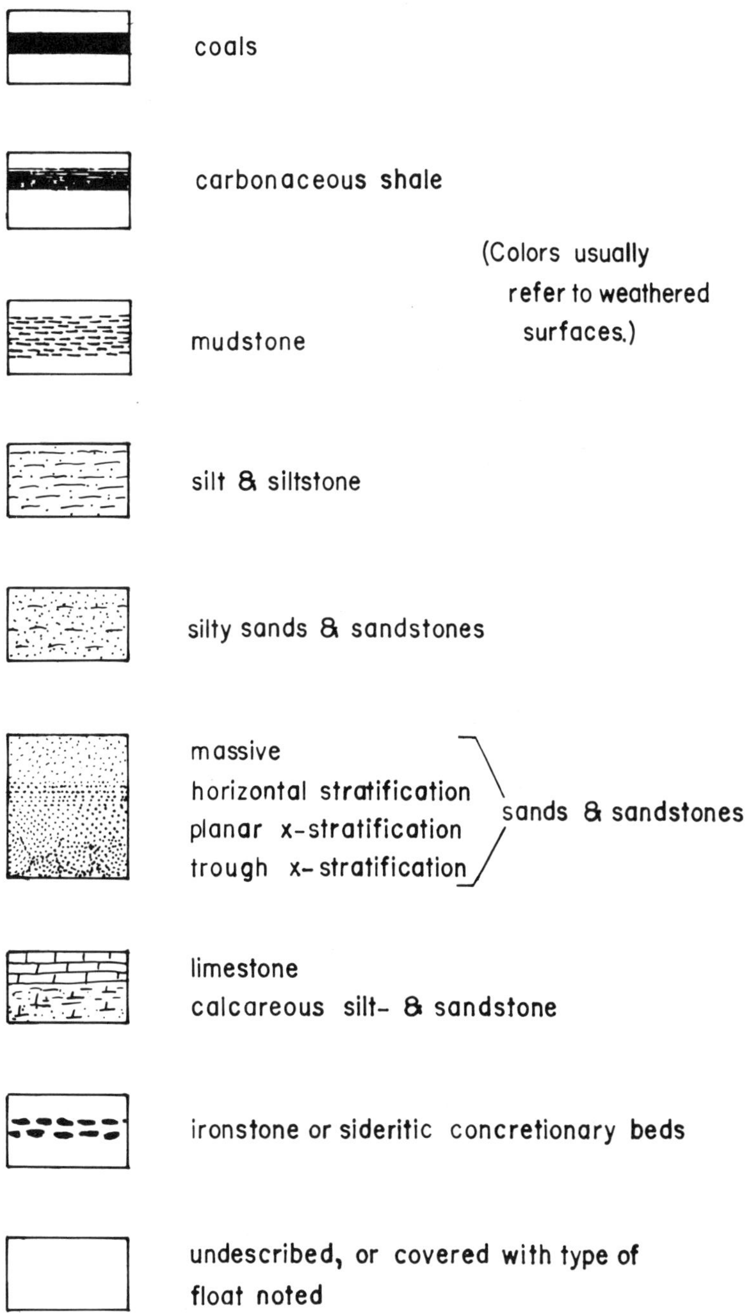

FIG. 3. Explanation of lithologic symbols used in all the stratigraphic sections (Figs. 5, 6, 8, and 9).

can be correlated to the main badlands exposures (see Baldy Butte, Fig. 4). The beds capping these two buttes appear to be remnants of the Hell's Hollow channel deposit (see discussion, below, of this deposit). Using these beds, the positions below the Hell Creek-Tullock contact of the three localities can be estimated as follows: Baldy Butte (V72215), 98 ft. (29.9 m); Baldy Butte SW–Blazing Saddle (V74116), 92 ft. (28.0 m); and Baldy Butte SW–Top Level (V74117), 82 ft. (25.0 m) (see stratigraphic section S5, Fig. 5). Another stratigraphically low Hell Creek faunal-facies locality, Clam Hollow (V76138), was found closer to the main badlands exposures in Hell's Hollow. This locality is in a gray mudstone extremely rich in freshwater molluscs and is about 115 ft. (35.0 m) below the Z coal.

A group of four Hell Creek faunal-facies localities north of the area mapped in Fig. 4 are probably stratigraphically lower than the localities described above. The stratigraphic position of these localities, Engdahl Anthills 1 (V72212) and Engdahl Anthills 2 through 4 (V73076–78), can only be crudely estimated by comparing elevations. This gives a figure of 150 to 200 ft. (45.7 to 61.0 m) below the Hell Creek–Tullock contact.

Two Hell Creek faunal-facies in the eastern portion of the mapped area in Fig. 4 have produced small mammalian samples. The first is Brownie Butte (V73097). This locality was discovered by Albert Silberling, whose collection is now at Princeton University (R. E. Sloan, pers. comm.). Collections from this locality were also made by the University of Minnesota, St. Paul Science Museum, and by the University of Michigan. Sloan and Van Valen (1965) mentioned this locality and indicated that it is 65 ft. (19.8 m) below the Tullock Formation. The second locality, Trumbo's Anthill (V72213), is 54 ft. (16.5 m) below the Tullock Formation.

The stratigraphically highest Hell Creek faunal-facies locality in the Hell Creek–Snow Creek area, and possibly in the whole county, occurs in Hell's Hollow. The exact position of this locality, Worm Coulee 5 (V75162), is uncertain; the figure given in Table 1 of 6 ft. (1.8 m) below the Hell Creek–Tullock contact is only an estimate. The locality occurs in the sandier portion of a gray siltstone. It has produced only a few mammalian tooth fragments but has abundant remains of lower vertebrates, including dinosaurs, common at Hell Creek faunal-facies localities. Immediately above the locality is a grass-covered slope composed of yellow silt and sand. This sand and silt is very similar to that seen in the Hell's Hollow channel deposit. As will be discussed below, this channel deposit is best considered to be within the Tullock Formation (not the Hell Creek Formation, as indicated by Archibald, 1979a). In places where the channel deposit is thickest it is below or at the level of the lower Z coal (see section S2 of Fig. 5). It is not clear whether the channel cut the lower Z or whether the coal is simply lacking in certain places. The lower Z coal could not be found above locality V75162, and the yellow sands and silts may be part of the Hell's Hollow channel deposit. However, the yellow sands and silts appear to be reworked. The most likely possibility is that they are Quaternary alluvia derived from the Hell's Hollow channel deposit. The estimated stratigraphic position of locality V75162 was obtained by a hand-level sighting across a 300 ft. (91.4 m)-wide coulee to good exposures of the lower Z coal and the Hell's Hollow channel deposit. A very remote possibility remains that this locality may be above the lower Z coal.

Subsequent to my work in Garfield County, UCMP crews under the direction of W. A. Clemens have extended mapping and exploration to the west of Hell's Hollow. One of several potentially important localities found by these crews is a Hell Creek faunal-facies locality, Hauso 1 (V77130). The locality occurs in a finely grained silty sandstone and is between 40 and 50 ft. (12.2–15.2 m) below the lower Z coal (Clemens, pers. comm.). Only a few atypical teeth of *Meniscoessus* and a few important teeth of *Essonodon* from this locality are described here.

Bug Creek faunal-facies localities are only questionably represented in Garfield County by two localities on the eastern edge of the Hell Creek–Snow Creek area. Rick's Place (V73098) produced two condylarth jaws (only one dentulous jaw). The locality is in a yellow-brown, silt-and-sandstone channel deposit capped by light gray siltstone 14 ft. (4.3 m) below the Z coal. The fossil-bearing strata are about 40 ft. (12.2 m) below the Z coal. Rick's Place 2 (V73099) produced one edentulous jaw that may belong to a condylarth. The stratigraphic position of this site was not determined.

In preliminary reports (Archibald, 1979a; Archibald and Clemens, 1977; and Clemens and Archibald, 1980) several localities, particularly Hell's Hollow Channel (V74110) and Worm Coulee 1 (V74111), were referred to the Bug Creek faunal-facies either by name or by inference. A reinvestigation of the stratigraphic data shows that some of the localities are best considered to be within the Tullock rather than the Hell Creek Formation. The time-stratigraphic placement (see chapter on Biostratigraphy) of these localities is still open to question, but they are considered here to be pre-mantuan Paleocene rather than Lancian (Bug Creek faunal-facies). These localities and their faunas provide an example of the confusion that can arise from failure to examine stratigraphic versus time-stratigraphic placement carefully. In view of this, it is worthwhile to examine more closely the correlations of the Z coal within the Hell Creek–Snow Creek area in conjunction with the descriptions of the pre-mantuan localities.

On the southern and eastern side of the area mapped in Fig. 4 (also see Fig. 5), the Z coal is composed of two lignites, a lower 1 to 3 ft. (0.3–0.9 m) lignite, and an upper 4 to 6 ft. (1.2–1.8 m) lignite, separated by 20 to 25 ft. (6.1–7.6 m) of gray and tan to brown variegated siltstones. In places, a middle carbonaceous shale or lignite is also present. On the western and northern side of the drainage and in Hell's Hollow, only one conspicuous 6 to 8 ft. (1.8–2.4 m) lignite is clearly present near the base of the Tullock Formation. Because of the intervening Hell Creek floodplain, no lignites can be traced with certainty between these two areas; however, based on the thicknesses of the lignites, relative elevations, and super- and subjacent lithologies, the single conspicuous lignite in Hell's Hollow correlates with the upper Z on the eastern side of the Hell Creek floodplain.

In earlier phases of this study it was believed that the lower Z seen on the eastern side of Hell Creek was absent on the western side in Hell's Hollow. The strongest argument for this was based on matching lithologies between the two areas. On the eastern side of Hell Creek, the lower Z coal is usually underlain by light gray siltstone with only occasional brown, tan, white, or greenish-gray bands. The lower Z coal is in turn overlain by 20 to 25 ft. (6.1–7.6 m) of gray and tan to brown variegated siltstones. The upper Z coal is superjacent to this unit. A similar sequence is present in Hell's Hollow except that no lower Z coal was found, which suggests that the lower Z coal is absent. The lower contact of the Tullock Formation in Hell's Hollow was placed at the base of the 15 to 20 ft. (4.6–6.1 m) tan and gray variegated siltstones. The assumption was made that these beds are the same as those which overlie the lower Z coal on the eastern side of the drainage. Using this definition, the important fossil-bearing Hell's Hollow channel deposit was believed to be within the Hell Creek Formation, because its channel capping lies about 25 ft. (7.6 m) below the tan and gray variegated siltstones (see, for example, Archibald, 1979a).

Subsequently, during the 1977–1978 summer field seasons, crews under the direction of W. A. Clemens found a series of small (measured in inches) lignite beds below the tan and gray variegated siltstones in Hell's Hollow. It was strongly suspected that the lowest of these lignites should be considered to be the base of the Tullock Formation. Based on a reexamination of Hell's Hollow in 1979, I concur with this finding. Although up to six minor lignites have been located below the upper Z coal in Hell's Hollow, only the lowest couplet of lignites will be discussed, as these are the lignites that are useful in redefining the Hell Creek–Tullock contact in Hell's Hollow.

The lowest lignite in Hell's Hollow ranges from 45 to 50 ft. (13.7–15.2 m) below the upper Z coal. It varies in thickness from 3 to 10 in. (7.6–25.4 cm) and in lithology from a dark gray mudstone to pure lignite. Strictly speaking, the name "lower Z coal" is a misnomer for this unit when it changes laterally from a lignite to a mudstone. Although such lateral variations should be noted, for stratigraphic work it is much more important to recognize the geographic extent of a bed than to become embroiled in discussions as to whether the informal name "lower Z coal" is appropriately applied to beds that are not lignitic throughout their lateral extent. A more striking example of the lateral variation in the lithology of a "lignite" is also found in Hell's Hollow. On the southern and western sides of Hell's Hollow, the upper Z coal has a very high lignitic content. In contrast, on the northern side the same bed is almost devoid of lignite, although the unit is no thinner than on the south side of the hollow.

It is sometimes difficult to locate the lower Z coal in Hell's Hollow, but the task is aided by the presence of a distinctive 1.5 to 5 ft. (0.5–1.5 m) whitish-gray mud-siltstone that immediately overlies the lower Z. In most places within Hell's Hollow this unit is in turn overlain by a 2 to 4 in. (5.1–10.2 cm) lignite or carbonaceous shale.

Following the criteria presented earlier in the discussion on determining the Hell Creek–Tullock formational contact, the lowest lignite (or even mudstone, in some places) in Hell's Hollow is considered to be the Hell Creek–Tullock contact. It should be clear from the preceding discussion that the lower Z coal may not be the same bed on the eastern as on the western side (Hell's Hollow) of Hell Creek. As was noted previously, the first work in Hell's Hollow suggested that the lignite called the lower Z coal on the eastern side of Hell Creek may be absent in Hell's Hollow. Even if this were shown to be the case, the Hell Creek–Tullock contact should still be based on the lowest lignite, as is being done in the present study. To repeat an earlier statement: there is no basis for assuming that a given coal bed is continuous over a broad region even though for convenience such beds have been given the same informal name (e.g., the "Z coal"). Scanning the stratigraphic sections in Fig. 5 and disregarding the informal names, it should be clear that the numbers, thicknesses, and stratigraphic placement of lignites vary considerably.

A final example demonstrating the need to recognize the discontinuity of the lignites also comes from the lowest lignite (the lower Z coal) in Hell's Hollow. In two separate instances, the lowest lignite (and the overlying whitish-gray mud-siltstone and second lowest lignite) are absent in Hell's Hollow. The first instance is straightforward. In 1978, a channel deposit was discovered just to the west of the area mapped in Fig. 4 and was dubbed the "Padgett channel deposit." This channel deposit has subsequently been traced laterally into Hell's Hollow. In one place (see stratigraphic section S1 in Fig. 5) the Padgett channel deposit clearly cuts through the lower Z coal. This type of relationship is not always clearly recognizable in the field, but once it has been found it is clearly seen as an example of cut-and-fill. The second instance has not been fully explained. Immediately adjacent to the measured section S2 (see Figs. 4 and 5), the lower Z coal and the overlying whitish-gray mud-siltstone and next higher lignite appear to have been cut by the other channel in Hell's Hollow, the Hell's Hollow channel deposit. When a fresh surface was exposed, it seemed more reasonable to suggest that the lower Z coal and the aforementioned superjacent units pinched out before the margin of the channel deposit was reached. Further examination in Hell's Hollow showed that the lower Z coal was absent in all places where the Hell's Hollow channel deposit thickened to form the channel proper. However, there was no instance where the channel clearly cut the lower Z coal. A reasonable explanation for this relationship may be that the channel began to form, or was forming, during the deposition of the lower Z coal in Hell's Hollow.

From a paleontological perspective, the Hell's Hollow channel deposit is the more important of the two channel deposits found in Hell's Hollow. The observed maximum thickness of the channel deposit is about 15 ft. (4.6 m) and in this portion the deposit is comprised of well-indurated, cross-stratified, red sandstone exposed in a series of outcrops running west to east in Hell's Hollow. As noted, the lower Z coal is absent around this portion of the channel deposit. Laterally, the channel deposit thins to as little as 1 ft. (0.3 m) of yellow sands and silts that can be traced for up to 0.5 mi. (0.8 km) on either side of the channel. The channel deposit appears to lie wholly within the Tullock Formation (*contra* Archibald, 1979a). The channel capping is about 39 ft. (11.9 m) below the upper Z coal and at its thickest portion the base of the channel coincides with or is slightly lower than the lowest Z coal. Two important mammal-producing localities have been discovered within the Hell's Hollow channel deposit, Hell's Hollow Channel (V74110) and Worm Coulee 1 (V74111). The mammals from these two localities have been included in the Hell's Hollow local fauna. Both localities are between 42 and 44 ft. (12.8–13.4 m) below the upper Z coal and 7 ft. (2.1 m) above the lower Z coal. Neither of the sections in which the localities occur appears to preserve the channel capping. However, this level can easily be found by lateral tracing and both localities are about 6 ft. (1.8 m) below the capping. The localities are probably Paleocene in age. The mammals suggest an age intermediate between Lancian and mantuan (as these Land Mammal "Ages" are presently defined) and the informal term *pre-mantuan* is used.

Another pre-mantuan locality, Herpijunk Promontory (V77129), was found in 1977 just west of Hell's Hollow. To date, only one identifiable condylarth dentary has been recovered, and thus the suggestion of a pre-mantuan age is very tentative. Unlike the pre-mantuan localities in Hell's Hollow, locality V77129 occurs in the uppermost Hell Creek Formation, 4.5 ft. (1.4 m) below the lower Z coal. The locality also appears to differ in that it occurs in a gray-brown siltstone with abundant lignite, plant impressions, and iron-stained stringers of yellow, silty sand rather than in a channel deposit as in other pre-mantuan localities. A more definite age assignment for locality V77129 may be determined based on a nearby locality now under study by W. A. Clemens. This new locality, Morales 1 (V77128), has produced a number of mammals and its sedimentologic setting and stratigraphic position are similar to those of locality V77129.

The second channel deposit within Hell's Hollow, the Padgett channel deposit, was originally discovered in 1978 west of Hell's Hollow. The lateral extent of this channel deposit is still being studied, but tentative stratigraphic relationships to the Hell's Hollow channel deposit have been established within Hell's Hollow (see sections S1 and S2 of Fig. 5). The Padgett channel deposit is capped 23 ft. (7.0 m) below the upper Z coal, 17 ft. (5.2 m) higher than the capping for the Hell's Hollow channel deposit. The Padgett channel deposit may also represent a larger, or at least deeper, stream, as it reaches a thickness of almost 50 ft. (15.2 m). It also cuts into the Hell Creek Formation to a depth of up to 25 ft. (7.6 m). With some difficulty, hand samples were used to differentiate the lithologies of the two channel deposits in the field. The samples from the Hell's Hollow channel deposit tended to be red or light yellow, in contrast to the white or yellow of the Padgett samples.

The sands from the Hell's Hollow channel deposit were more mature: fine-grained, better-rounded, with fewer ferro-magnesium minerals and possibly a higher silt content. The sands from the Padgett were less mature: slightly coarser-grained, less rounded, with more ferro-magnesium minerals and possibly a lower silt content. Only a few lower vertebrates and no mammals have been recovered from the Padgett channel deposit so faunal comparisons are not possible.

With the exception of a few localities, most of the fossil material found in the Tullock Formation of Garfield County has come from the headwaters in the valley of Hell Creek. Also, almost without exception these localities are restricted to large channel-fill deposits. Two stratigraphically superposed channel-fill deposits containing mammals have been recognized in the valley of Hell Creek: the lower of these is the Garbani channel deposit, and the higher is the Farrand channel deposit.

Systematic and biostratigraphic research on the mammals from these channel deposits is being conducted by W. A. Clemens. Based on his work (pers. comm. and Clemens, 1974), the mammals from the Garbani and Farrand channel deposits can tentatively be referred to the Puercan and Torrejonian Mammal "Ages," respectively (approximately early and middle Paleocene).

Outcrops of the Garbani channel deposit are present in six sections (see Fig. 4). In most exposures, the deposits have an obvious cut-and-fill relationship with the surrounding overbank and floodplain deposits. The channel usually cuts into the uppermost lignite of the Y coal. In all places examined it cuts through the X coal. A capping for the channel deposit was found only near the center of Sec. 13, T20N, R26E (see stratigraphic section S13 in Fig. 5). The capping is approximately 23 ft. (7.0 m) below the W coal. The Garbani Quarry (which includes 19 UCMP localities) and the MacDonald Locality (V72201) are located near the base of the channel deposit. Both sites are 90 to 95 ft. (27.4–29.0 m) above the base of the Tullock Formation and, by correlation, are approximately 80 ft. (24.4 m) below the channel capping. The sediment at the Garbani Quarry is composed of poorly sorted fine- to medium-grained sand and silt. Most of the bedding persists laterally for no more than 1 to 2 ft. (0.3–0.6 m), but well-indurated beds of siltstones or stringers of lignite can occasionally be traced further. No cross-stratification is readily apparent, but occasionally some of the beds (e.g., at Harley's High, V73080) show curious "tilting" of as much as 45°. This may be a primary sedimentary structure, but other non-primary causes cannot be dismissed. The lignite stringers are interpreted here as being scours that have been filled by plant detritus. Carbonized plant fragments and seed casts also occur. Vertebrate fossils at this locality range in size from mammal, fish, and other teeth measured in millimeters to limb fragments of crocodilians and champsosaurs up to 15 cm long. Most of the larger material is in the 5 cm range. Much of the postcranial material is worn, indicating transport and abrasion. Some smaller material, such as the mammal teeth, shows postmortem abrasion, but to a lesser degree than the large material. Dentulous jaw fragments of mammals are not uncommon, but only two maxillae have been recovered. This great range in size of fossil material indicates poor sorting, as was noted above for the surrounding sediment.

The fossils and sediment at Biscuit Springs (V74122) indicate a much higher degree of sorting. This locality, as well as the Biscuit Butte (V73096) and Yellow Sand Hill localities (V74123–28, V75193–95), appear to be slightly higher in the channel deposit than the Garbani Quarry or the MacDonald locality. All these localities are approximately 115 to 120 ft. (35.1-36.6 m) above the base of the Tullock and are 50 ft. (15.2 m) below the channel capping.

The sediment at and around Biscuit Springs was observed in some detail. Based on these observations, the Biscuit Springs locality is interpreted as having been deposited at the top of a point bar sequence (see stratigraphic section S14 in Fig. 5).

Below Biscuit Springs, a cliff-forming sandstone is exposed in a natural amphitheater. The lowest 15 to 20 ft. (4.6–6.1 m) are composed of a relatively clean, tan-colored, massive sandstone. At one place, a sharp contact can be seen between the channel sandstone and the abutting overbank deposits. Above the massive sandstone there is a maximum of 8 ft. (2.4 m) of similarly colored sandstone formed into large-scale trough cross-stratification; each set measures about 3 ft. (0.9 m) in thickness. Two strikes and dips of these beds are N83W 21° NE and N30W 25° NE.

At the top of these beds are northeasterly-trending loglike "concretions" about 2 to 3 ft. (0.6–0.9 m) in diameter, with somewhat flattened upper surfaces. Exposed portions of these bodies are up to 20 ft. (6.1 m) long. As was discussed in the description of the Hell Creek Formation, these are not considered to be true concretions. Four of these bodies examined in the Biscuit Springs sequence all show low, angled, planar cross-stratification. Each loglike body appears to be composed of only one set of this type of cross-stratification. The bases of these bodies are not well exposed, so it is unknown whether the lower erosional surfaces are slightly curved. Strikes and dips of these sets measured on three of the bodies are N55W 18° NE, N50W 10°NE, N25W 9° NE, and N53W 14° NE. Very worn ripple marks can be seen on the tops of some of these bodies. At one place where they are less worn, they are straight-crested and probably asymmetrical. Their strike is east-west and the flow was probably to the north. The stratigraphic position, shape, and type of cross-stratification suggest that these loglike bodies are chute bar deposits.

Several other types of sedimentary features are found above and lateral to the chute bar deposits. An approximately 2 ft. × 2 ft. (0.6 × 0.6 m) exposure of a sandstone remnant preserves a co-set of small-scale trough and planar cross-stratification. Each set is wedge-shaped and averages 6 in. (15.2 cm) in thickness. Orientations were difficult to measure, but the planar sets appear to be striking about N15E, with a 7° northwesterly dip. The trough sets appear to be striking N80E, with a 15° southeasterly dip.

Laterally from this outcrop are less well-indurated gray sands, silts, and clays which include the Biscuit Springs locality. Most of these finer-grained sands and silts do not preserve primary sedimentary structures; however, up to 1 cm thick by 20 cm long lignite stringers suggest scouring and subsequent filling by plant detritus. For the most part the stringers are horizontally bedded. Occasionally, clay balls as much as 1 cm in diameter are present.

The majority of fossils collected at this locality (by underwater screening), including thousands of mammal teeth, are under 1 cm in diameter. The smaller size range of fossils collected at Biscuit Springs than that of those collected at the Garbani Quarry indicates a lower flow regime at Biscuit Springs. The large fossils, particularly limb fragments, from the Garbani Quarry have been abraded by transport. Such material is rare at Biscuit Springs. However, an almost complete lower jaw of a crocodilian measuring about 1 ft. (30.1 cm) in length was recovered from this locality. Although damaged by compaction and collecting, it does not show stream abrasion. No larger fossils as complete as this have been recovered from the Garbani Quarry.

Both their physical geology and type of fossil preservation indicate that the Garbani Quarry was deposited near the base of a stream channel in a relatively high flow regime, while Biscuit Springs was deposited near the top of a point bar sequence in a relatively lower flow regime. The few paleocurrent indicators present at Biscuit Springs suggest an easterly or northeasterly flow.

The Garbani channel deposit can be traced to the western edge of Sec. 17, T20N, R37E. Additional portions of the channel deposit were probably present to the east or northeast but have been removed by erosion. Portions could also be present below the W coal in the badlands to the south and east. Several other localities in this deposit have produced small samples of lower vertebrates and, occasionally, mammals.

The overlying Farrand channel deposit was discovered in 1975, but mammals suggesting a Torrejonian or middle Paleocene age (Clemens, pers. comm.) were not discovered until 1976. In some places the channel overlies the W coal and in other places the W coal is cut by the channel. The channel deposit is capped by the V coal (see stratigraphic section S15 in Fig. 5). The base of the U coal, which defines the lower limit of the Lebo Formation, is 18 ft. (5.5 m) above the channel capping. The Farrand channel deposit has not been as thoroughly examined as the Garbani channel deposit, but much of the sandstone in the former unit appears to be coarse to very coarse sand. This contrasts with the usually fine- to medium-grained sand and sandstone in the underlying Garbani channel deposit. The coarser-grained sand is often yellowish or white in color and is probably arkosic. The two localities from which mammal teeth and dentaries have been recovered occur in this type of sandstone. The productive layer also contains clay balls, iron concretions, and, occasionally, lignite stringers. The two localities, Mosquito Gulch (V76170) and Two Mosquitos (V76171), are both 48 ft. (14.6 m) below the channel capping. Planar cross-stratified sandstone was observed lateral to the localities. The sets vary from 4 to 12 in. (10.2–30.5 cm) in thickness. The cross-stratified sandstone often contains laminae of carbonaceous material. The strike of the sets varies from N65W to N75W, with dips of 10° to 15° W.

Billy Creek–McKeever area. This area is about 10 mi. (16.1 km) northwest of the Hell Creek–Snow Creek area. During my initial work in Garfield County in 1973 through 1976, detailed mapping was not conducted and only one stratigraphic section was measured along Billy Creek. Subsequently, crews under the direction of W. A. Clemens have undertaken an extensive mapping program for this area. Several Lancian Hell Creek faunal-facies localities occur in or near this area. They have produced only a few mammalian specimens and these are not noteworthy, especially since their stratigraphic positions are not known. However, one possible pre-mantuan locality and two mantuan localities deserve comment.

The possible pre-mantuan locality, Billy Creek 1 (V72211), was discovered in 1965 by H. Garbani. Collections made by him and by later UCMP–LACM field parties recovered fragmentary dinosaurian remains as well as other lower vertebrate material from the locality. The locality was first catalogued as LACM 7213. In 1975, the field party recovered a dentulous condylarth jaw from the locality. Tentative assignment to the pre-mantuan is based on this condylarth. In a careful examination of the same strata, no dinosaurian remains were recovered. It is probable that the earlier dinosaur remains were recovered in surrounding sediments but not at locality V72211. Therefore, UCMP locality V72211 and LACM 7213 do not entirely represent the same locality. Originally it was believed that locality V72211 was in the uppermost Hell Creek Formation. However, a reexamination strongly suggests the locality is in the lowermost Tullock Formation (see Fig. 6). The locality is in a channel deposit situated on an isolated hill separated from the main badlands exposures by a narrow road. By plane tabling, L. Dingus has correlated the lithologic units in the isolated hill to the main badlands exposures, but he regards the correlations as tentative (Dingus, field notes). On the isolated hill as much as 25 ft. (7.6 m) of section, including what is here regarded as the lower Z coal, have been cut out by the channel deposit. In the main badlands exposures, the same channel deposits do not breach the lower Z coal. In the main exposures, a channel capping of carbonaceous shale is present. Five lignites or carbonaceous shales are present above the channel capping. It is not certain which, if any, of the informal letter designations should be applied to these beds, so in Fig. 6 no informal names are given to these beds. In the main badlands exposure, the channel capping is approximately 17 ft. (5.2 m) above the lower Z coal and about 16 ft. (4.9 m) below the next higher lignite. Within this exposure, the channel deposit is about 17 ft. (5.2 m) thick, but probably would have thickened to 35 to 40 ft. (10.7–12.2 m) on the isolated hill where locality V72211 occurs. Locality V72211 is near the base of the channel deposit in poorly indurated, yellow sand and silt. Higher in the deposit the color changes to red and the sediments are better indurated.

The two mantuan localities, McKeever Ranch 1 (V72210) and McKeever Ranch, Harley's Point (V77087), are located south of Billy Creek 1 (V72211). McKeever Ranch 1 is in a channel-fill deposit that rests on a thin coal. During the summer of 1977, W. A. Clemens (pers. comm.) was able to determine that the coal, although thin, is laterally continuous for at least a mile (1.6 km). He has tentatively identified this as the lower Z coal. It is not possible at the present time to determine if the channel-fill deposit containing McKeever Ranch 1 is capped by the upper Z coal, which is on the order of 30 to 50 ft. (9.1–15.2 m) above the lower Z. Nevertheless, the locality is within the Tullock Formation. McKeever Ranch, Harley's Point, appears to be a continuation of the channel that includes McKeever Ranch 1. It also rests directly on the lower Z coal and is thus at the very base of the Tullock Formation.

A simplistic correlation based solely on stratigraphic position would place the younger, mantuan localities at McKeever Ranch lower (immediately above the lower Z coal) than the older, pre-mantuan localities in Hell's Hollow [about 7 ft. (2.1 m) above the lowest Z coal]. It must be emphasized that both groups of localities occur within channel-fill deposits, as do many localities in the Hell Creek and Tullock formations, and thus the stratigraphic positions of these localities cannot be directly correlated with stratigraphic levels in the surrounding non-channel deposits. Without a detailed analysis of a channel deposit, it must be assumed that in such deposits the level of the fossil occurrences approximates the lowest possible stratigraphic level, while the channel capping is the highest possible stratigraphic level. Unless there are geologic factors complicating the correlation between Hell's Hollow and McKeever Ranch (e.g., incorrectly correlated lignites, significantly time-transgressive lignites, or multiple cycles within one or both of the channel deposits), the faunal evidence predicts that the channel capping should be higher at McKeever Ranch than at Hell's Hollow.

Flat Creek area. The Flat Creek area is in the northeastern quadrant of Garfield County, about 30 mi. (48.3 km) east of the Hell Creek–Snow Creek area. The most important locality in this area, Flat Creek 5 (V73087), is Lancian in age and belongs to the Hell Creek faunal-facies. Not only is the locality the richest of this faunal-facies discovered so far in Montana, but at 17 ft. (5.2 m) below the Z coal, it is also one of the stratigraphically higher localities of this faunal-facies. The sediment in which this locality occurs is composed largely of gray siltstone with stringers of fine- to medium-grained sand and smaller stringers of lignite. Other than the stringers, which suggest scour filling, no primary sedimentary structures are obvious. The exposure is about 10 ft. (3.0 m) long and about 2 to 3 ft. (0.6–0.9 m) thick. The only difference between this and the majority of the surrounding sediment is the slight increase in sand content at the locality. Without a more complete analysis, it is difficult to describe accurately the sedimentological environment. The evidence available suggests that this exposure, like many of the other Hell Creek faunal-facies localities, was produced in a small, possibly ephemeral stream.

Four other Flat Creek localities, Flat Creek 1 through 4 (V73083–86), are included with Flat Creek 5 in the Flat Creek local fauna. All but Flat Creek 4 (V73086) have produced at least one mammalian species. Flat Creek 2 (V73084) has produced only one mammalian specimen, an edentulous jaw of ?*Pediomys,* but is unique in the type of fossil accumulation. Most larger vertebrate specimens, usually dinosaurs, are found as single specimens or in concentrations scattered laterally along a single level. Flat Creek 2 (V73084) is the

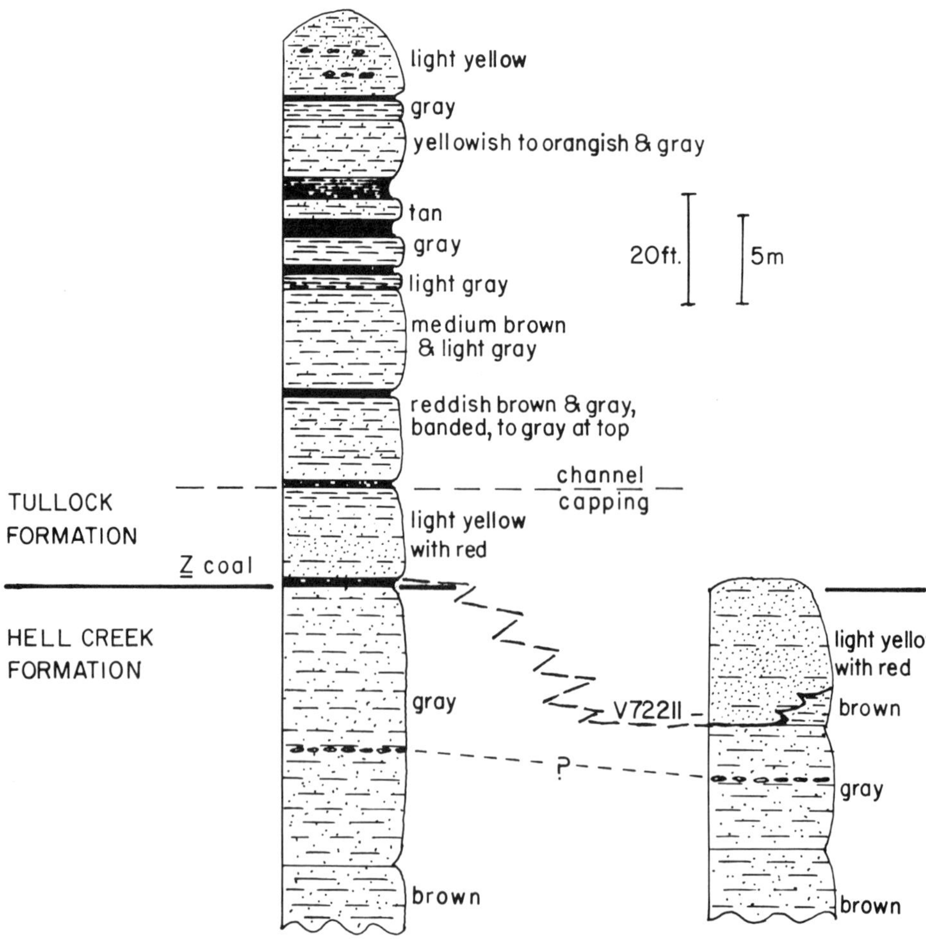

FIG. 6. Stratigraphic section measured along Billy Creek, Sec. 35, T22N, R34E. (See Fig. 3 for explanation of lithologic symbols.)

only locality discovered in the Hell Creek Formation that resembles a "logjam" of bones. The exposure is about 20 ft. (6.1 m) long and 4 to 6 ft. (1.2–1.8 m) wide. Larger elements such as limb bones, vertebrae, etc., are closely packed and in some instances overlap one another. Smaller specimens such as fish vertebrae and gar pike scales are scattered throughout the exposure. Unlike most other sites in this formation preserving large elements, this locality distinctly suggests that moving water was the transporting and depositing agent.

Approximately 3 mi. (4.8 km) southeast of the Flat Creek area, three localities, Kerr Butte 1–3 (V75196–98), have produced small samples of Paleocene mammals. The mammals are possibly Puercan in age. The localities are at the base of a 20+ ft. (6.1+ m) well-indurated sandstone channel deposit. The base of the channel is about 210 ft. (64.0 m) above the base of the lower Z coal. No attempt has been made to apply the informal coal

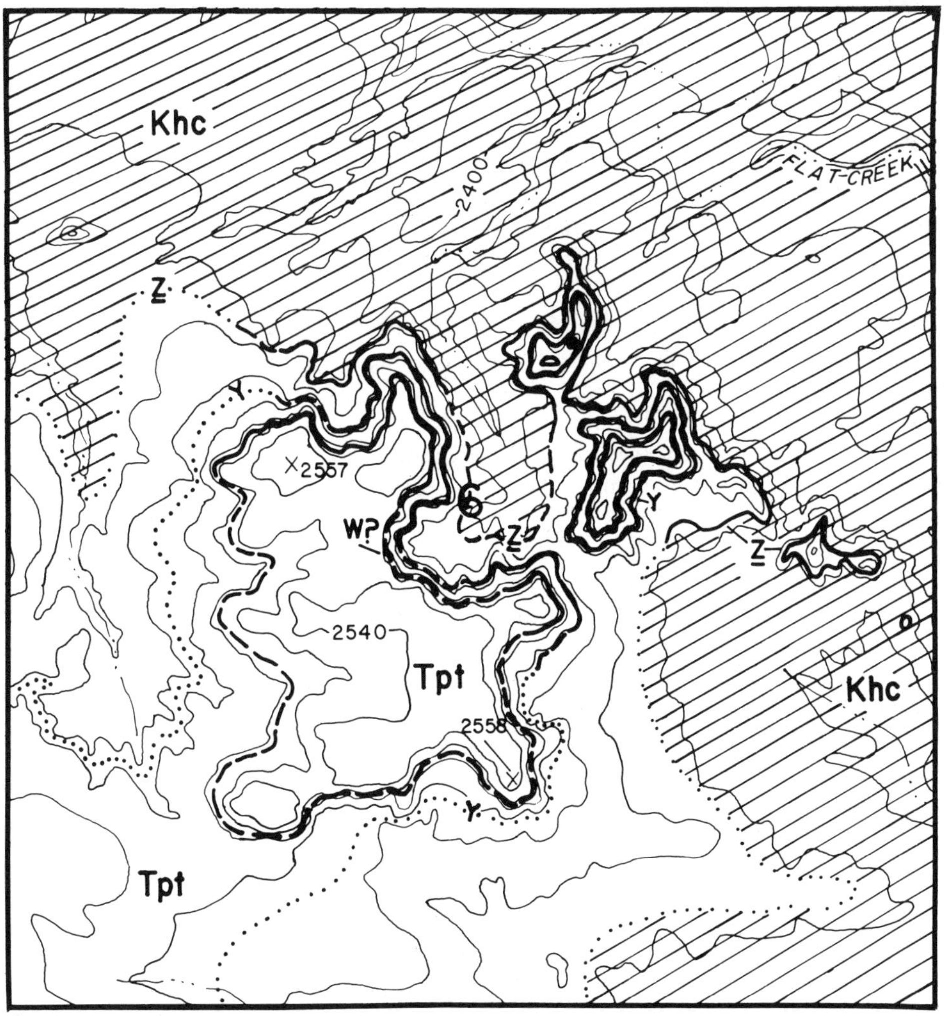

FIG. 7. Geologic map of Sec. 6, T20N, R42E, along Flat Creek in the vicinity of the Flat Creek local fauna. Map symbols are the same as those in Fig. 4. Scale: section line equals 1 mi. (~1.61 km).

nomenclature to the lignites intervening between the lower Z and the channel deposit. The localities are probably within the Tullock Formation, but this is not known with certainty.

Horsethief Canyon area. The Bliss Ranch (V72214) and Horsethief Canyon (V73094–95 and V75192) localities are approximately 20 mi. (32.2 km) southwest of Hell Creek. Tentative identification of the mammals suggests a middle Paleocene (Torrejonian) age (Clemens, pers. comm.). The localities occur in a sand- and siltstone channel that is under- and overlain by coals (Fig. 9). The identification of these coals has been somewhat problematic. The overlying coal is questionably called the U coal; this identification places the locality within the Tullock Formation. The coal underlying the channel deposit is tentatively identified by W. Rohrer of the U.S.G.S. as the UW? coal (written comm.). This could also be the W or some other lower coal.

Other areas and geologic aspects of the faunal-facies. As can be seen in Table 1, the greatest number of localities that are not in or near the four areas just described are referred to the Hell Creek faunal-facies of the Lancian. Most of these localities are isolated occurrences or occur as clusters of localities within the northeastern quadrant of Garfield County. These localities are not worthy of detailed discussion, particularly since the geology associated with them has not been studied in detail. Collectively, however, they strengthen two lithologic-stratigraphic observations that are useful in differentiating between the Hell Creek and Bug Creek faunal-facies.

First, by examining the data from these additional localities plus those described in more detail, it is clear that Hell Creek faunal-facies localities occur throughout the stratigraphic extent of the Hell Creek Formation, except for possibly the uppermost 10 ft. (3.0 m) or so. In marked contrast, the stratigraphically lowest Bug Creek faunal-facies locality, Bug Creek Anthills, is only 80 ft. (24.1 m) below the Hell Creek–Tullock formational contact (Sloan and Van Valen, 1965).

Second, Bug Creek faunal-facies localities are limited to the larger, sandier channel-fill deposits. Hell Creek faunal-facies localities are more nearly ubiquitous, being found in a variety of sedimentological regimes. However, they do tend to be found most often in small lenses of poorly sorted sand and silt representing small, possibly ephemeral streams and overbank deposits.

These stratigraphic and lithologic criteria should be regarded as being preliminary, although they are now useful in separating the two faunal-facies. Continued research in Garfield County and elsewhere will undoubtedly complicate this simple, dual faunal-facies pattern. As is discussed in the chapter on Biostratigraphy, there is also a faunal separation between the two faunal-facies. The apparent faunal, stratigraphic, and lithologic differences between these two faunal-facies permit a first attempt at sorting the various ecologic and evolutionary events that occurred at the close of the Cretaceous in northeastern Montana.

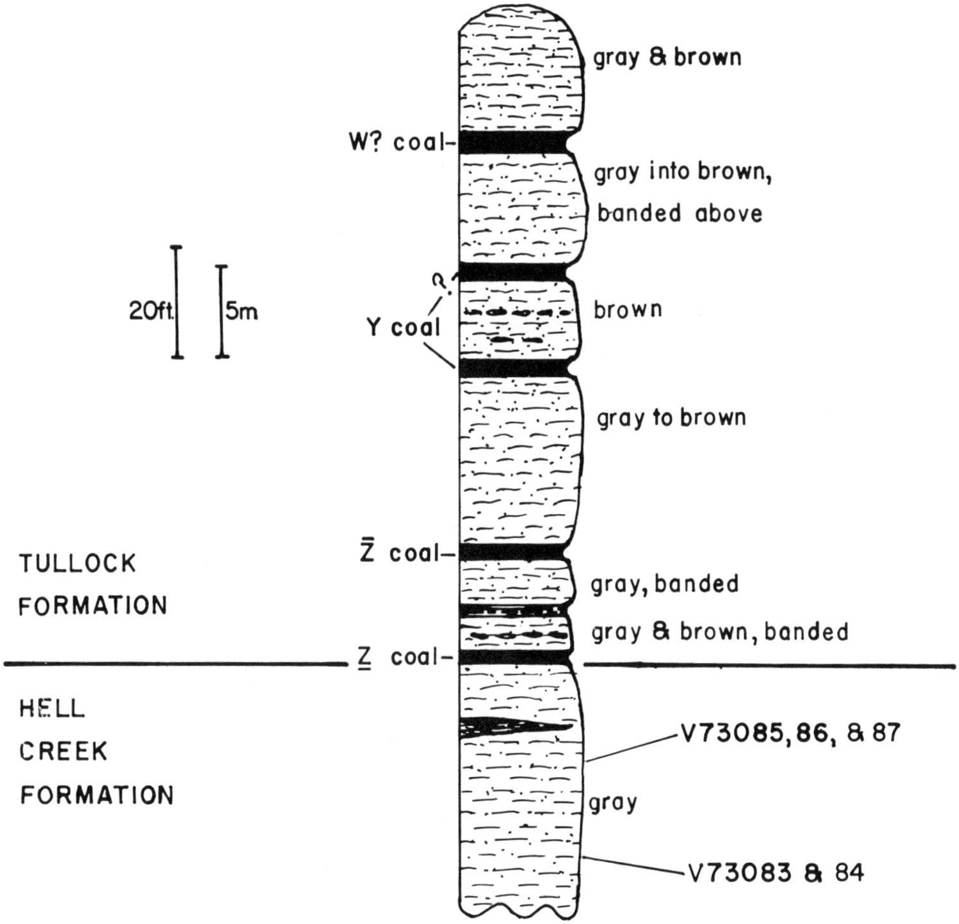

FIG. 8. Composite stratigraphic section measured in Sec. 6, T20N, R42E, along Flat Creek. See Fig. 3 for explanation of lithologic symbols.

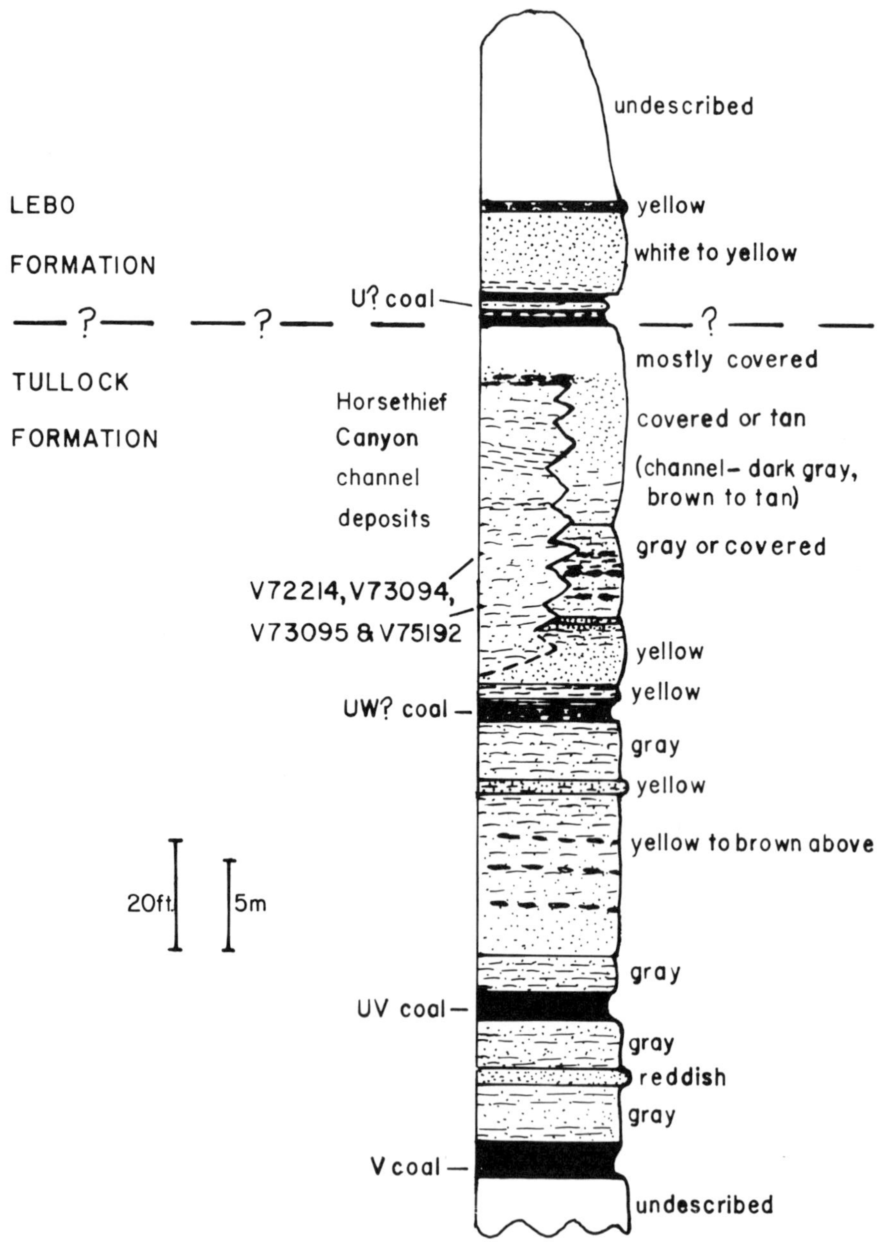

FIG. 9. Composite stratigraphic section measured in Sec. 28, T17N, R34E, in Horsethief Canyon. (See Fig. 3 for explanation of lithologic symbols.)

SYSTEMATICS

Class MAMMALIA
Subclass ALLOTHERIA
Order MULTITUBERCULATA

The multituberculates from at least the Cretaceous portion of the Hell Creek Formation, Garfield County, Montana, are known only from the Hell Creek faunal-facies. The only probable Bug Creek faunal-facies locality in Garfield County (V73098) has not yielded multituberculates. Pre-mantuan localities known from the lowermost Tullock Formation and the uppermost Hell Creek Formation have yielded multituberculates.

The Hell Creek faunal-facies, best represented by UCMP loc. V73087, Flat Creek local fauna, is most similar to the Trochu local fauna (Lillegraven, 1969) and to the Lance local fauna (Clemens, 1964, 1973) in its representatives of the Multituberculata. In most aspects the Flat Creek local fauna is intermediate between the Trochu and Lance local faunas in the multituberculate (and other mammalian) taxa present. This is not surprising, considering the geographically intermediate position of the Flat Creek local fauna.

The pre-mantuan localities, best represented by UCMP loc. V74111, Hell's Hollow local fauna, have yielded a number of multituberculates not found in the Hell Creek faunal-facies but with genera in common with the Bug Creek faunal-facies.

One of the most vexatious problems in considering the systematics of the multituberculates from Garfield County is the lack of teeth associated in jaws. Lillegraven (1969) faced a similar problem in dealing with the multituberculates from the Trochu local fauna. He partially solved this problem for the case of *Mesodma* by constructing bivariate plots of M^2_2. This idea is used and expanded here, again for *Mesodma*.

Misallocation of isolated teeth to a particular genus or species of multituberculate is nevertheless a real possibility in the following systematic descriptions. The specific instances of possible misallocation are considered under each species (or genus) description. Table 2 is a summary of the number of teeth and jaws assigned to each specific taxon, with the possible instances of misallocation. The taxa are arranged according to the classification used here. It should be noted that several changes to the classifications most recently used by Sloan and Kielan-Jaworowska (various papers) have been made. No new taxa above the generic level have been proposed, although a number of species have been removed from currently recognized families and suborders of multituberculates. Most importantly, the Cimolomyidae has been removed from the Taeniolabidoidea, and *Cimexomys* has been removed from the Ptilodontoidea and a new genus named on the basis of several species currently included in this genus. Although it might be possible to name new taxa at or above the familial level, it would certainly be premature to do so before comparisons have been made between the Late Cretaceous multituberculates of Asia and North America.

TABLE 2. Frequency of occurrence of teeth and jaws of multituberculates, Hell Creek and lowermost Tullock formations, Garfield County, Montana. Numbers before species are given as a superscript in the table where it is believed that the species may be represented at that tooth site of the second species. None of the species occurs both at Hell Creek faunal-facies and pre-mantuan localities *in Garfield County,* although many do occur together at Bug Creek faunal-facies localities in McCone County (H = Hell Creek faunal-facies; p = pre-mantuan).

	P_3 (in jaw)	P_4	M_1	M_2	P^4	M^1	M^2	Teeth found in association
Ptilodontoidea								
Neoplagiaulacidae								
H 1 *Mesodma hensleighi*	0	3	$^{2,14}11$	$^{2,14}4$	2	$^{2}2$	$^{2,14}15$	M_{1-2}
H 2 *Mesodma* cf. *M. formosa*	0	$^{1}1$	$^{5}1$	$^{5}1$	$^{5}4$	$^{5}6$	$^{5}9$	0
H 3 *Mesodma thompsoni*	0	3	3	$^{5}1$	7	6	1	0
p 4 *Mesodma garfieldensis*, n. sp.	2	15	$^{14}55$	$^{14}44$	35	43	$^{14}35$	P_{3-4}
H 5 ?*Neoplagiaulax burgessi*, n. sp.	0	5	0	0	0	0	0	0
?Neoplagiaulacidae								
H 6 Gen. et sp. indet.	0	1	0	0	0	0	0	0
Cimolodontidae								
H 7 *Cimolodon nitidus*	0	4	2	4	8	6	5	0
Taeniolabidoidea								
Eucosmodontidae-Eucosmodontinae								
p 8 *Stygimys* aff. *S. kuszmauli*	—	1	5	7	4	4	11	0
Eucosmodontidae-Microcosmodontinae								
p 9 *Acheronodon garbani*, n. gen. et n. sp.	?	1	0	0	0	0	0	0
Taeniolabididae								
p 10 *Catopsalis* cf. *C. foliatus*	—	0	0	0	0	1	0	0
Suborder, *incertae sedis*								
Cimolomyidae								
H 11 *Cimolomys gracilis*	0	0	0	3	2	0	2	0
H 12 *Meniscoessus robustus*	0	0	3	6	$^{13}8$	7	8	I^2; $P^{1,2,4}$; M^{1-2} I_1; P_4; M_{1-2}
?Cimolomyidae								
H 13 *Essonodon browni*	?	0	2	1	0	4	2	P^4; M^{1-2}
Suborder and Family, *incertae sedis*								
p 14 *Cimexomys minor*	0	2	0	1	0	3	0	0
p 15 *Cimexomys hausoi*, n. sp.	1	2	6	6	4	5	7	P_{3-4}; M_{1-2}
Suborder and Family, *incertae sedis*								
H 16 *Paracimexomys priscus*, n. gen.	0	0	4	3	0	5	3	0

Suborder PTILODONTOIDEA (Gregory and Simpson, 1926)
Sloan and Van Valen, 1965
Family NEOPLAGIAULACIDAE Ameghino, 1890

Mesodma Jepsen, 1940

Comments. In characterizing the family Ectypodidae (now Neoplagiaulacidae), Sloan and Van Valen (1965, p. 221) noted: "These small multituberculate mammals are more common and usually pose a more difficult taxonomic problem at the species level than other mammals in any fauna." After studying *Mesodma* from Garfield County, I agree with this assessment.

In the Hell Creek faunal-facies *Mesodma hensleighi* and *M. thompsoni* are represented, although P_4's of *M. thompsoni* are larger than those known from the Lance (Clemens, 1964) or Trochu (Lillegraven, 1969) local faunas. *M. formosa* is questionably represented in the Hell Creek faunal-facies, but the presence of a new neoplagiaulacid complicates the referral of molars and P^4's to the former species. A single, or at most two, species of *Mesodma* are represented at pre-mantuan localities in Garfield County. This conclusion is based on analysis of premolars and molars and tends to support Novacek and Clemens' (1977) thesis that only one species of *Mesodma* is present in the Bug Creek Anthills. *Mesodma garfieldensis* from the pre-mantuan, Garfield County, is a new species, but until comparative data on the molars of *Mesodma* from Bug Creek Anthills, McCone County, has been published, closer comparisons of this latter form and *M. garfieldensis* will not be possible.

Clemens (1964) was able to separate *Mesodma* from the Lance local fauna into two species, *M. formosa* and *M. thompsoni,* largely on the basis of P_4 morphology and size. Lillegraven (1969) used finer mesh screens for his screen-washing operations and he was able to recognize a third, small species, *M. hensleighi,* from the Trochu local fauna, Alberta. Subsequently, Clemens (1973) determined that *M. hensleighi* occurs in the Lance local fauna. Unlike the Lance local fauna, the Trochu local fauna does not contain a moderately large sample of P_4's. However, Lillegraven (1969, p. 19 and fig. 9) recognized three size groupings of *Mesodma* based on a bivariate plot (length vs. width) of M_2's and M^2's. The Hell Creek and lowermost Tullock formations, Garfield County, similarly contain few P_4's of *Mesodma,* making it necessary to utilize molars for recognition of species. Bivariate plots of length vs. width were constructed for M^1_1's and M^2_2's of *Mesodma* from UCMP locs. V73087 and V74111. Similar plots were made for the same molars from the most fossiliferous Lance localities, UCMP locs. V5620 and V5711, based on Clemens' original unpublished measurements. By this method, size groupings were easily recognized. Fig. 10 is a summary of these bivariate plots. Since molar length contributed most to the recognition of these groups, it is the dimension shown in Fig. 10, even though width was also used in constructing this figure. In the compilation of the dimensions given in appendices 1, 2, and 3, Lillegraven's data (1969, from table 1) are included. With the exception of the loc. V74111 sample, the molars consistently form three groups for almost every tooth site. As the diagonal lines in Fig. 10 indicate, there is some overlap between the groupings. Also, as might be expected, in some instances in Fig. 10 the observed ranges do not correlate well between localities.

There are a number of reasons for the overlap of groups at one molar site for a given locality and for the apparently poor correlation of some of the size groupings between localities. First, although relatively rare, *Cimexomys minor* is known from UCMP locs. V5620 and V5711 (Clemens, 1973, p. 79), and V74111. Based on isolated molars, only the M^1's of *C. minor* can be distinguished from *Mesodma*. Therefore, the other molars (M_1, M_2^2) of *C. minor* (close to the size of *Mesodma hensleighi)* are probably included in the samples from these localities.

Second, Clemens (1964, p. 29) noted that UCMP loc. V5711 is a wind deflation pocket. The winnowing action of wind probably accounts for the recovery of more large fossils at loc. V5711 than at UCMP loc. V5620. This difference in size range is clearly shown in Fig. 10: for any given molar site except possibly the M_1 of *M. formosa,* the upper limit for the observed range is larger from loc. V5711 than from loc. V5620. It is doubtful whether there is any biological significance to larger teeth being found at loc. V5711, because this phenomenon affects taxa besides multituberculates. It is possible that larger representatives of the various taxa may be present at loc. V5620, but unlike loc. V5711 the former

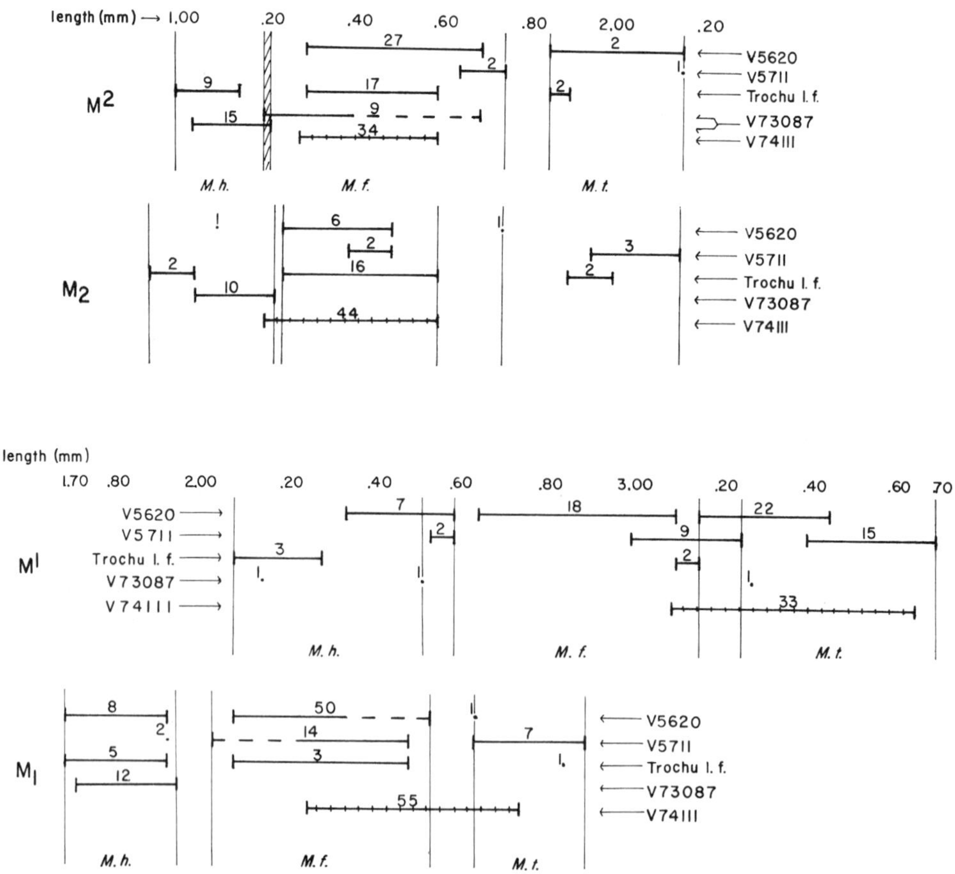

FIG. 10. Comparison of *Mesodma* molar lengths from various Lancian localities. (See text for explanation.)

locality has not undergone any extensive winnowing by wind that would tend to accumulate the larger fossils. In addition, the original sedimentological regime may have favored the size groupings now found at the two localities, so that the wind may only have accentuated this difference at loc. V5711.

A third factor concerns the differences in screen grid used in the various studies represented in Fig. 10. Clemens used window screen with 18 × 16 openings per in. (= approx. 7.1 × 6.3 openings per cm) in the Lance study. Lillegraven used screen with 40 openings per in. (= approx. 15.7 openings per cm). Screen with 24 openings per in. (= approx. 9.5 openings per cm) was used in the collection of the Garfield County specimens. On the basis of P_4's Clemens (1973) recognized *M. hensleighi* at locs. V5711 and V5620. However, no M^2's and only one M_1 referable to *M. hensleighi* were recovered from these two localities, suggesting loss through the washing screen (Lillegraven, 1969; Novacek and Clemens, 1977).

A fourth consideration is sample size. It can be seen in Fig. 10 that the observed range generally increases with an increase in sample size.

The fifth factor concerns the recognition of group boundaries in some of the bivariate plots. Occasionally, teeth lay equidistant between recognizable groups, making their allocation to one group depend arbitrarily on size criteria, especially since no differences in morphology could be detected. The uncertainty of these group boundaries is shown in Fig. 10 by dashed lines.

Even with the vagaries discussed above, plus others probably not realized, the molars at all but UCMP loc. V74111 form remarkably consistent tripartite groups. These data corroborate Lillegraven's (1969) and Clemens' (1973) recognition of three species of *Mesodma* in the Trochu and Lance local fauna, respectively. The occurrence of a new neoplagiaulacid at UCMP loc. V73087, which somewhat complicates the recognition of three *Mesodma* species at this locality, is discussed in greater detail on the following pages. The bivariate plots suggest the presence of only one (or two?) species of *Mesodma* at loc. V74111 that is not referable to any of the described species and is considered separately in the following descriptions.

Mesodma hensleighi Lillegraven, 1969

(Table 3 & Appendix 1)

Mesodma hensleighi Lillegraven, 1969, p. 16

Type. UA 3596, isolated left P_4 (Lillegraven, 1969, fig. 6, la-b).

Type locality. Loc. KUA-1, Scollard Formation, Alberta.

Referred specimens. Mandible fragment with M_{1-2}, UCMP 115960; mandible fragment with M_2, UCMP 115961; 3 P_4's; 11 M_1's; 3 M_2's; 2 P^4's; 2 M^1's; 15 M^2's from UCMP loc. V73087.

Locality. UCMP loc. V73087.

Distribution. Scollard Formation, Alberta; Lance Formation, Wyoming; Hell Creek Formation, Montana (all Lancian).

Diagnosis. Lillegraven, 1969, p. 17.

Description. Only additions to the description of Lillegraven (1969) are given.

Mandibles. Two mandible fragments from UCMP loc. V73087 preserve teeth (UCMP 115960 and 115961), but neither preserves the ventral margin. The portions present indicate that the jaw is relatively deeper than in *Mesodma ambigua* and probably *M. formosa.* As in *M. formosa,* the anteroventral border of the ascending ramus of the dentary merges with the corpus below the posterior half of M_1 (Clemens, 1964, p. 31). UCMP 115961, a right dentary, preserves M_2 with a cusp formula 3:2. UCMP 115960, a left dentary, preserves M_{1-2}. The M_1 has a cusp formula 6:4 and the M_2 has a cusp formula ?3:2. The M_1 to M_2 length ratio for UCMP 115960 is 1.74. (See Table 3 for measurements of these molars.)

P_4. Of the three P_4's referred to *Mesodma hensleighi,* one has nine serrations and two have ten. The highest point on the blade is usually at the fourth serration (two teeth) but can also be at the fifth (one tooth). All three teeth have seven internal and external ridges.

M_1. There are twelve M_1's from loc. V73087, including the dentary UCMP 115960. Unworn teeth have a cusp formula 6:4. There is no clear indication of bimodality in the dimensions. For all twelve M_1's the CV for length is 4.04. The CV for width is 9.58 for the same teeth. Although the CV for width is high, the observed ranges for length and width agree well with those given by Lillegraven (1969, table 1 and repeated here in Appendix 1) for *Mesodma hensleighi.* The observed ranges of *M. hensleighi* from UCMP loc. V5620 are also in agreement (Appendix 1 and Fig. 10). The three M_1's Lillegraven assigned to *M. formosa* are all longer than those from loc. V73087. Clemens (1964, p. 35) identified two M_1's in dentaries associated with P_4's as *M. formosa.* The lengths of these M_1's, 2.2 and 2.4, fall within the OR, 2.10–2.50, observed by Lillegraven. All twelve M_1's from loc. V73087 are referred to *M. hensleighi* on the basis of size and cusp formula.

M_2. As with the M_1's, there is no clear indication of bimodality in the sample of ten M_2's (including UCMP 115960 and 115961) from loc. V73087. In a bivariate plot of length vs. width, all the M_2's from loc. V73087 cluster at the lower end of the graph. Three of the ten teeth fall between the ranges of *Mesodma hensleighi* and *M. formosa* indicated by Lillegraven (1969, fig. 9 & table 1). There are no morphological differences between the M_2's except that the cusp formula tends to increase from 3:2 in the smaller teeth to 4:2 in the larger teeth. The teeth are assigned to *M. hensleighi* because of their small size and the lack of bimodality in the sample.

P^4. Only two complete P^4's from loc. V73087 are referable to *Mesodma hensleighi.* UCMP 115927 has a cusp formula 1:6:0, with the penultimate medial cusp highest, and UCMP 115928 has a cusp formula 2:6:0, with the penultimate and ultimate medial cusps of equal height. These two teeth (Table 3) are within the size range given for *M. hensleighi* by Lillegraven (1969, table 9: OR 1.60–1.90, N = 3).

M^1. Of the ten M^1's in the *Mesodma* sample from loc. V73087, only three are complete anteriorly. However, there is a strong suggestion of trimodality in the sample based on width. The three groups agree well with the width OR's given for M^1's in Lillegraven (1969, table 9). Two of the ten teeth belong to *M. hensleighi.* Only one is complete (UCMP 115934), and has a cusp formula 6:8:5, plus a small accessory cusp at the anterolabial corner next to the first cusp in the external row. The internal cusp row is greater than 50 percent of the tooth length and terminates at the sixth cusp from the rear in the medial row.

M^2. As in the Trochu local fauna, M^2's are the most common teeth of *Mesodma* (other than P^x's). The size ranges of teeth from loc. V73087 are slightly different from those given by Lillegraven (1969, table 1) but are bimodal. (*M. thompsoni* is not represented by M^2's at loc. V73087.) The fifteen M^2's of *M. hensleighi* have a cusp formula 1:3:3. The anterior cusp in the medial row is smaller and incompletely divided from the second cusp.

TABLE 3. Measurement of isolated P_4^4's and dentaries of *Mesodma hensleighi* from UCMP loc. V73087

		M_1			M_2		
		L	*W*		*L*	*W*	
UCMP 115960		1.86	0.94		1.07	0.96	
UCMP 115961					1.06	0.86	

		L1	*L*	*H*	*W*	*L1/L*	*H/L*
	$\bar{x}$	1.31	2.79	1.25	1.17	0.47	0.45
P_4	OR	1.24 –1.39	2.66 –2.89	1.04 –1.39	1.11 –1.21	0.45 –0.49	0.39 –0.48
	N	3	3	3	3	3	3

		L	*W*
	$\bar{x}$	1.76	1.07
P^4	OR	1.73 –1.79	1.07 –1.10
	N	2	2

Discussion. *Mesodma hensleighi* is the most common species at loc. V73087. Although samples are small, it can be differentiated from *M. thompsoni* on the basis of its smaller size and the more arcuate shape of P_4. In the sample from loc. V73087 only one P_4 may be referable to *M. formosa.* It is near the lower end of the size range for P_4's of this species and could also belong to *M. hensleighi.*

Mesodma cf. *M. formosa* (Marsh, 1889b) Clemens, 1964
(Table 4 & Appendix 2)

Halodon formosus Marsh, 1889b, p. 179
(non) *Cimolomys formosus* (Marsh), Hay, 1902, p. 560

Type. YPM 11812, left P_4 (Marsh, 1889b, pl. VIII, figs. 36–39).

Type locality. Peterson's quarry of Lull (1915), Lance Formation, Wyoming.

Referred specimens. 1 P_4; 4 P^4's; 4 M^1's; 9 M^2's from UCMP loc. V73087. Isolated teeth from various localities.

Localities. UCMP locs. V72213, V73087, V73090, and V74116.

Distribution. Lance Formation, Wyoming; Scollard Formation, Alberta; Hell Creek Formation, Montana (all Lancian).

Revised diagnosis. Clemens, 1964, p. 31, with the addition of Lillegraven's comment "that only the specimens in the higher size ranges of Clemens' diagnosis are now included in the taxon." (1969, p. 19).

Description. Only one P_4 from the Hell Creek Formation of Garfield County (loc. V73087) resembles those of *Mesodma formosa.* A new neoplagiaulacid (also from loc. V73087), slightly larger than *M. formosa,* is represented by five P_4's. Isolated P^4's and molars from loc. V73087 could belong to *M. formosa* or to this new neoplagiaulacid. Because of this uncertainty the isolated teeth, other than P_4's, are referred to *Mesodma* cf. *M. formosa,* and only the five P_4's are assigned to the new neoplagiaulacid.

P_4. The only P_4 possibly referable to cf. *M. formosa* (UCMP 115931) is small (Table 4), but within the OR's for P_4's of *M. formosa* given by Clemens (1973, table 7) and Lillegraven (1969, table 1). It has eleven serrations and seven internal and eight external ridges. This tooth is distinguished from the three P_4's of *M. hensleighi* by its slightly larger size, lower L1/L ratio (0.42), and presence of one additional serration.

M_1. All M_1's at loc. V73087, because of their small size and the lack of bimodality in the sample, have been assigned to *Mesodma hensleighi.* Some of the larger M_1's from this locality could belong to cf. *M. formosa.* UCMP 116585, UCMP loc. V72213, is an M_1 of cf. *M. formosa* (length 2.45; width 1.02; cusp formula 6:4 or 5).

M_2. Although cf. *M. formosa* is possibly present at loc. V73087, there is no clear indication of bimodality in the M_2 sample, and all M_2's from this locality have been referred to *M. hensleighi.* UCMP 116586, UCMP loc. V73090, might be an M_2 of cf. *M. formosa* (length 1.58; width 1.34; cusp formula 3:2).

P^4. Four P^4's from loc. V73087 are referable to cf. *M. formosa.* Three of the teeth have a cusp formula 1:6:0 and one has a formula 2:7:0. These P^4's could belong to the new neoplagiaulacid described later, although their morphology agrees with that given by Clemens (1964, pp. 35-36) for *M. formosa.*

M^1. Only one complete M^1 from loc. V73087 (UCMP 115935) could belong to cf. *M. formosa.* The cusp formula of this tooth is ?7:8:4. Three incomplete M^1's from loc. V73087 are also referred to cf. *M. formosa* based on their width (Appendix 2). UCMP 116594, loc. V74116, (length 3.16; width 1.37; cusp formula 7:9:4/5) is also an M^1 of cf. *M. formosa.*

M^2. Lilligraven (1969, fig. 9) was able to differentiate three size groupings of *Mesodma* M^2's and M_2's. The sample of M^2_2's from loc. V73087 does not follow this tripartite pattern as closely. However, the CV's for length and width of the total sample of M^2's from loc. V73087 are high (CV for length, 14.41; CV for width, 14.16), suggesting the presence of more than one species. Fig. 10 and Appendices 1–3 show that the M^2's follow the groupings recognized by Lillegraven more closely than do the M_2's from loc. V73087. Of the nine M^2's referred to cf. *M. formosa* from loc. V73087, only the two largest are substantially outside the range observed by Lillegraven. It is possible that these two teeth belong to *M. thompsoni* or the new neoplagiaulacid described below. All nine M^2's have a cusp formula 1:3:3. There is no perceptible difference in morphology between the smaller M^2's belonging to *M. hensleighi* and those belonging to cf. *M. formosa.*

Discussion. Based on P_4's, *Mesodma formosa* is not a common mammal in the Hell Creek Formation of Garfield County, although *Mesodma* is the most abundant multituberculate in the Hell Creek and Bug Creek faunal-facies. However, if the P^4's and isolated molars described above do belong to *M. formosa,* then it is a fairly common mammal. The presence of a new neoplagiaulacid similar in size to *M. formosa* at loc. V73087 complicates the referral of isolated teeth to the latter species.

TABLE 4. Measurements of P_4^4's of *Mesodma* cf. *M. formosa* from UCMP loc. V73087.

	L1	*L*	*H*	*W*	*L1/L*	*H/L*
P_4	1.29	3.09	1.44	1.21	0.42	0.47
	(N = 1 for all parameters)					

		L	*W*
	$\bar{x}$ =	2.27	1.24
P^4	OR =	2.18 −2.42	1.19 −1.28
	N =	4	4

Mesodma thompsoni Clemens, 1964

(Table 5 & Appendix 3)

Mesodma thompsoni Clemens, 1964, p. 39

Type. UCMP 47217, an isolated left P_4 (Clemens, 1964, fig. 10).

Type locality. UCMP loc. V5620, Lance Formation, Wyoming.

Referred specimens. 1 P_4; 1 M_1; 3 P^4's; 4 M^1's from UCMP loc. V73087. 1 P_4 from UCMP loc. V72212. 1 P_4 from UCMP loc. V74116. Isolated teeth from various localities.

Localities. UCMP locs. V72212, V73076, V73078, V73087, V73090, V74116, and V75157.

Distribution. Lance Formation, Wyoming; Scollard Formation, Alberta; Hell Creek Formation, Montana (all Lancian); St. Mary River Formation, Alberta (Edmontonian).

Diagnosis. Clemens, 1964, p. 39 (although size range has been extended to include longer P_4's).

Description. *P_4.* Three complete P_4's from the Hell Creek Formation, Garfield County, belong to *Mesodma thompsoni.* All three teeth are longer than any P_4's of *M. thompsoni* reported from the Lance and Trochu local faunas by Clemens and Lillegraven, respectively (Table 5). However, they do not exceed the OR for width of *M. thompsoni* from the Lance local fauna. One P_4 (UCMP 108053) has twelve serrations and the other two (UCMP 116604 and UCMP 116605) each have thirteen serrations. All three P_4's have ten external and nine internal ridges. On UCMP 108053 and UCMP 116605 the last external and internal ridges are not well developed. UCMP 116605 is worn, but like the other two P_4's it possesses a well-developed posterolabial basal ridge.

M_1. Only one M_1 fragment of *Mesodma thompsoni* occurs in the sample from loc. V73087. Two complete M_1's belonging to this species are from two different localities (Table 5). UCMP 116607 (loc. V73090) has a cusp formula 6:5. UCMP 116608 (loc. V74116) has a cusp formula 7:5. The anterolabial cusp on the latter M_1 is small.

M_2. A single M_2, UCMP 108983, from loc. V73076 may belong to *Mesodma thompsoni.* There are two large internal cusps. The external cusp row is worn, but there appear to have been four cusps. The labial grooves are not as well developed as the lingual grooves separating the third and fourth external cusps. The longitudinal basin separating the external and internal cusp rows has numerous small ridges, giving the basin a reticulate appearance.

P^4. Three P^4's from loc. V73087 belong to *Mesodma thompsoni.* One very worn specimen, UCMP 116610, is assigned to *M. thompsoni* on the basis of size. The other two P^4's, UCMP 116609 and UCMP 116611, are unworn and both have a cusp formula 2:7:0. In UCMP 116609 the medial row of cusps does not increase greatly in height anteriorly to posteriorly and the individual cusps are also not high. As in UCMP 116609, the individual cusps of UCMP 116611 do not increase in height. However, in UCMP 116611 the medial cusp row increases in height posteriorly, with the penultimate cusp taller than the last cusp.

Four additional P^4's from four different localities (Table 5) are referred to *Mesodma thompsoni.* Three of these have seven medial cusps. In the fourth (UCMP 116614) the apices of the posterior cusps have been worn flat, but constrictions of the pulp cavity suggest that six medial cusps were present. This apical wear is not unique in ptilodontoids; Clemens noted this pattern in *Cimolodon* and *Mesodma,* in addition to that previously known for the type of *Kimbetohia campi.* He (1964, p. 55) suggested that this wear pattern resulted when the posterior end of the P^4 occluded with the M_1 in older individuals. The number of cusps in the anteroexternal row varies from two to five in these four P^4's, with two teeth having three cusps.

The three P^4's from loc. V73087 and the four other P^4's all fall within the OR for length given by Clemens (1964, table 2). However, five of the teeth exceed the greatest width given by Clemens for *Mesodma thompsoni.* One P^4 in particular does not group well with the others because of its great length and width. This is the P^4 (UCMP 116613, loc. V75157) that has five anteroexternal cusps.

M^1. Two complete and four fragmentary M^1's are referred to *Mesodma thompsoni* (Table 5). UCMP 116616, loc. V73087, has a cusp formula 7:9:4. UCMP 116619, loc. V73078, has a cusp formula 8:9:6. The anterior and posterior cusps of the internal row in UCMP 116619 are distinct but extremely small. In both M^1's the last cusp in the external row is small and incompletely separated from the preceding cusp.

M^2. Only one M^2, UCMP 108982, from UCMP loc. V73076 belongs to *Mesodma thompsoni.* It has a cusp formula 1:3:4. The cusps in the external row are small and not distinctly separated from one another.

Discussion. The three P_4's of *Mesodma thompsoni* from the Hell Creek Formation approach or exceed all the size parameters of *M. ambigua* given by Novacek and Clemens (1977, table 4) except for that of width. The number of serrations on these three P_4's (mode = 13) agrees with *M. thompsoni* (Clemens, 1964) and is lower than *M. ambigua* (mean = 13.7). These P_4's are not preserved in jaw fragments, so it cannot be determined if the jaw was deep, as in *M. thompsoni*, or shallow, as in *M. ambigua*. The P_4's are referred to *M. thompsoni* because of their serration count, but their *M. ambigua*-like size is another indication of a possible ancestor-descendant relationship between the two species.

Mesodma garfieldensis, n. sp.

(Figs. 11–14 & Table 6)

Etymology. Named for Garfield County, Montana.

Type (holotype). UCMP 116622, right dentary with P_{3-4}.

Paratype. UCMP 116621, right dentary with P_{3-4}.

Type locality. UCMP loc. V74111, both type and paratype, lowermost Tullock Formation, Montana.

Referred specimens. 15 P_4's; 55 M_1's; 44 M_2's; 35 P^4's; 43 M^1's; 35 M^2's from UCMP loc. V74111. A small sample of isolated teeth from UCMP loc. V74110.

TABLE 5. Measurements of premolars and molars of *Mesodma thompsoni.*

P_4's from various UCMP localities

UCMP Number	*UCMP Locality*		*L1*	*L*	*H*	*W*	*L1/L*	*H/L*
116604	V73087		2.15	4.61	1.96	1.55	0.47	0.43
108053	V72212		1.95	4.55	2.21	1.55	0.43	0.49
116605	V74116		1.82	4.58	1.97	1.56	0.40	0.43
		$\bar{x}$ =	1.97	4.58	2.05	1.55	0.43	0.44

P^4's

UCMP loc. V73087			Other localities			
	L	*W*	*UCMP Number*	*UCMP Locality*	*L*	*W*
$\bar{x}$	2.54	1.45	116614	V75178	2.47	1.58
			116612	V73103	2.60	1.65
OR	2.51 –2.56	1.36 –1.58	116613	V75157	3.11	1.88
			108981	V73076	2.72	1.65
N	3	3				

Molars from various UCMP localities
(see App. 3 for UCMP loc. V73087)

UCMP Number	*UCMP Locality*	*Tooth Site*	*L*	*W*
116607	V73090	M_1	2.86	1.11
116608	V74116	M_1	2.77	1.26
108983	V73076	M_2	1.81	1.64
116619	V73078	M^1	3.19	1.55
108982	V73076	M^2	2.00	1.79

Localities. UCMP locs. V74110 and V74111.

Distribution. Lowermost Tullock Formation, Montana (pre-mantuan), possibly extending into uppermost Hell Creek Formation (pre-mantuan and Lancian).

Diagnosis. P_4 variable, similar to *Mesodma thompsoni* in length, greatest height to length ratio of any species of *Mesodma* (except possibly *M. senecta),* eleven to fourteen serrations (mode 12 or 13, probably 13); M^1 and M_1 longer relative to M^2 and M_2, respectively, than in any Cretaceous species of *Mesodma* (relationship unknown in mantuan, *M. ambigua*); cusp formula averages higher on M^1 and M_1 than in other species of *Mesodma* (where formulae are known). Cusp formulae: M_1, 6–7:4–5; M_2, 3–4:2; P^4, 1–4:7–8; M^1, 6–8:8–11:4–5(+); M^2, 1:3:3–4.

Comments. Since dentary fragments preserving the P_4 and molars are lacking, the tradition of naming a P_4 (the type also preserves P_3) as the type specimen has been followed. However, as will be discussed, there is some suggestion of bimodality, especially in the P_4's. Therefore, a type specimen was chosen that could unambiguously be placed in one of the two size groups (a larger P_4 was chosen rather arbitrarily) and a paratype (a smaller P_4) was chosen from the other size group.

Description. In their description of the Bug Creek local faunas, Sloan and Van Valen (1965) recognized *Mesodma formosa* and *M. thompsoni* as being present. Novacek and Clemens (1977) believed that only one species of *Mesodma* was present at Bug Creek Anthills. Their study was based on a comparison of P_4's and P^4's of all named species of *Mesodma.* A new species at Bug Creek Anthills was not named, but they suggested that this sample was most similar to *M. thompsoni* but was not remarkably different from *M. formosa.* Their study did not consider the molars of *Mesodma.* The *Mesodma* sample from UCMP loc. V74111 cannot be allocated with certainty to any of the named species of *Mesodma* or to the sample from Bug Creek Anthills. Its greatest number of dental similarities is with *M. thompsoni.* However, enough differences are present to show that the loc. V74111 sample does not belong to *M. thompsoni.*

Mandible. Two mandibular fragments are present in the sample from loc. V74111. Both specimens (UCMP 116622, type, and UCMP 116621, paratype) have the P_3 and P_4, but no molars are preserved. The P_3's agree with the description given by Clemens for *Mesodma formosa* (1964, p. 33).

P_4. Of the seventeen complete, or nearly complete, P_4's referred to *Mesodma garfieldensis,* the serration count can be determined on only ten and it ranges from 11 to 14. One each of the P_4's has a count of 11 and 14, three have 12 serrations, four have 13, one has 11 or 12, and two have 12 or 13. The modal value is between 12 and 13. The number of external and internal ridges is also hard to determine because of wear, but ten is the most common number of external ridges and the number of internal ridges is almost equally divided between 9 and 10. The profile of the blade is quite variable. It ranges from a moderately high, relatively symmetrical arc as seen in *M. formosa* (as in the paratype, UCMP 116621) to a relatively lower, longer arc as seen in *M. thompsoni* (as in the type, UCMP 116622). The fourteen most complete teeth can be divided into two groups of seven each based on their profiles (Fig. 11) and to a lesser degree on their size. However, the ratios that express the shape of the blade (L1/L and H/L, Novacek and Clemens, 1977) do not suggest alignment of these two groups with *M. formosa* and *M. thompsoni.* Novacek and Clemens (1977) found the average L1/L ratio of *M. formosa* to be 0.45 and that of *M. thompsoni* to be 0.42. These ratios are almost the same for the two groups of P_4's referred to *M. garfieldensis,* but there they are reversed. The larger, lower-profiled blades average 0.45, while the smaller, higher-profiled blades average 0.41. For the H/L ratio Novacek and Clemens arrived at mean values of 0.46 for *M. formosa* and 0.41 for *M. thompsoni.* Although the average H/L values calculated for the two groups within *M. garfieldensis* are higher (small-sized group, 0.49; large-sized group, 0.46) than for *M. formosa* and *M. thompsoni,* they support a decrease in relative height with increase in length of the blade.

In the continuous variables the total sample of seventeen P_4's (Table 6) compares most favorably with *Mesodma thompsoni* (Novacek and Clemens, 1977, table 1), although the absolute height for the total sample is closer to *M. ambigua.* In an approximate t-test comparing the P_4 lengths of the *M. garfieldensis* sample with *M. thompsoni* from loc. V5620, Lance Formation (Novacek and Clemens, 1977, table 1), the null hypothesis that the sample means are equal was not rejected at $\alpha = 0.05$. (An approximate t-test was used because the sample variances based on an F-test were significantly different.) A similar t-test between the *M. garfieldensis* sample and the sample from Bug Creek Anthills (Novacek and Clemens, 1977, table 1) resulted in rejection of the null hypothesis at $\alpha = 0.05$. This suggests that the size (length) of the P_4's in the total sample referred to *M. garfieldensis* more closely approximates that of *M. thompsoni* (from Lance). However, other features of the dentition discussed below (especially the molars) do not favor referral of the *M. garfieldensis* sample to *M. thompsoni.*

M_1. Fifty-five isolated M_1's from loc. V74111 are identified as *Mesodma garfieldensis.* The number of cusps in the external row is almost equally divided between six and seven. The number in the internal row is similarly divided between four and five. This compares to the very stable cusp count of 6:4 that Clemens (1964, p. 45) found for *Mesodma* from the Lance Formation. Like the P_4's but unlike the other molars and P^4's, there is some suggestion of bimodality in the M_1 sample of *M. garfieldensis.* This is not expressed in the relatively low CV's for length (4.57) and width (7.06) of the M_1's. In slightly less than half the M_1's, the anterior margin of the tooth ends in a blunt point or is slightly

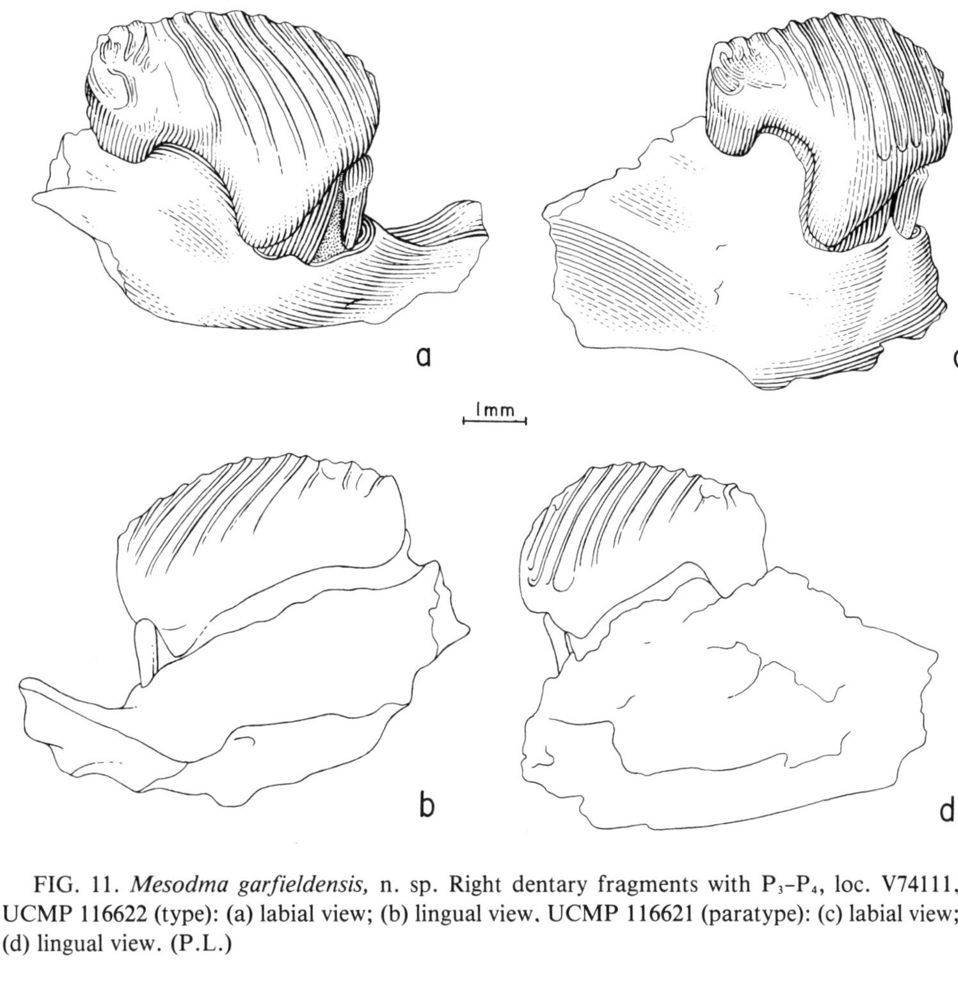

FIG. 11. *Mesodma garfieldensis,* n. sp. Right dentary fragments with P_3–P_4, loc. V74111, UCMP 116622 (type): (a) labial view; (b) lingual view. UCMP 116621 (paratype): (c) labial view; (d) lingual view. (P.L.)

squared (Fig. 12b). Almost 100 percent of these M_1's have a cusp count of 6:4. The second group, comprising slightly more than half the sample, has an anterior margin that is more tapered in outline (Fig. 12a). Almost 100 percent of these M_1's have a cusp count of 7:5. The tapered outline is apparently due to the presence of small external and internal cusps at the anterior margin of the tooth.

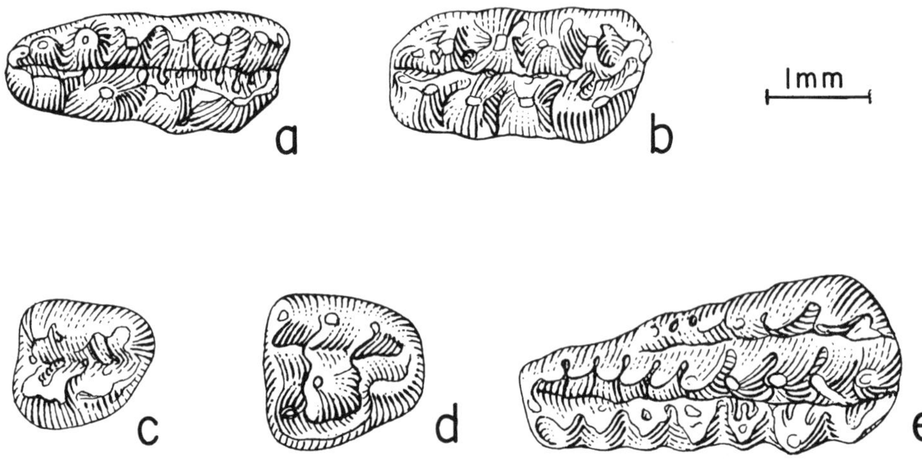

FIG. 12. *Mesodma garfieldensis*, n. sp. Occlusal view of isolated molars, loc. V74111: (a) right M_1, UCMP 116643; (b) right M_1, UCMP 116642 (note the difference in shape between the two M_1's); (c) right M_2, UCMP 116704; (d) right M^2, UCMP 116838; (e) right M^1, UCMP 116790. (P. L.)

The lack of these cusps in the bluntly pointed M_1's, although relatively shortening these teeth, is not paralleled by an equal reduction in width. In a bivariate plot of length vs. width (Fig. 13) the greater relative width of the blunt M_1's is apparent. The more plausible explanations for this variation are: (1) two species of *Mesodma* are present; (2) one species of *Mesodma* is present, showing variation in M_1 morphology; or (3) the bluntly pointed M_1's with cusp formula 6:4 belong to a third genus and species, possibly *Cimexomys minor.* Although showing some variation in cusp formula, the M^1's do not display any comparable bimodality in width-length relationships. This factor, plus the low CV's for length and width seen in all other molars and P^4's and the absence of bimodality in bivariate plots of length vs. width of these teeth, suggests alternative (2) as the most plausible. It is possible that some of the smaller, bluntly pointed M_1's with a cusp formula 6:4 belong to *Cimexomys minor.*

M_2. Slightly more than 50 percent of the M_2's have an external cusp formula of three. The remainder have four external cusps, and a few show the incomplete separation of the third and fourth external cusps. All the M_2's have two internal cusps. There is no indication of bimodality in the sample and, although slightly high, the CV's for length (6.79) and width (6.49) are within the expected range for a single fossil species.

P^4. Of the thirty-five P^4's of *Mesodma garfieldensis,* twenty have two anteroexternal cusps. Of the remainder, three have one cusp, seven have three, and five have four cusps.

Thirty of the thirty-four P^4's have seven cusps in the medial row; the remaining four have eight cusps. The mean values of length and width of the P^4's are noticeably larger than those given by Novacek and Clemens (1977, table 2) for *M. thompsoni* and the Bug Creek *Mesodma* sample. The CV's for length and width (7.24 and 8.06, respectively), although slightly high, do not suggest the presence of more than one species.

Novacek and Clemens (1977) noted that the *Mesodma* sample from Bug Creek shows very little variability in the anteroexternal cusp formula. Ninety-five percent of the specimens have two cusps in this region. *M. garfieldensis* from loc. V74111 also has a modal value of two for the number of anteroexternal cusps. The percentage having this number is lower (57%), but the next most common number of anteroexternal cusps, three, comprises only 20 percent of the sample. In contrast, the modal value for the number of anteroexternal cusps in *M. formosa* is three and in *M. thompsoni* is four (Clemens, 1964).

M^1. The cusp formula ranges as follows: 6–8:8–11:4⁺–5⁺. Eighty percent (28 of 34) of the M^1's have seven cusps in the external row. The number of cusps in the medial row is equally divided between nine and ten, with only four of five teeth having eight cusps and two having eleven cusps. The modal value of external cusps (mode = 7) for *Mesodma garfieldensis* is the same as that determined for M^1's from the Lance (Clemens, 1964). The mode for the number of medial cusps from the Lance was 9 (Clemens, 1964). Examination of the unpublished data sheets kindly provided by W. A. Clemens showed that only one M^1 has a medial cusp formula of 10. This contrasts to 50 percent of the M^1's of *M. garfieldensis* possessing ten medial cusps. Clemens (1964, p. 47) noted that determination of

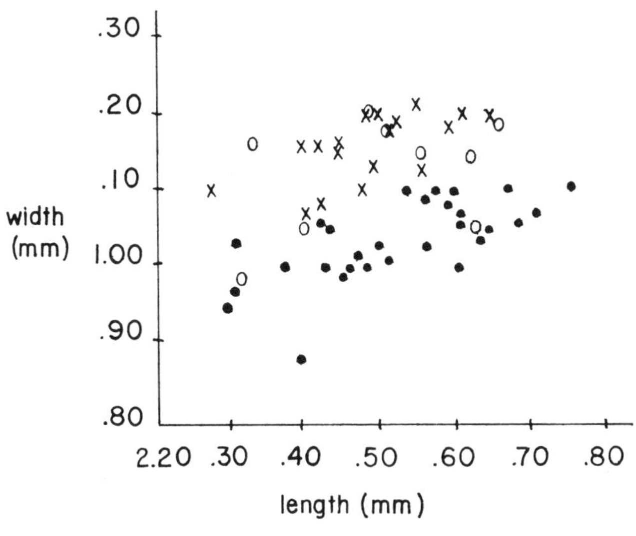

FIG. 13. Bivariate (width vs. length) plot of M_1's of *Mesodma garfieldensis*, n. sp., loc. V74111. x = bluntly pointed M_1's; ∘ = intermediates; • = pointed M_1's. (See text for explanation.)

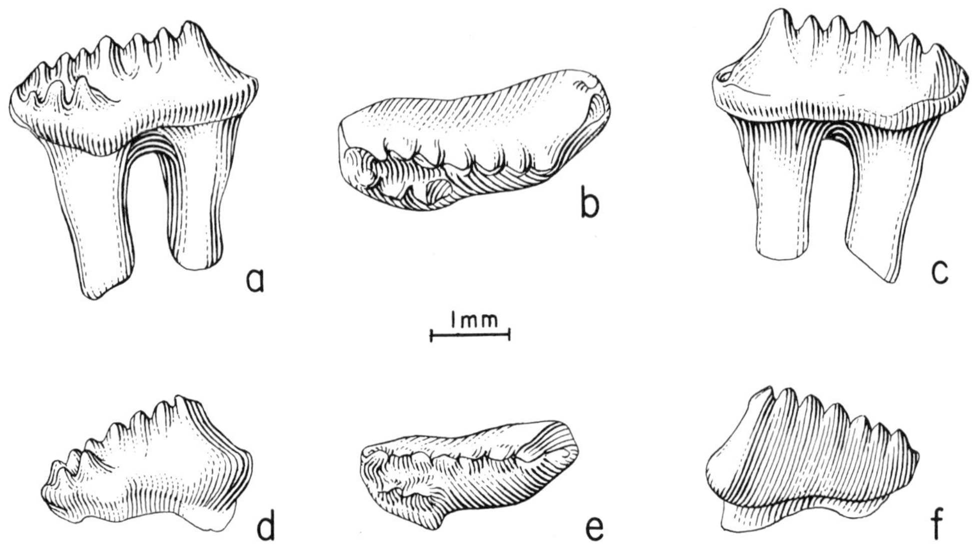

FIG. 14. *Mesodma garfieldensis*, n. sp. Isolated right P^4's, loc. V74111. UCMP 116751: (a) labial view; (b) occlusal view; (c) lingual view. UCMP 116750: (d) labial view; (e) occlusal view; (f) lingual view. (P. L.)

internal cusp count is rather subjective because of the transition from smaller anterior cusps to larger posterior cusps. In the *M. garfieldensis* sample most of the M^1's have from four to five larger internal cusps, but many of the teeth also have several small anterior cuspules.

The M^1's of *Mesodma garfieldensis* are closest in size to *M. thompsoni.* When length and width are considered together, the M^1's of *M. garfieldensis* show the least amount of dispersion of any of the molars, indicated by low CV's (length 4.73; width 4.99). Of the molariform teeth, the M^1's of *Mesodma* are the only isolated teeth that can be separated morphologically from isolated M^1's of *Cimexomys minor. C. minor* is relatively rare at loc. V74111, but the rather high CV's for molars of *M. garfieldensis* other than M^1's suggest that teeth of *C. minor* may be included in the samples of *M. garfieldensis* M_1's, M_2's, and M^2's.

M^2. Of the thirty-five M^2's only one (or possibly two or three) has any other cusp formula than 1:3:3. The few M^2's that do differ have four cusps in their internal row. There is no suggestion of bimodality in the sample.

Discussion. This group of teeth can clearly be assigned to *Mesodma.* However, their allocation to any named species is not warranted. If the sample from loc. V74111 represents only one species of *Mesodma*, the following characters suggest the naming of a new species. (1) Although the P_4's are very similar in size to *M. thompsoni,* the H/L ratio suggests that the P_4's from loc. V74111 are proportionally higher. (2) The modal value of 2 for anteroexternal cusps on the P^4's most closely approximates that seen in the sample of

TABLE 6. Measurements of teeth of *Mesodma garfieldensis,* n. sp., UCMP loc. V74111.

		L1	*L*	*H*	*W*	*L1/L*	*H/L*
	$\bar{x}$	1.74	3.98	1.89	1.44	0.43	0.47
P_4	s	0.234	0.338	0.165	0.136	0.034	0.043
	CV	13.45	8.49	8.73	9.44	7.95	9.21
	OR	1.36 −2.06	3.50 −4.57	1.64 −2.13	1.23 −1.66	0.35 −0.48	0.42 −0.57
	N	17	17	17	14	16	16

		$\bar{x}$	*s*	*CV*	*OR*	*N*
P^4	L	2.82	0.204	7.24	2.42 −3.24	35
	W	1.42	0.114	8.06	1.13 −1.63	35
M_1	L	2.50	0.114	4.57	2.27 −2.76	55
	W	1.09	0.077	7.06	0.88 −1.22	55
M_2	L	1.41	0.096	6.79	1.22 −1.66	44
	W	1.18	0.077	6.49	1.03 −1.39	44
M^1	L	3.35	0.158	4.73	3.09 −3.65	33
	W	1.49	0.074	4.99	1.31 −1.65	43
M^2	L	1.46	0.079	5.42	1.28 −1.60	34
	W	1.42	0.060	4.20	1.32 −1.57	35

Mesodma from Bug Creek Anthills, although the P^4's (and P_4's) from Bug Creek Anthills are markedly smaller. (3) The M_1 cusp formula is almost evenly split between 6:4 and 7:5. In the Lance sample of *Mesodma,* Clemens (1964) found fewer than 10 percent of the M_1's had five or seven external cusps. This percentage remains about the same when the M_1's from the Lance are separated into species, as is done in Fig. 10 and Appendices 1, 2, and 3. (4) The number of medial cusps on the M^1's from loc. V74111 is evenly divided between nine and ten. In the Lance sample only one M^1 has ten medial cusps. (5) In Fig. 10 it can be seen that the M_2^2's from loc. V74111 conform in length most closely to *M. formosa,* while the M_1^1's are intermediate in size between *M. formosa* and *M. thompsoni.* This indicates that the length ratios of M^2 to M^1 and M_2 to M_1 are lower for the loc. V74111 sample of *Mesodma.* The increase in number of cusps on the M_1's and M^1's from loc. V74111 supports the suggestion that these lower length ratios are real.

Most of the evidence suggests the presence of only one species of *Mesodma* in the Hell's Hollow local fauna, although this is by no means certain. Assuming that the sample does represent a single species, *M. garfieldensis,* it is most similar to *M. thompsoni,* particularly in the size and morphology of P_4. There are some, but fewer, resemblances to the *Mesodma* sample from Bug Creek Anthills (particularly in the number of P^4 antero-external cusps). Some resemblances (particularly absolute, but not relative, height of P_4) can also be seen to *M. ambigua* but like *M. thompsoni, M. garfieldensis* is smaller and has fewer P_4 serrations (probably a modal value of 13 for both later species compared to a mean value of 13.7 for *M. ambigua). M. garfieldensis* differs most from previously named Cretaceous species of *Mesodma* in the relative length of M^1_1. The samples of isolated molars of *M. garfieldensis* are large enough to show that M^1_1 is lengthened relative to M^2_2 mostly through the addition of cusps. The molar relationships have not been studied in the *Mesodma* sample from Bug Creek Anthills and only one molar (an M_1) has been recorded for *M. ambigua*. However, it would not be surprising to find that in at least *M. ambigua* the M^1_1's are relatively longer. As will be discussed later, there is a general trend to lengthen and add more cusps to M^1_1 in Tertiary ptilodontoids. This trend is already obvious in *M. garfieldensis* and thus might also be present in the probably younger (mantuan) and probably related *M. ambigua.*

Neoplagiaulax Lemoine, 1880

?*Neoplagiaulax burgessi,* n. sp.

(Figs. 15 & 16 & Table 7)

Etymology. Named for the Burgess family, on whose land these specimens were collected.

Type. UCMP 116896, an isolated left P_4.

Type locality. UCMP loc. V73087, Hell Creek Formation, Montana.

Referred specimens. 4 isolated P_4's from UCMP loc. V73087.

Locality. UCMP loc. V73087.

Distribution. Hell Creek Formation, Montana (Lancian).

Diagnosis. Only P_4's known, which are slightly larger than those of *Mesodma formosa*; P_4 with eleven or twelve serrations (mode = 11); serrate edge high, arcuate, and symmetrical; highest point on blade at greater than 50 percent of total P_4 length; anterior margin of P_4 slopes posteriorly, forming smooth transition into serrate edge.

Description. Five P_4's from UCMP loc. V73087 are morphologically distinct from any P_4's previously reported from the Cretaceous of North America.

Of the five P_4's, three have eleven serrations and two have twelve serrations. Three of the teeth have ten internal ridges and the remaining two have eight and nine internal ridges. The number of external ridges are distributed as follows: three have nine external ridges, one has ten, and one has eight. The P_4's are slightly larger than those of *Mesodma formosa* (Table 7). The serrate edge of the blade is high and almost semicircular in profile. The anterior edge of the blade slopes posteriorly, forming a smooth transition from the anterior edge to the serrate crown. This is in contrast to the larger species of *Mesodma,* where the anterior edge forms a slightly convex, almost vertical margin noticeably delimited from the serrate portion of the blade. The highest point on the blade is at the fifth serration in four of the five teeth. The last one to two serrations have no internal or external ridges, and tend to be slightly rounded and better separated by deeper notches.

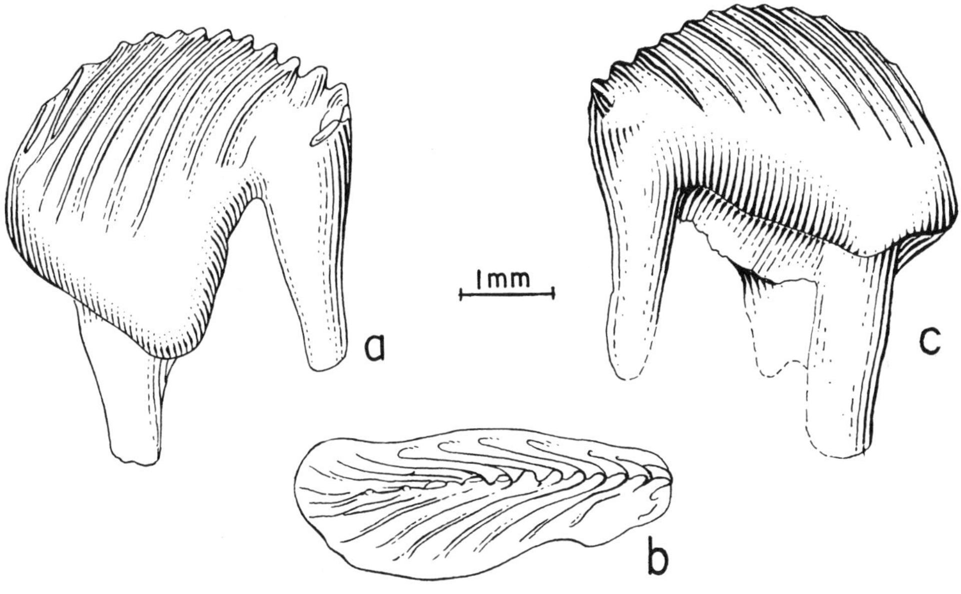

FIG. 15. ?*Neoplagiaulax burgessi*, n. sp., type. Isolated left P_4, UCMP 116896, loc. V73087: (a) labial view; (b) occlusal view; (c) lingual view. (a, b, P.L.; c, A.L.)

All five P_4's are relatively unworn, but three of them show the beginnings of small wear facets on their extreme posterolabial side. The wear facet tends to bevel the labial side of the last two to three serrations and accentuates a small posterolabial basal ridge which is present on all five P_4's.

The first, weak internal ridge is intersected near its base by the second and third internal ridges. The external ridges are generally parallel throughout their length, except that in two of the P_4's the second and third external ridges join toward the base of the crown.

Of the two roots, the anterior is much larger. Both roots are ovoid in cross-section. The anterior root is grooved on its anterior surface and the root is greatly overhung by the enamel-lined anterior edge of the crown, indicating that a P_3 had been present. All five teeth have a continuous interradicular crest joining the anterior and posterior roots.

TABLE 7. Measurements of P_4's of ?*Neoplagiaulax burgessi*, n. sp.

UCMP Number		*L1*	*L*	*H*	*W*	*L1/L*	*H/L*	
116894		1.87	3.53	1.68	1.24	0.53	0.48	
116895		1.93	3.53	1.67	1.35	0.55	0.47	
116893		2.07	4.01	1.96	1.49	0.52	0.49	
116896		a2.09	3.81	1.91	1.57	0.55	0.50	(type)
116897		1.99	3.82	1.83	1.53	0.52	0.48	
	$\bar{x}$ =	1.99	3.74	1.81	1.44	0.53	0.48	

a Approximate measurement.

Discussion. The unique feature of these P_4's compared to those of *Mesodma* is the shape of the blade, which is high and almost semicircular in profile. The highest point on the blade occurs at slightly more than half the crown length. This is seen in the L1/L ratio, which has an average of 0.53 and an OR of 0.52–0.55. The L1/L ratio was calculated for twelve samples of *Mesodma*. Nine of the mean values of L1 and L were taken from Novacek and Clemens (1977, table 1). Three of these samples are from Garfield County. The OR of L1/L for these twelve samples is 0.42–0.50, and the mode is 0.45. Fig. 16 shows that the shape of the blade in ?*Neoplagiaulax burgessi* does not follow the P_4 shape trend of *Mesodma*, as expressed by the L1/L ratio. The regression line for the twelve *Mesodma* samples is $Y = -0.2164 + 0.5087X$, and the correlation coefficient is 0.9643. Both the slope and correlation coefficient are significantly different from 0 at $P < 0.001$. The 95 percent and 99 percent confidence limits for this line are plotted in Fig. 16. The mean value for the L1/L ratio of the ?*Neoplagiaulax burgessi* sample lies well beyond the 99 percent confidence limits of the *Mesodma* trend. This suggests that the five P_4's of ?*Neoplagiaulax burgessi* follow a shape trend other than that expressed by *Mesodma*.

Among North American Lancian multituberculates the only species similar to ?*Neoplagiaulax burgessi* in the shape of the P_4 is *Cimolodon nitidus*. As in the former species, the anterior margin of the P_4 in *C. nitidus* slopes posteriorly and forms a gradual transition to the serrate edge. The blade in *Cimolodon* is also similarly high, arcuate, and nearly symmetrical. However, in comparison to the three P_4's of *C. nitidus* described below, the height of the P_4 of ?*N. burgessi* (as expressed by the H/L ratio) is lower. In addition, the P_4's of ?*N. burgessi* are significantly smaller than those of *C. nitidus*. (A t-test comparing the means of the P_4 lengths would be superfluous, since the OR's of the two species do not overlap.)

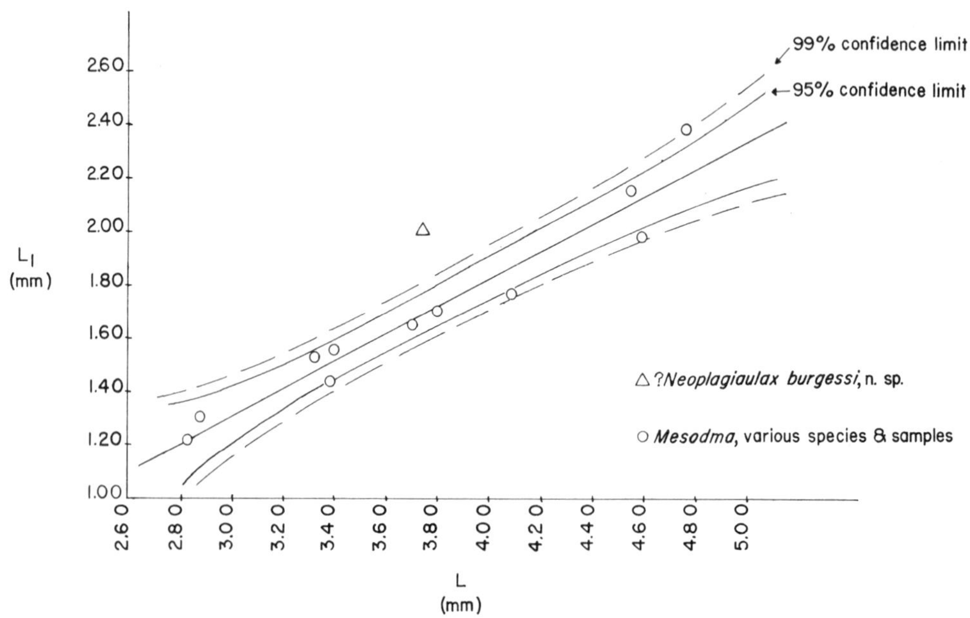

FIG. 16. Regression analysis of various species of *Mesodma*, using mean values of P_4's (L1/L), and ?*Neoplagiaulax burgessi*, n. sp. (See text for explanation.)

The basis upon which this new species is erected is somewhat tenuous, as only five P_4's can definitely be referred to it. However, all five blades share a distinctive morphology markedly different from any previously described Lancian multituberculate. The specimens could represent a new genus, but because of the nature of the material the naming of a new genus is not warranted, at least at this time. In addition, these P_4's are quite similar to several unnamed species of neoplagiaulacids from the Paleocene currently being studied by R. E. Sloan. Dr. Sloan generously provided outline drawings and average measurements for these unnamed species. It is largely upon his data that I have referred this new species questionably to *Neoplagiaulax*. After more complete descriptions of these new neoplagiaulacid species have been published it may be necessary to refer ?*Neoplagiaulax burgessi* to another genus, or, less likely, to erect a new genus for it.

Finally, as noted in the description of *Mesodma* cf. *M. formosa,* there is a possibility that molars allocated to that species may actually belong to ?*Neoplagiaulax burgessi.* It would not be surprising to find the molars of these two multituberculates to be quite similar, especially if Van Valen and Sloan (1966) were correct in suggesting that *Neoplagiaulax* was derived from *M. formosa.*

Family ?NEOPLAGIAULACIDAE

Gen. et sp. indet
(Fig. 17)

Description. One isolated P_4, UCMP 110626, from loc. V72208 cannot be referred to any recognized genus and species of Lancian multituberculate. The continuous variables for this P_4 are as follows; length, 4.15; width, 1.76; L1, 1.93; and height, 2.16. The H/L ratio is 0.52 and the L1/L ratio is 0.47. The serrate edge is high and arcuate, and was probably symmetrical when unworn. There are twelve serrations and nine or ten external and internal ridges. The anterior margin of the P_4 forms a long (greater than one-third of the total length), straight border distinct from the serrate edge. The first serration bears short, indistinct external and internal ridges, plus a more distinct ridge on its anterior margin. All but the first external and internal ridges parallel each other throughout their length. If the second through fourth external and internal ridges continued anteroventrally, they would intersect the first external and internal ridges. A small, posterolabial ridge is accentuated by a wear facet between it and the serrate edge. The wear facet is triangular in shape, with its narrowest portion ending dorsally on the labial side of the sixth from last serration. The posterior root is missing but appears to have been roughly circular in cross-section. The anterior root is larger and ovoid in cross-section. There is a large medial accessory root. The anterior margin of the crown does not greatly overhang the anterior root and the anterior margin of this root is not grooved. However, the presence of an enamel-lined anterobasal cup indicates that a P_3 was probably present.

Discussion. Of the Lancian multituberculates, UCMP 110626 most closely resembles *Mesodma,* ?*Neoplagiaulax burgessi,* and *Cimolodon nitidus.* The relative height of the blade decreases from *C. nitidus* to ?*N. burgessi* to *Mesodma.* In *C. nitidus* and ?*N. burgessi* the highest point on the blade is usually at or slightly behind the midpoint of the crown, while in *Mesodma* it is slightly in front of the midpoint. UCMP 110626 is intermediate between *C. nitidus* and ?*N. burgessi* in relative height of the blade and is similar to *Mesodma* in relative position of the highest point on the blade. Also as in *Mesodma* but not the other two genera, the anterior margin of UCMP 110626 forms a relatively straight border distinct from the serrate edge. However, the anterior margin of UCMP 110626 is longer than in P_4's of *Mesodma*. In its combination of twelve serrations, long, straight anterior margin, high blade, and highest point of the blade less than 50 percent of the total P_4 length, UCMP 110626 resembles Tertiary neoplagiaulacids such as *Neoplagiaulax* and *Parectypodus*.

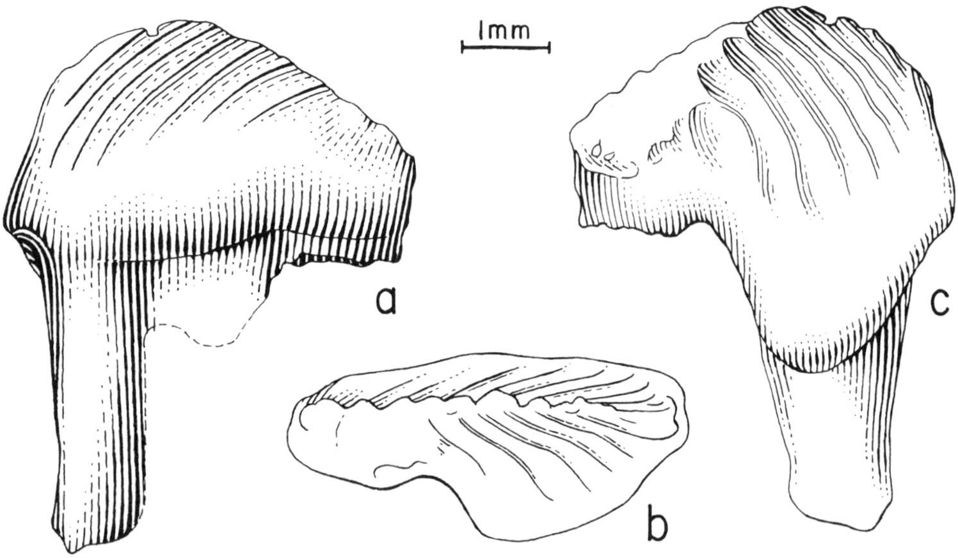

FIG. 17. ?Neoplagiaulacidae, gen. et sp. indet. Isolated worn right P_4, UCMP 110626, loc. V72208: (a) lingual view; (b) occlusal view; (c) labial view. (a, A. L.; b, c, P. L.)

Family CIMOLODONTIDAE Marsh, 1889a

Cimolodon Marsh, 1889a

Cimolodon nitidus Marsh, 1889a

(Figs. 18–20 & Table 8)

Cimolodon nitidus Marsh, 1889a, p. 84
(See Clemens, 1964, p. 56, for synonymies.)

Type. YPM 11776, left M_1 (Marsh, 1889a, pl. II, figs. 5–8).

Type locality. Mammal locality no. 1 of Lull (1915), UCMP loc. V5003, Lance Formation, Wyoming.

Referred specimens. 4 P_4's; 2 M_1's; 4 M_2's; 8 P^4's; 5 M^1's; 4 M^2's from various localities.

Localities. UCMP locs. V72208, V72212, V72213, V73077, V73083, V73085, V73087, V73093, V74116, V75157, and V75161.

Distribution. Lance Formation, Wyoming; Scollard Formation, Alberta; Hell Creek

Formation, Montana (all Lancian); St. Mary River Formation, Alberta (Edmontonian).

Revised diagnosis. Clemens, 1964, p. 56.

Comments. Clemens (1964, 1973) noted that the teeth of *Cimolodon nitidus* from the Lance are quite variable in size, but concluded: "*Cimolodon nitidus* from the type Lance Formation is probably a species exhibiting a relatively high variability in the dentition but might be composed of two closely related species" (1973, p. 83). In contrast, Lillegraven (1969, p. 29) found that *C. nitidus* from the Scollard Formation formed a "closely knit group." The Hell Creek Formation collection resembles the Lance Formation sample in

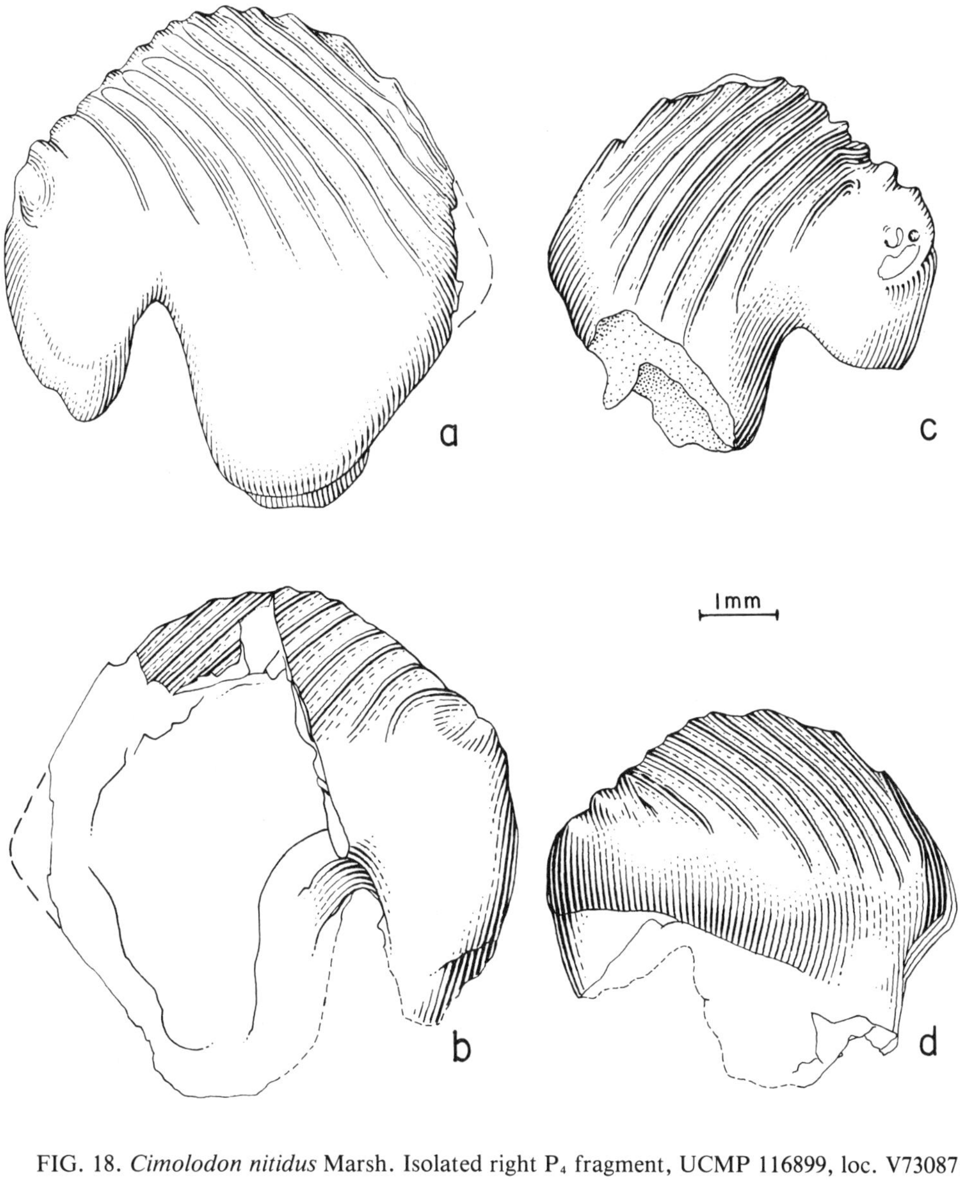

FIG. 18. *Cimolodon nitidus* Marsh. Isolated right P_4 fragment, UCMP 116899, loc. V73087: (a) labial view (anterior edge missing); (b) lingual view. Isolated left P_4, UCMP 116901, loc. V73093: (c) labial view; (d) lingual view. (a, c, P. L.; b, d, A. L.)

its great amount of variability. In Table 8, tooth sites that show a great amount of variability in size are given both as total sample and as individual size groupings. The presence of two morphs of Lancian *Cimolodon* in the Hell Creek and Lance formations receives some corroboration from the presence of two Aquilan morphs (*C. electus* and *C. similis*) in the Upper Milk River Formation (Fox, 1971 and pers. comm.).

Description. P_4. The four P_4's separate into two size classes. The first group (Group A in Table 8) are the largest P_4's and are represented by UCMP 116900, loc. V75161, and UCMP 116899, loc. V73087. UCMP 116900 is a posterior half of a P_4. UCMP 116899 is most of the labial side of a P_4 but is missing the most anterior portion (Fig. 18a, b). The latter specimen preserves fourteen serrations and could have had one more on its anterior edge. The differences between this P_4 and those described by Clemens (1964, pp. 59–61) are related to an increase in height and posterior shift of the highest point of the serrate edge in UCMP 116899 (Fig. 19). The profile of UCMP 116899 is very close to that of *Anconodon gidleyi* and is higher than the other two P_4's from the Hell Creek Formation referred to *Cimolodon nitidus* (Figs. 18c & d & 19 & Group B in Table 8). The latter two

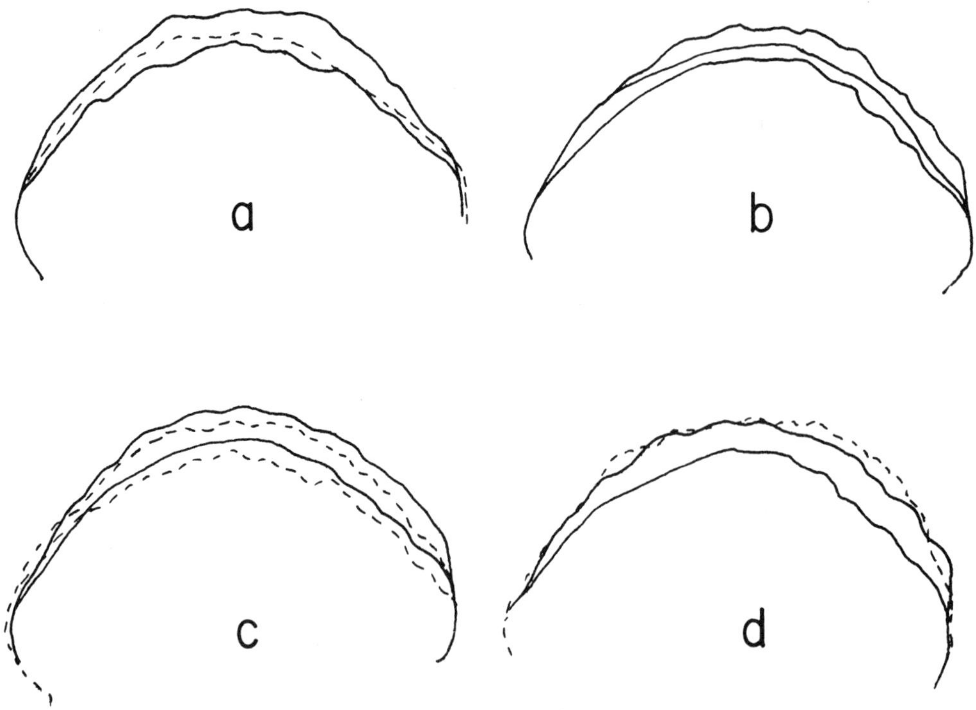

FIG. 19. *Cimolodon nitidus* Marsh. Profile of P_4's drawn to same scale: (a) observed range of variation, Lance Formation, Wyoming, shown by solid lines, after Clemens (1964, fig. 24) and UA 2333, Scollard Formation, Alberta (based on cast) shown by dashed lines; (b) Hell Creek Formation, Montana, from lowest to highest UCMP 116901, UCMP 116898, UCMP 116899; (c) observed ranges of variation, Lance Formation, Wyoming, shown by dashed lines, and Hell Creek Formation, Montana, shown by solid lines; (d) observed range of variation, Hell Creek Formation, Montana, shown by solid lines, and *Anconodon gidleyi,* type, USNM 9763, shown by dashed line, after Jepsen (1940, fig. 10).

P_4's have 12–13 serrations and conform closely to the shape of P_4's from the Lance Formation. These differences in profile can also be seen by comparing the H/L and L1/L ratios of UCMP 116899, 0.60 and 0.55 (both estimates), respectively, to the average H/L and L1/L ratios of the other two Hell Creek specimens, which are 0.55 and 0.51, respectively. The morphology of UCMP 116899 strengthens the possibility of a phylogenetic relationship between *Cimolodon* and *Anconodon*, as was suggested by Clemens (1964, pp. 70–71, table 8) and Van Valen and Sloan (1966, fig. 4).

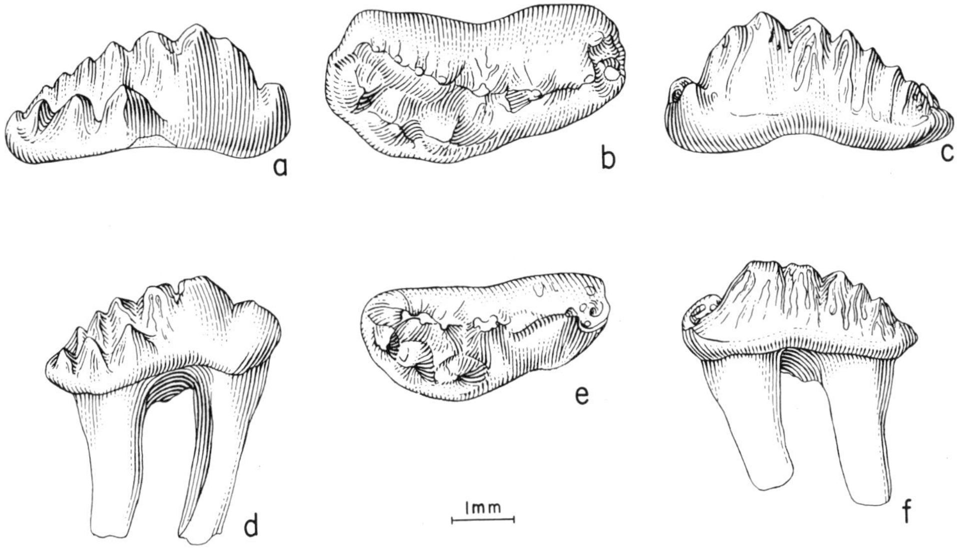

FIG. 20. *Cimolodon nitidus* Marsh. Isolated right P^4's, UCMP 116909, loc. V75178: (a) labial view; (b) occlusal view; (c) lingual view, UCMP 116906, loc. V73087: (d) labial view; (e) occlusal view; (f) lingual view. (P. L.)

M_1. The two M_1's referable to *Cimolodon nitidus,* UCMP 116902 and UCMP 116903, have a cusp formula 6 or 7:4 and agree with the description given by Clemens (1964, p. 61).

M_2. All four M_2's have two subcrescentic internal cusps. The three relatively unworn M_2's have four smaller external cusps. On all three, the fourth cusp is incompletely subdivided into two cusps. There are numerous small ridges on the opposing faces of the internal and external cusps and in the longitudinal valley.

P^4. Like the P_4's, the P^4's can be divided into two groups based on size (Table 8). The two larger P^4's (Group A in Table 8) have a cusp formula 3–?4:7:2. The remaining five P^4's (Group B) have a cusp formula 3-4:5–?7:1–2. Clemens (1964, pp. 63–64) noted modal values for the medial and anteroexternal cusp rows as 6 and 4 respectively. In Group B, three P^4's have six medial cusps, one has five cusps, and one may have had seven medial cusps when unworn. The same group has three P^4's with three anteroexternal cusps and two P^4's with four. Most of the Hell Creek samples (Groups A and B) have one posterolingual basal cusp. In the cusp formula for P^4, Clemens did not indicate the presence of any basal cusps. However, in his description he did mention the presence of

basal cusps on the P^4's, as is noted for the Hell Creek sample. Unlike the P^4's of other Lancian multituberculates, the P^4's of *Cimolodon nitidus* are difficult to align for measurement. In most of the P^4's of *Cimolodon*, especially the larger ones, the medial cusp row is curved in occlusal view, with the concave side lingual. This is another character separating *Cimolodon* from other Lancian multituberculates.

M^1. The cusp formula for the four most complete M^1's is 5–6:?7–8:?4–6. The morphology of these M^1's agrees well with the description given by Clemens (1964, p. 65).

TABLE 8. Measurements of *Cimolodon nitidus*.

				P_4's				
UCMP Number	*UCMP Locality*	*L1*	*L*	*H*	*W*	*L1/L*	*H/L*	*Size Group*
116899	V73087	3.05	5.88	3.79				A actual
		3.47	6.30	3.79		0.55	0.60	estimated
116898	V73087	2.66	a5.22	a2.88		0.51	0.55	B
116901	V73093	2.51	5.11	2.83	2.26	0.49	0.55	B

UCMP Number	*UCMP Locality*	*Tooth Site*	*L*	*W*	*Size Group*
116903	V75157	M_1	4.48	2.11	
116902	V73087		4.53	1.97	
116904	V74116	M_2	2.49	1.74	
116905	V72213		2.78	2.38	
108962	V73083		3.11	a2.28	
109010	V73085		2.95	2.23	
116907	V73087	P^4	4.54	2.79	A
116906	V73087		3.72	2.33	B
108947	V73077		3.70	2.16	B
116909	V75178		4.56	2.87	A
116910	V74116		3.49	2.40	B
116911	V74116		3.40	2.20	B
116912	V74116		3.57	2.24	B
116913	V72208	M^1	4.48	2.52	
110621	V72208		—	2.64	
116914	V73087		4.34	2.52	
116915	V73087		3.66	1.88	
116916	V73087		4.31	2.32	
116919	V73087	M^2	2.42	2.02	
116917	V73087		3.07	—	
116918	V73087		—	a2.61	
109032	V73087		3.57	2.70	
108957	V72212		3.10	2.64	

a Approximate measurement.

M^2. Four M^2's have a cusp formula 1-3:4:4–6 and are similar to the M^2's from the Lance Formation, Wyoming. The M^2's cannot be divided into two size groups, but they vary considerably in size.

Suborder TAENIOLABIDOIDEA (Granger and Simpson, 1929) Sloan and Van Valen, 1965

Family EUCOSMODONTIDAE (Jepsen, 1940) Sloan and Van Valen, 1965

Subfamily EUCOSMODONTINAE Jepsen, 1940

Stygimys Sloan and Van Valen, 1965

Stygimys aff. *S. kuszmauli* Sloan and Van Valen, 1965

(Fig. 21 & Table 9)

Stygimys kuszmauli Sloan and Van Valen, 1965, p. 224

Type. UMVP 14, left lower jaw fragment with I and alveoli for P_4 and M_1, shown as part of restoration in fig. 4 of Sloan and Van Valen (1965).

Type locality. Bug Creek Anthills, Hell Creek Formation, Montana.

Referred specimens. 2 P_4's; 5 M_1's; 7 M_2's; 4 P^4's; 4 M^1's; 11 M^2's from UCMP loc. V74111.

Locality. UCMP loc. V74111.

Distribution. Lowermost Tullock Formation, Montana (pre-mantuan), probably also uppermost Hell Creek Formation, Montana (Lancian).

Diagnosis. Sloan and Van Valen, 1965, p. 224, but see discussion.

Description. *Stygimys* was named by Sloan and Van Valen (1965) on the basis of material from the Bug Creek Anthills local fauna. Besides the type species, *S. kuszmauli,* they referred four other species to this genus, including *S. "Eucosmodon" gratus* Jepsen, 1930b. They report *S. gratus* as occurring at Harbicht Hill, Montana, and Mantua lentil, Wyoming. *S. kuszmauli* occurs at the stratigraphically lower Bug Creek Anthills and Bug Creek West localities.

The specimens from UCMP loc. V74111 have been referred to aff. *Stygimys kuszmauli* after comparison with the UCMP Bug Creek Anthills and Harbicht Hill collections and with Jepsen's descriptions of *S. gratus* (1930b, 1940). The prefix *aff.* is used because the loc. V74111 sample most closely resembles *S. kuszmauli* but may, along with the Harbicht Hill sample, represent a new species. A comparison of P_4's from the above samples (see discussion) also suggests that *S. kuszmauli* is not distinguishable, based solely on P_4 size, from *S. gratus*. Most of the following comparisons are made with *S. gratus*, because *S. kuszmauli*, except for the P_4's, has not been described in detail.

P_4. Only two P_4's of this species were recovered from loc. V74111. Like *Stygimys kuszmauli* and *S. gratus*, they have eleven serrations. The length of one P_4, UCMP 116920, is similar to the size of *S. kuszmauli* given by Sloan and Van Valen (1965). This P_4 has ten external ridges. Only the last serration does not bear an external ridge. The first four external ridges (one for each serration) are short and parallel each other. The next six ridges are longer and set at a lower angle to the horizontal than are the preceding four. There are nine internal ridges. The more anterior three to four internal ridges are shorter and set at a higher angle than the more posterior ridges. The more posterior internal ridges are more undulatory than their external counterparts. The last internal ridge (on the ninth serration) is very short and weakly developed. There are short vertical striations on the lingual side of P_4 immediately below serrations nine and ten. A small posterolabial basal

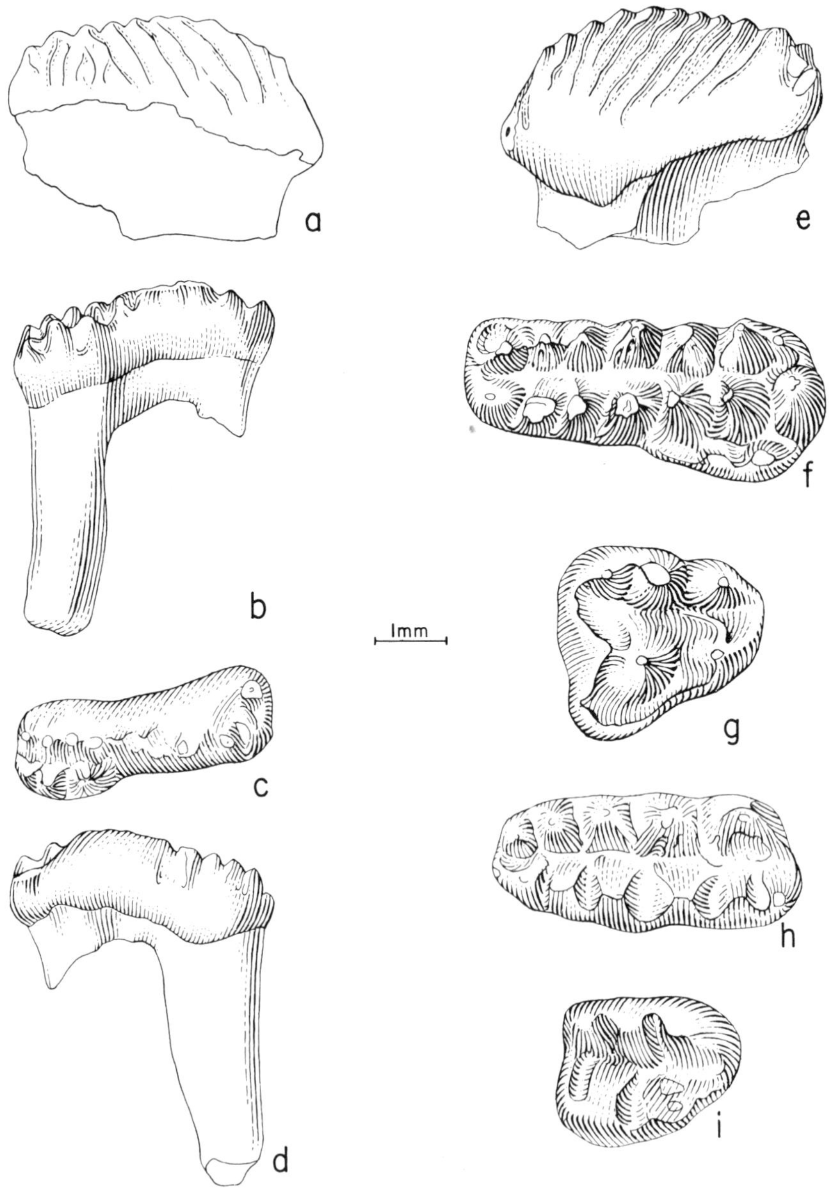

FIG. 21. *Stygimys* aff. *S. kuszmauli* Sloan and Van Valen. Isolated teeth, loc. V74111. Left P_4, UCMP 116920: (a) lingual view; (e) labial view. Right P^4, UCMP 116934: (b) labial view; (c) occlusal view; (d) lingual view. Left M^1, UCMP 116938: (f) occlusal view. Right M^2, UCMP 116942: (g) occlusal view. Left M_1, UCMP 116922: (h) occlusal view. Right M_2, UCMP 116927: (i) occlusal view. (P. L.)

expansion is present below serrations 10 and 11. A small wear facet is developed between the posterolabial basal expansion and the last three serrations. Not much of the two roots is preserved, but the anterior one is slightly larger and an inter-radicular crest connects the two roots. The second P_4, UCMP 116921, differs only slightly from the morphology of UCMP 116920. The first serration on UCMP 116921 is not very pronounced and the posterior external and internal ridges are more muted, probably due to wear. However, UCMP 116921 is distinctly larger than UCMP 116920. It is also larger than the two P_4's of *S. gratus* reported by Jepsen (1940, table 3). The statistics (P_4 length, $\bar{x} = 4.6$, $s = 0.3$) given by Sloan and Van Valen (1965, p. 224) for S. *kuszmauli* show that this P_4 is greater than two standard deviations from the mean of the P_4 length of this species (see discussion).

M_1. The four complete M_1's from loc. V74111 all have a cusp formula 7:5. The five internal cusps increase in size and height posteriorly. The more posterior cusps become subcrescentic. The last internal cusp bears a thickened ridge extending from its apex down to the midline of the tooth. The more posterior internal cusps have shallow grooves on their labial surface. The external cusps are conical to quadrate in outline and increase in size posteriorly. The notches separating the external cusps become shallower posteriorly; the penultimate and ultimate cusps are incompletely separated. On three of the four M_1's the notch separating the fourth and fifth external cusps is distinctly larger. In two of these three M_1's there is a noticeable labial bulge next to this enlarged notch and near the posterior end of external cusp four. The length and number of cusps of these M_1's from loc. V74111 is very similar to that given by Jepsen (1940, table 3) for one M_1 of *Stygimys gratus*. The width of this single M_1 of *S. gratus* is greater than that of the M_1's from loc. V74111.

M_2. Six M_2's from loc. V74111 have a cusp formula 3–4:2. The number of external cusps varies from three (one tooth), to the third cusp being incompletely separated into a third and fourth cusp (two teeth), to four cusps (three teeth). Even when four external cusps are present the third and fourth cusps are separated only by a shallow lingual groove. The external cusps are conical and much smaller than the two internal cusps. The internal cusps are subcrescentic and bear ridges and shallow grooves on their labial side. Although the one M_2 Jepsen (1940, table 3) assigned to *Stygimys gratus* is similar in cusp formula and width to the M_2's from loc. V74111, the length of the latter M_2's is greater.

P^4. Although Sloan and Van Valen (1965) illustrated a P^4 of *Stygimys kuszmauli*, they did not give its cusp formula or describe it. To my knowledge no P^4's of either *Eucosmodon* or *Stygimus* have previously been described in detail.

Like some other multituberculates, *Stygimys* has medial basal cusps on P^4 posterior to the main medial cusp row. Unlike those of other multituberculates from the lowermost Tullock and Hell Creek formations, the two posteromedial basal cusps in *Stygimys* are not clearly delimited from the cusps in the medial row. Therefore, they have been included as part of the medial cusp row.

The four P^4's consistently have three anteroexternal cusps that bear vertical striations on their sides. All four P^4's also have one basal cusp on the posterolingual side of the tooth. The number of medial cusps varies from six(?) to ten, including the posteromedial cusps. The crown height increases gradually to the largest cusp above the posterior root. The greatest variation in the number of medial cusps occurs on the anterior half of the P^4. There are from four to six small, equal-sized conical cusps in this region. They all have vertical striations on their lingual and labial surfaces. In the highest portion of the crown there are one, two, or three(?) closely spaced cusps which probably do not bear striations on their flanks when unworn. Posteriorly there are two smaller (and lower) well-separated conical cusps. A low posterior ridge connects the last of these one or two small medial cusps with the single posterolingual basal cusp. In unworn P^4's a weak ridge leads from this basal cusp up the lingual side of the highest medial cusp. A wear facet develops on the lingual surface below the highest medial cusps. In addition, even the least worn P^4's show wear on the apices of the highest medial cusps. The P^4's have two roots, circular in cross-section, and the anterior root is the largest.

M^1. Four M^1's from loc. V74111 belong to *Stygimys*. Of these four only two are complete and only one is sufficiently unworn to determine the cusp formula 7:7:2–3. The seven external cusps are conical and increase in size posteriorly except for the last cusp, which is the smallest and is positioned on a ridge directed down the posterior side of the penultimate cusp. The first external cusp is noticeably deflected medially. Anterolabial to this cusp is a low cuspidate ridge running from the anterior midline of the M^1 around the first cusp and joining the second external cusp. The larger external cusps bear striations on their lingual sides. The seven medial cusps gradually increase in size toward the back of the M^1 and change from conical to subcrescentic in shape. There are two to three small internal cusps. A small ridge is directed anteriorly from the first internal cusp and merges with the M^1 between the fourth and fifth medial cusps. The internal cusp row is less than 50 percent of the total M^1 length. The larger anterior root is ovoid in cross-section and the smaller posterior root is circular in cross-section. Jepsen (1940, table 3) recorded two M^1's of *S. gratus*. Both are noticeably larger (length and width) than the M^1's from loc. V74111. The number of external and medial cusps is the same for *S. gratus* as for *S. kuszmauli*. However, *S. gratus* has many more internal cusps (5–8) than the M^1's from loc. V74111.

M^2. The eleven M^2's have a cusp formula 1–3:3:3. The external cusp row bears one

TABLE 9. Measurements of teeth of *Stygimys* aff. *S. kuszmauli*, UCMP loc. V74111.

	UCMP Number	*L1*	*L*	*H*	*W*	*L1/L*	*H/L*
P_4	116920	2.23	4.70	1.92	1.94	0.47	0.41
	116921	2.33	5.48	2.22	2.22	0.42	0.41

		$\bar{x}$	*OR*	*N*
P^4	L	3.76	3.60–3.85	4
	W	1.95	1.81–2.10	4
M_1	L	4.44	4.17–4.98	4
	W	1.87	1.82–1.93	4
M_2	L	2.64	2.46–2.72	6
	W	2.03	1.97–a2.14	6
M^1	L	5.01	4.94–5.08	2
	W	2.37	2.35–2.38	2
M^2	L	2.87	2.72–a3.08	9
	W	2.61	2.45–2.74	9

a Approximate measurement.

elongate, ridgelike cusp which may have one to two small cuspules behind it. The internal row has three small, equal-sized, conical cusps. A small ridge is directed down the anterior face of the first internal cusp toward the midline of the tooth. The last internal cusp, like its larger counterpart in the medial row, is separated from the preceding two cusps by a larger notch. The first, small medial cusp is closely appressed to the second cusp, and the former is actually a ridge across the front of the tooth joining the external and internal cusp rows. The second medial cusp is conical, and the larger third cusp is also conical but distinctly canted anteriorly.

Discussion. Sloan and Van Valen (1965) recognized *Stygimys kuszmauli* as occurring at Bug Creek Anthills and Bug Creek West and the supposedly larger *S. gratus* as occurring at Harbicht Hill. The former species was diagnosed on the basis of the P_4. Each species has eleven serrations. Sloan and Van Valen (1965) gave the P_4 length of *S. kuszmauli* as $\bar{x} = 4.6$ and $s = 0.3$. The sample size of the P_4's was not given. A sample of twenty-seven P_4's of *S. kuszmauli* in the UCMP Bug Creek Anthills collection has a similar mean and standard deviation ($\bar{x} = 4.68$, $s = 0.308$; CV = 6.58).

Although the lengths of the two P_4's of *Stygimys gratus* (4.8 and 4.9 mm) given by Jepsen (1940) are greater than either of the means given above for *S. kuszmauli*, they fall within one standard deviation of these means for *S. kuszmauli*. *S. kuszmauli* and *S. gratus* are not separable solely on the size of P_4. No specimens of *S. gratus* were examined in this study, but based on the illustrations of Jepsen (1940, pl.I, fig. 1) the P_4 of *S. gratus* appears to have a more symmetrical blade with the highest point at or near the midpoint (i.e., the L1/L ratio is approximately .50). In contrast, in a sample of twenty-five P_4's of *S. kuszmauli* from the UCMP Bug Creek Anthills collection the highest point on the blade was anterior to the midpoint (i.e., the L1/L ratio average = .44; $s = 0.024$; CV = 5.45). Also, based on Jepsen's illustrations (pl. I, fig. 1), the height of the blade in *S. gratus* probably approaches 50 percent of the total length (i.e., H/L = approx. .50). The H/L ratio in a sample of twenty-two P_4's of *S. kuszmauli* averaged only .42 ($s = 0.02$; CV = 5.00). Thus, utilizing only the P_4 morphology, *S. kuszmauli* is distinguishable from *S. gratus* on the shape of the blade but not on size (if the small sample, N = 2, of *S. gratus* is close to the mean for that population).

In view of these differences the sample of *Stygimys* from UCMP loc. V74111 is of some interest. Unfortunately, as for *S. gratus*, this sample contains only two P_4's. The two P_4's vary considerably in size (Table 9). The P_4's could represent different species; however, the small sample of other teeth does not suggest the occurrence of more than one species of *Stygimys* at loc. V74111. The L1/L and H/L ratios for these two P_4's (Table 9) most closely approach those of *S. kuszmauli*, although if the two teeth are assumed to represent the range of size variation for the P_4's from loc. V74111, then the mean for this sample is larger than either *S. gratus* or *S. kuszmauli*.

The three P_4's of *Stygimys* from the UCMP Harbicht Hill collection are also larger than *S. kuszmauli* or *S. gratus*. The mean length is 5.16 and $s = 0.10$. The L1/L and H/L ratios for this sample are, respectively, $\bar{x} = .47$, $s = 0.10$, CV = 2.13, and $\bar{x} = .44$, $s = 0.012$, CV = 2.73. These values suggest that the P_4 of *Stygimys* from Harbicht Hill is intermediate in shape between *S. kuszmauli* and *S. gratus*.

The samples of *Stygimys* from UCMP loc. V74111, Harbicht Hill, and Mantua lentil are too small to make any meaningful statistical comparisons with each other or to the larger sample from Bug Creek Anthills. Nevertheless, these samples suggest the following: (1) *S. kuszmauli* and *S. gratus* are similar in size but are distinct species, based on differences in P_4 shape. (2) The sample of *Stygimys* from Harbicht Hill could belong to *S. gratus*, but the probable differences in P_4 shape and the larger size of the Harbicht Hill P_4's make it possible that this sample represents a distinct species. (3) The sample from loc. V74111 is the most problematic. The shape of the P_4 blade suggests assignment of this sample to *S. kuszmauli*, while the larger size of the P_4 suggests a similarity to the Harbicht Hill sample. If it can be established that a new species of *Stygimys* is present at Harbicht Hill, I favor the inclusion of the loc. V74111 sample in this species.

It must be kept in mind that the preceding considerations are based solely on the P_4. At least one other feature appears to distinguish *Stygimys gratus* from the UCMP Bug Creek Anthills collection and the loc. V74111 sample of *Stygimys*. Jepsen (1940) notes that the two M^1's of *S. gratus* from the Mantua lentil have five to eight cusps in the internal row. The anterior cusps are very small and near its midline the internal row is interrupted by a vertical groove. In addition, M^1 has three roots. Most of the M^1's from Bug Creek Anthills and loc. V74111 have only two to three internal cusps, occasionally with an additional one or two small anterior cuspules. Infrequently there is evidence that a very small rootlet was present anterolabial or anterolingual to the posterior root. Of the two M^1's of *Stygimys* from the UCMP Harbicht Hill collection only one is well enough preserved to show five internal cusps, of which the posterior three are larger.

Subfamily MICROCOSMODONTINAE Holtzman and Wolberg, 1977

Acheronodon, n. gen.

Etymology. *Acheron*, L., river in Hades; the first river crossed in Dante's *Inferno*. The allusion is the proximity of the type locality to the Hell Creek–Tullock contact, *odon*, Gr., tooth; following Jepsen's names for the other genera now in this subfamily.

Genotypic species. *Acheronodon garbani*, new, and only known, species.

Distribution. Lowermost Tullock Formation, Montana (pre-mantuan).

Diagnosis. Smallest microcosmodontine; P_4 with six cusplike serrations, last serration distinct and separate, posterolabial basal ledge prominent with two cuspules, external and internal ridges relatively long and distinct.

Acheronodon garbani, n. sp.

(Fig. 22 & Table 10)

Etymology. Named for Harley Garbani, in recognition of his important contributions to this project.

Type. UCMP 116953, isolated right P_4. Only known specimen.

Type locality. UCMP loc. V74111, lowermost Tullock Formation, Montana.

Distribution. Same as for genus.

Diagnosis. Same as for genus.

Description. UCMP 116953, an isolated P_4 from loc. V74111, cannot be referred to any recognized multituberculate species. It is extremely small (L, 1.65; L1, 0.85; H, 0.77; max. W, 0.63) and has a distinctive morphology. It has six cusplike serrations. The much lower, vertically oriented first serration is well separated from those following. The last two serrations are well-developed conical cusps, slanting posteriorly. There are four external ridges from the second through the fifth serration. The first external ridge, arising from the second serration, is short and weakly developed. The external ridges parallel each other except for the lower portion of the last ridge, which is convex in a dorsoanterior direction. The external ridges are set at a 45° angle to the vertically oriented first serration. The internal ridges are more weakly developed but parallel their external counterparts in all other features, including the dorsoanterior flexure in the lower part of the fourth external ridge. There is a very prominent posterolabial basal shelf bearing two small cuspules below the last two serrations. The shelf is proportionally wider than in any other premantuan or Lancian multituberculate. Because of the presence of this shelf, the posterior half of the P_4 is wider (W = 0.63) than the anterior half (W = 0.54). Since the lower portion of the crown and the roots are not preserved, it is impossible to determine whether a P_3 was present.

Discussion. Among Late Cretaceous multituberculates the closest similarities are to P_4's of *Cimexomys*. Like the P_4 of *Cimexomys*, *Acheronodon* has a small number of

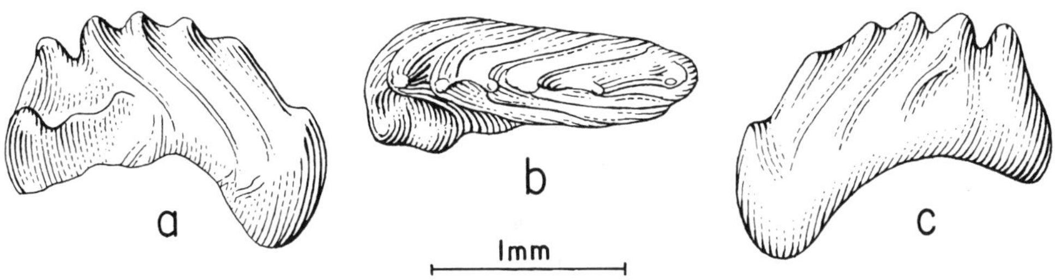

FIG. 22. *Acheronodon garbani*, n. gen. et n. sp., type. Isolated right P_4, UCMP 116953, loc. V74111: (a) labial view; (b) occlusal view; (c) lingual view. (P. L.)

TABLE 10. Comparison of microcosmodontine P_4's. Information on *Pentacosmodon* and *Microcosmodon conus* from Jepsen, 1940, and on *M. woodi* from Holtzman and Wolberg, 1977.

	Pentacosmodon	*Microcosmodon*		*Acheronodon*
	pronus	*conus*	*woodi*	*garbani*
Sample size	2	6	4	1
Length (mean)	1.9	1.7	3.05	1.65
Width (mean)	0.95	0.5	1.03	0.63
Serration count	5	5	5-6	6
External and internal ridges	Long and distinct	Short and indistinct	Short(?) and indistinct	Long and distinct
Posterolabial basal ledge	Weak	Prominent	Reduced	Very prominent
First serration lower and separate	Yes	Yes	Yes	Yes

relatively widely spaced serrations; the posterior serrations are cusplike; and the serrate edge forms a smooth arc, with the first serration distinctly lower and separated from the more posterior serrations. The small number of serrations (6) in *Acheronodon* reflects its extremely small size. However, even in the largest species of *Cimexomys* (*C. hausoi*, n. sp., described below) there are only nine to ten serrations, whereas the smaller *Mesodma thompsoni* has twelve to fourteen serrations.

Acheronodon garbani resembles most closely the microcosmodontines *Pentacosmodon pronus, Microcosmodon conus,* and *M. woodi* (Jepson 1930b, 1940; Holtzman and Wolberg, 1977) known from various Tiffanian localities. Table 10 compares the characters of these microcosmodontines and *A. garbani*. The lateral profile of *A. garbani* forms a smoother arc—as in *M. woodi*—than that in *M. conus* and *P. pronus*, due to the presence of an additional posterior cusp in *A. garbani* and in most specimens of *M. woodi*. Jepsen (1940) and Holtzman and Wolberg (1977) did not give height measurements of the P_4's, but based on their illustrations, at least the relative heights of P_4's in *A. garbani* and *M. woodi* appear to be less than in *P. pronus* and *M. conus*. Holtzman and Wolberg (1977) did note that the P_4 of *M. woodi* is lower than in *M. conus*. In overall shape *A. garbani* most closely resembles *M. woodi*. In size and in the presence of a prominent posterolabial basal ledge, *A. garbani* resembles *M. conus*. As will be discussed later, the presence of a posterolabial basal ledge is regarded as primitive for multituberculates; however, the markedly large posterolabial basal ledges in *M. conus* and *A. garbani* probably represent an increase in size over the primitive condition. *M. conus* and *M. woodi* have reduced external and internal ridges on the P_4. In contrast, in *P. pronus* and *A. garbani* the presumably more primitive condition of distinct P_4 ridges has been maintained.

The overall similarities and the presence of several, possibly shared, derived characters suggest that *Acheronodon* is most closely related to *Microcosmodon* and *Pentacosmodon*, and it is therefore referred to the Microcosmodontinae. This constitutes a range extension for the subfamily back into the earliest Paleocene. Although by no means a proof, this extension supports the idea of Holtzman and Wolberg that the microcosmodontines and eucosmodontines have a long separate history. As noted earlier, there are similarities between *Acheronodon* and *Cimexomys*. Such similarities have also been noted between *Microcosmodon* and *Cimexomys* by Holtzman and Wolberg (1977) and Krause (1977). Like Holtzman and Wolberg, I regard these similarities as primitive retentions and thus of little phyletic significance. These authors also raised the possibility that microcosmodontines do not share a common ancestor with eucosmodontines. Based on their analysis, I agree that this is indeed a possibility, but like them I prefer to retain the taxon as a subfamily of the Eucosmodontidae (although only tentatively).

The naming of a new genus and species based on a single specimen is not a practice with which I generally agree. However, UCMP 116953 is so distinctive that to do so seems justified in this instance. *Acheronodon* is most similar to *Microcosmodon*, and the recovery of additional material of *Acheronodon* may show that it is congeneric with *Microcosmodon*. Although about six million years may separate the known occurrences of *Acheronodon* and *Microcosmodon*, this in itself is not grounds for generic separation.

A final possibility is that UCMP 116953 is a deciduous P_4. This is suggested by the lack of roots, the exceedingly small size, and the unique morphology of the P_4. This must remain an unlikely possibility until it is demonstrated by better material that multituberculates replaced the P_4.

Family TAENIOLABIDIDAE Granger and Simpson, 1929

Catopsalis Cope, 1882b

Catopsalis cf. *C. foliatus* Cope, 1882b

Description. The posterior half of a right M^1 from loc. V74111 and a left M^2 and an M_2 from loc. V74110 are referable to *Catopsalis.* The M^1 fragment preserves three cusps in each of the three rows. The maximum width of the preserved portion is 5.27 mm. Based on its width and length of the last three cusps in the medial row as compared to M^1's of *C. joyneri* in the UCMP Bug Creek Anthills collection, the M^1 (UCMP 116954) was approximately 10 mm long when complete. The cusp morphology is similar to that found in *C. joyneri* Sloan and Van Valen (1965), but the individual cusps are distinctly larger.

The as yet uncatalogued left M^2 from loc. V74110 was called to my attention by W. A. Clemens. He reported (written comm.) that the cusp formula is 1:3:4, the length is 6.0 mm, and the width is 5.3 mm. The uncatalogued M_2 from loc. V74110 was discovered during the summer of 1979 and was not available for inclusion in this study.

Discussion. Catopsalis joyneri Sloan and Van Valen (1965) is the only North American species of this genus in which M^1 has been described. The statistics for width of a sample of M^1's of *C. joyneri* in the UCMP Bug Creek Anthills collection are: $\bar{x} = 4.26$; $s = 0.161$; $CV = 3.79$; $OR = 4.00–4.46$; $N = 9$. In a t-test comparing the width of the single M^1 fragment from loc. V74111 to the above sample, the $P < .001$ that the single M^1 could have come from the same statistical population. This strongly suggests that the M^1 does not belong to *C. joyneri.* However, since the UCMP Bug Creek Anthills sample size is relatively small and the CV (3.79) is quite low, the full range of M^1 widths may not be represented in this sample.

The next largest species of *Catopsalis* is *C. foliatus* from the Puercan (Kielan-Jaworowska and Sloan, 1979). Granger and Simpson (1929) report the M_1 in the type specimen of *C. foliatus* to be 10.7 mm long and 4.9 mm wide. The width of the M^1 fragment (UCMP 116954) is greater than the width of the M_1 in *C. foliatus* and thus it may belong to this species. However, the estimated length of the M^1 is closer to 10 mm. Further, comparison of M^1 and M_1 widths of *C. joyneri* in the UCMP Bug Creek Anthills collection and examination of the illustration of Sloan and Van Valen (1965, fig. 5), indicate that an M^1 of *C. foliatus* would be close to 6 mm in width, rather than 5.27 mm as in UCMP 116954.

The uncatalogued left M^2 from loc. V74110 also suggests an animal larger than *Catopsalis joyneri.* In their diagnosis of *C. joyneri,* Sloan and Van Valen (1965) report the length of M^2 as 5.2 mm, although there is no indication whether this figure is a mean or a single measurement. A sample of four M^2's in the UCMP Bug Creek Anthills collection has the following measurements: length, $\bar{x} = 5.5$, $OR = 5.0–5.8$; width, $\bar{x} = 4.6$, $OR = 4.5–4.8$ (Clemens, written comm.). These measurements are all smaller than those noted previously for the uncatalogued M^2. In addition, the cusp formula for M^2's of *C. joyneri* is 1:3:3, whereas the uncatalogued M^2's formula is 1:3:4. M^2's, like M^1's, have not been described for the larger *C. foliatus.*

Although the upper dentition of *Catopsalis foliatus* has not been described, it seems likely that the specimens from the Hell's Hollow local fauna are more closely allied with this species than with *C. joyneri.* Based on these comparisons, these specimens are tentatively (indicated by "cf.") referred to *C. foliatus.*

M. Middleton is currently studying a sample of *Catopsalis* from an earliest Paleocene locality near Denver, Colorado. From his sample he kindly provided width measurements of six M^1's, with a mean of 5.35 mm and an ordered range of 4.95 to 5.50 mm. Although he originally thought that his specimens were referable to *C. foliatus*, he now feels that his sample represents a new species intermediate in size between *C. joyneri* and *C. foliatus*. One dentary belonging to his sample was figured in Kielan-Jaworowska and Sloan (1979) and they also noted its small size compared to the type of *C. foliatus*. Middleton (pers. comm.) believes that the sample from Hell's Hollow local fauna may belong to his undescribed new species.

Suborder, *incertae sedis*
Family CIMOLOMYIDAE (Marsh, 1889b) Sloan and Van Valen, 1965
Cimolomidae Marsh 1889b, p. 177

Referred genera. Cimolomys Marsh, 1889a; *Meniscoessus* Cope, 1882c; and possibly *Essonodon* Simpson, 1927a.

Revised diagnosis. Dental formula $\frac{?2:0:4:2}{1:0:?1\text{–}2:2}$; lower incisor completely enamel-covered; eight to eleven serrations on P_4, with posterior two to three serrations distinct and facing posteriorly; profile of P_4 variable from low to high and arcuate; posterolabial basal ridge on P_4; serrate edge of P_4 at or below level of M_{1-2}; P_4 to M_1 length ratio (0.70–1.19) low; M_1 with variable posterolabial ridge with cuspules; internal cusp row on M^1 equal to or greater than 50 percent of M^1 length; number of molar cusps similar to ptilodontoids and greater than in *Cimexomys* and *Paracimexomys;* molar cusps pyramidal to crescentic (to ridgelike in *Essonodon*); M^1_1 tend to develop multiple accessory roots; M^2_2 long relative to M^1_1 (except *Essonodon*); P^4 morphology variable, but with 0?–2 anteroexternal cusps and usually with posteromedial and posterolingual basal cusps (except *Essonodon*?).

Discussion. Marsh (1889b, p. 177) named the Cimolomyidae (= Cimolomidae) including only *Cimolomys*. Sloan and Van Valen (1965, p. 222) resurrected the family and added *Meniscoessus* Cope, 1882c. Sahni (1972, pp. 370–371) was the first to define the family. He followed the lead of Sloan and Van Valen in including the cimolomyids in the Taeniolabidoidea. The most important feature Sloan and Van Valen used to characterize the Taeniolabidoidea was the restriction of the enamel to the ventrolateral surface of the lower incisor. On lower incisors of *M. robustus* (the only cimolomyid in which lower incisors are known), enamel encircles the entire tooth. On this basis the family has been removed from the Taeniolabidoidea. In the Cimolomyidae the shape and number of molar cusps resemble those of ptilodontoids. However, in the ptilodontoids the P_4 becomes enlarged in comparison to the molars, which is not the case in cimolomyids. Therefore, the Cimolomyidae has not been included in the Ptilodontoidea. This suite of characters is strong evidence for aligning *Meniscoessus* and *Cimolomys* (and *Essonodon*?), but it must be remembered that only *Meniscoessus* and *Essonodon*) is known from more than isolated teeth. Until *Cimolomys* (and *Essonodon*) is better known the concept "Cimolomyidae" should be used only tentatively.

Cimolomys Marsh, 1889a

Cimolomys gracilis Marsh, 1889a

(Fig. 23 & Table 11)

Cimolomys gracilis Marsh, 1889a, p. 84
Selenacodon brevis Marsh, 1889b, p. 177

Type. YPM 11835, right M^1 (Marsh, 1889a, pl. II, figs. 1–4).

Type locality. Mammal locality no. 1 of Lull (1915), UCMP loc. V5003, Lance Formation, Wyoming.

Referred specimens. 3 M_2's from loc. V72208, 2 P^4's from loc. V73087, 1 M^2 from loc. V75178, 1 M^2 fragment from loc. V77208.

Localities. UCMP locs. V72208, V73087 and V75178.

Distribution. Lance Formation, Wyoming; Scollard Formation, Alberta; Hell Creek Formation, Montana (all Lancian).

Revised diagnosis. Clemens, 1964, p. 76.

Description. *M_2.* Three M_2's from loc. V72208 may belong to *Cimolomys gracilis.* The range of the cusp formula is 4–6:2. Their morphology agrees with the description given by Clemens (1964, p. 78). However, when the samples of isolated M_2's of *Cimolodon nitidus* and *Cimolomys gracilis* from the Lance were examined it could be seen

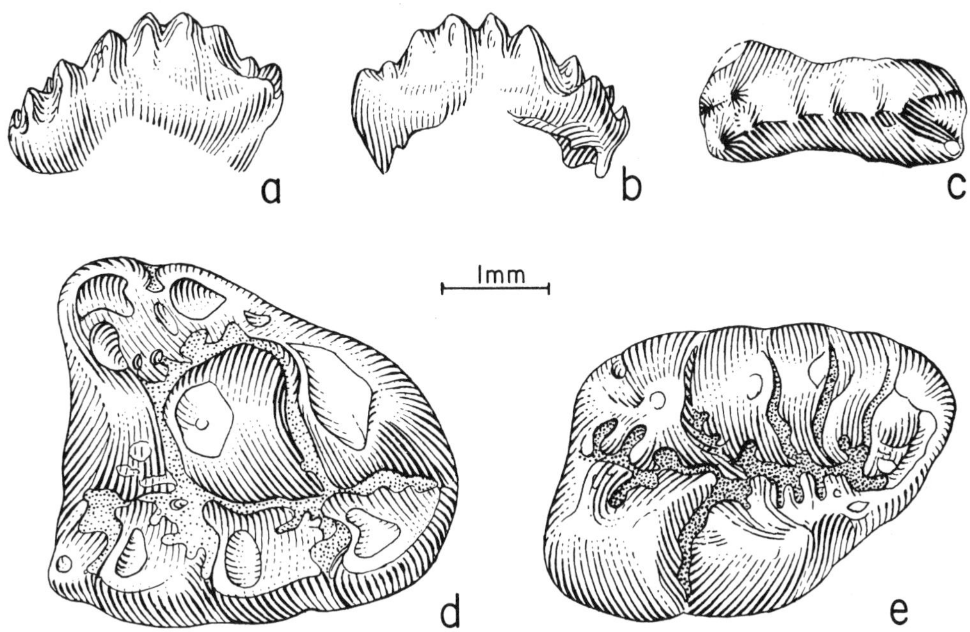

FIG. 23. *Cimolomys gracilis* Marsh. Isolated teeth. Left P^4, UCMP 116956, loc. V73087: (a) lingual view; (b) labial view; (c) occlusal view. Left M^2, UCMP 116957, loc. V75178; (d) occlusal view. Right M_2, UCMP 110639, loc. V72208: (e) occlusal view. (a, b, d, e, P. L.; c, A. L.)

that the M_2's of these two taxa overlap in size and morphology. The only basis for assigning the three M_2's from loc. V72208 to *Cimolomys gracilis* is the presence of other slightly smaller M_2's in the Hell Creek which resemble the M_2's of *Cimolodon nitidus*.

P^4. Only two P^4's from loc. V73087 can be referred to this species with any confidence. These teeth are smaller than the P^4's of *Cimolomys gracilis* described by Clemens (1964) and Lillegraven (1969). However, both Clemens' and Lillegraven's samples were small (Lance, N = 6; Trochu, N = 1). One of the P^4's, UCMP 116956, has six medial cusps (cusp formula for the two P^4's is 1–2:5–6:2–3), with the penultimate cusp highest. UCMP 116956 has two posteromedial basal cusps which are smaller, lower, and slightly separated from the preceding six cusps. The other P^4, UCMP 116955, has five main medial cusps and one lower, smaller posteromedial basal cusp. It also has a small cuspule in the anterolingual corner of the crown. Both P^4's also have one small posterolingual basal cusp linked to the medial row by a ridge. Clemens (1964, pp. 78–79) noted that unlike *Mesodma thompsoni, C. gracilis* has a large area of the crown posterior to the last medial cusp. Not including the smaller posteromedial basal cusps, this is true of the two P^4's from loc. V73087, but to a lesser degree. Also, the medial cusps in these two P^4's are more widely spaced and bear small striations on their sides. These differences may be great enough to suggest that the two P^4's belong to another taxon. These teeth are the only ones from loc. V73087 that are referable to *C. gracilis*. *Paracimexomys priscus* (n. gen.) also occurs at loc. V73087 and P^4's are not known for this species. The two P^4's could belong to this taxon.

M^2. A complete M^2, UCMP 116957, from loc. V75178 and an M^2 fragment, UCMP 110640, from loc. V72208 could belong to *Cimolomys gracilis*. Both have a cusp formula 2:3:4. The morphology of these teeth agrees well with the description given by Clemens (1964, pp. 79–80). These two M^2's are also referred to *C. gracilis* because they are larger and have fewer medial and internal cusps than other M^2's from the Hell Creek Formation probably belonging to *Cimolodon nitidus*.

TABLE 11. Measurements of teeth of *Cimolomys gracilis*.

UCMP Number	*UCMP Locality*	*Tooth Site*	*L*	*W*
116955	V73087	P^4	2.88	1.46
116956	V73087	P^4	2.68	1.50
110622	V72208	M_2	3.55	2.60
110638	V72208	M_2	3.50	2.44
110639	V72208	M_2	3.95	2.66
110640	V72208	M^2	3.38	2.95
116957	V75178	M^2	3.73	3.25

Discussion. *Cimolomys gracilis* is a rare animal in the Hell Creek Formation of Garfield County. The only other tooth of this multituberculate from Garfield County is an isolated M_1, AMNH 21986, from B. Brown's Crooked Creek locality (identified by Simpson, 1927a, fig. 3C, as *Cimolodon* sp. but referred to *Cimolomys gracilis* by Clemens, 1964, p. 78).

Meniscoessus Cope, 1882c

Meniscoessus robustus (Marsh, 1889a) Osborn, 1891

(Figs. 24–30, pl. 1, & Tables 12 & 13)

Dipriodon robustus Marsh, 1889a, p. 85
Dipriodon lunatus Marsh, 1889a, p. 85
Tripriodon coelatus Marsh, 1889a, p. 86
Selenacodon fragilis Marsh, 1889a, p. 86
Halodon sculptus Marsh, 1889a, p. 87
Oracodon anceps Marsh, 1889b, p. 178
Oracodon conulus Marsh, 1892, p. 256
Meniscoessus borealis Simpson, 1927a, p. 3

Type. YPM 11234, right M_2 (Marsh, 1889a, pl II, figs. 13–15).

Type locality. Mammal locality no. 2 of Lull (1915), UCMP loc. V5815, Lance Formation, Wyoming.

Referred specimens. UCMP 107405, dentary with I, P_4, M_{1-2}, UCMP loc. V73051; UCMP 116960, dentary with P_4 fragment, M_1, M_2 fragment, UCMP loc. V76145; LACM 24524, crushed skull fragment, left I^2, P^{1-2}, P^4, M^{1-2} fragments, right I^2, alveoli for P^{1-2}, M^{1-2} fragments, LACM loc. 3075; LACM 55753, left maxilla with P^{1-2}, P^3 roots, LACM loc. 3485; 2 I^1's; 2 I^2's; 8 P^4's; 8 M^1's; 8 M^2's, 3 M_1's; 6 M_2's.

Localities. Various UCMP and LACM localities. (See Table 13 for listing of localities by specimen.)

Distribution. Hell Creek Formation, Montana and South Dakota; Lance Formation, Wyoming (all Lancian); and St. Mary River Formation, Alberta (Edmontonian).

Revised diagnosis. Largest species of *Meniscoessus*; compared to known molars of *M. intermedius, M. major,* and *M. ?conquistus*, cusp formula is lower, the M^2 to M^1 and M_2 to M_1 length ratios are greater, and the internal cusp row on M^1 is longer and usually with more cusps. (*M. ferox* may be included in the above list; the only known molar, M^2, was mentioned but not described by Fox, 1971.)

Description. The teeth of this species have already been described in detail by Clemens (1964 and 1973) and only additions to his description will be noted here.

Mandible, associated teeth, and incisors. Three mandibular fragments belonging to this species have been recovered from the Hell Creek Formation, Garfield County. Two of them contain only one measurable tooth (M_1) and are included with the isolated teeth in Table 13. The third specimen (UCMP 107405) preserves the lower incisor (not in contact with the jaw), P_4, and M_{1-2}. In addition to these two specimens, a third jaw (CCM 68–104) from the Hell Creek Formation, Fallon County, Montana, is described through the generosity of Mr. Marshall Lambert, Carter County Museum. Measurements of these specimens are in Table 12. Specimen CCM 68–104 is a left mandible which preserves the lower incisor (erupting), P_{3-4}, and M_1; this is the first association of the incisor with cheek teeth. The lower incisor is protruding only slightly above the alveolus and is laterally compressed. Two ridges emanate from the lingual surface of the incisor tip. The dorsal edge bears four small cuspules behind the incisor tip. More cusps may be present but the rest of the incisor is covered by matrix. On the lingual side of the jaw below M_1 a small piece of bone is missing, revealing the lingual side of the incisor. The posterior edge of the exposed incisor is broken. The incisor shows no constriction at this point; it probably extended posteriorly below M_2. The incisor found with UMCP 107405 is similar to that

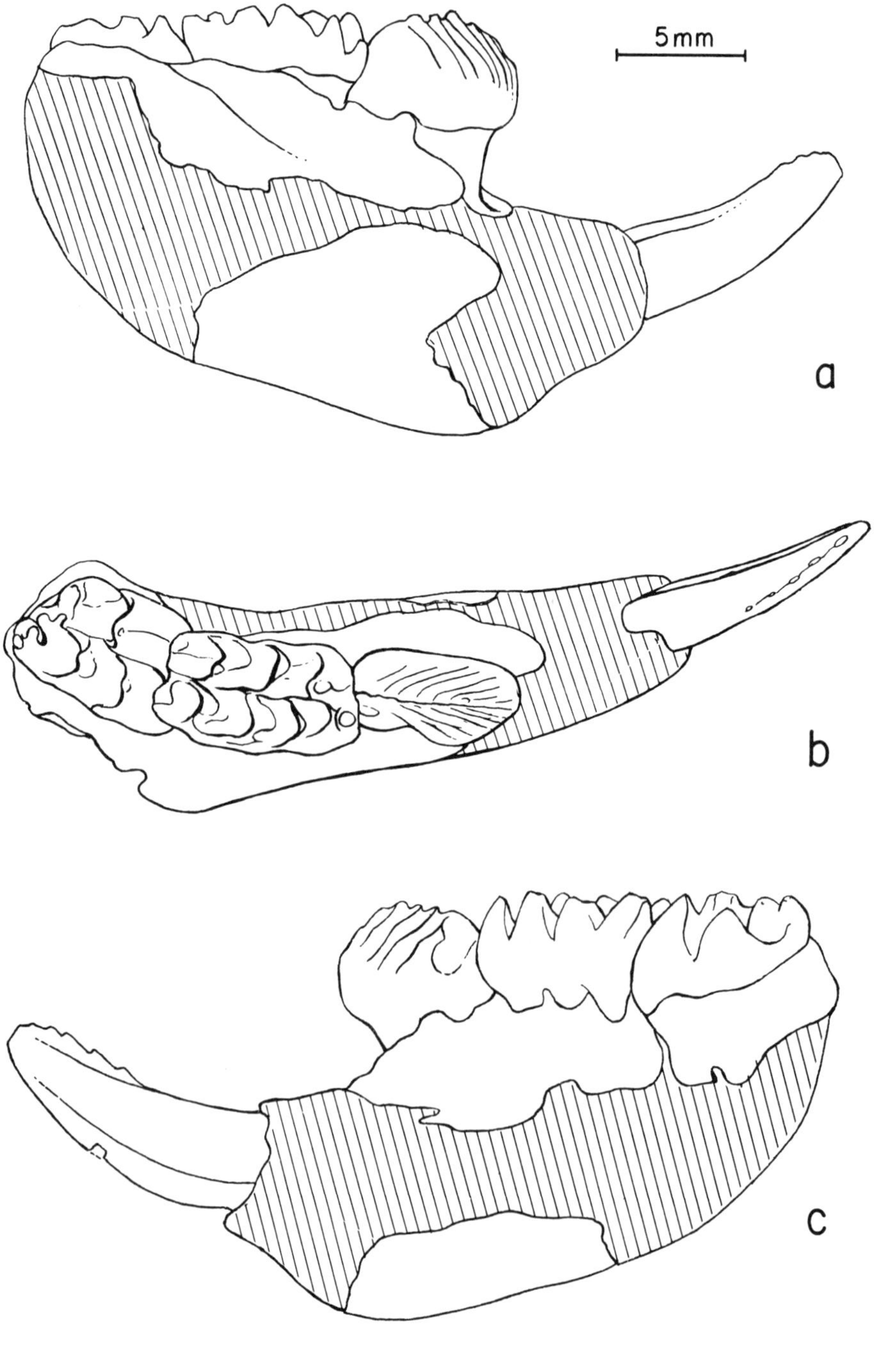

FIG. 24. *Meniscoessus robustus* (Marsh). Right dentary with I, P_4, M_{1-2}, UCMP 107405, loc. V73051: (a) labial view; (b) occlusal view; (c) lingual view. Lines indicate plaster. (P. L.)

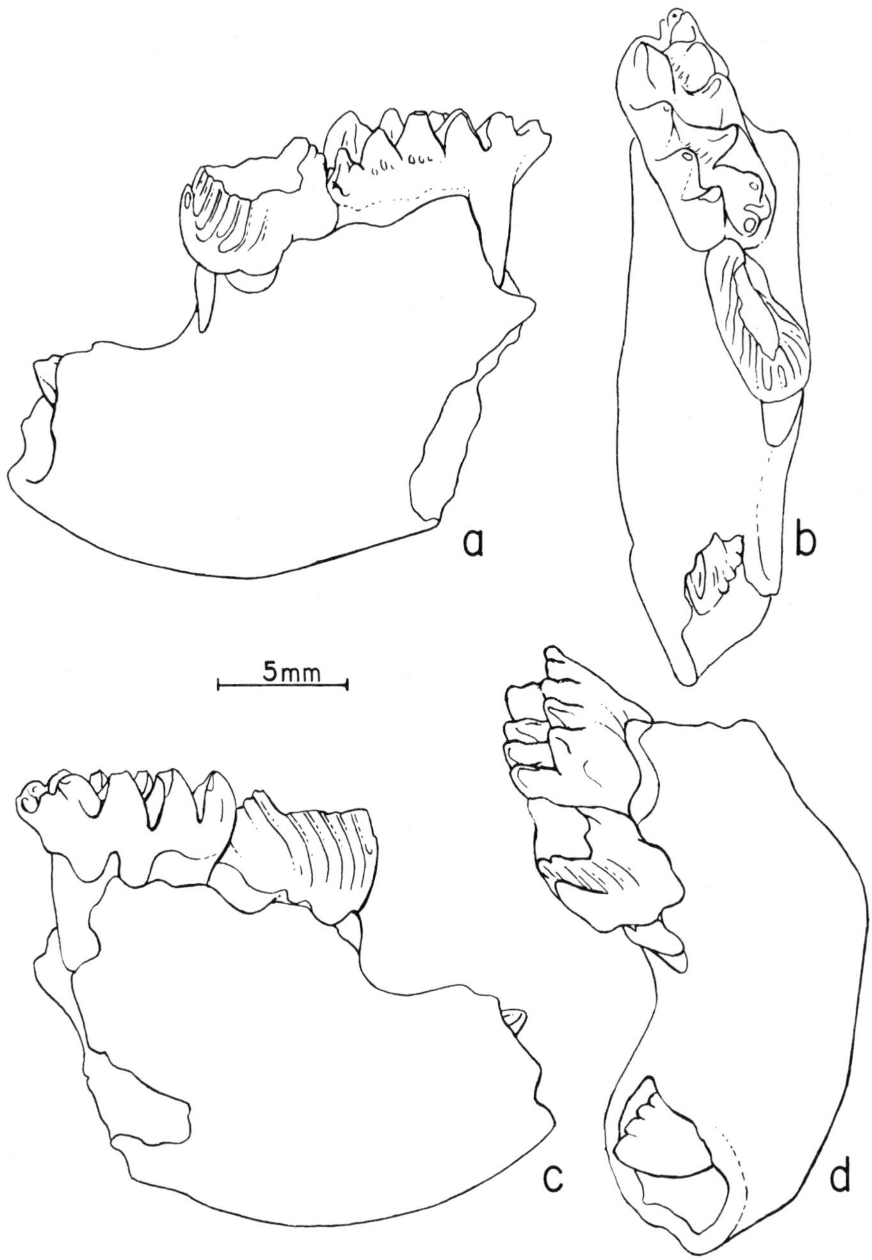

FIG. 25. *Meniscoessus robustus* (Marsh). Left dentary with erupting I, P_{3-4}, M_1, CCM 68-104, Fallon County, Montana; (a) labial view; (b) occlusal view; (c) lingual view; (d) anterolabial view, showing erupting incisor. (P. L.)

found in CCM 68–104. The former was not found in place in the jaw, but it was nearby. The similarity in preservation also indicates that the jaw and incisor belong together. The incisor is laterally compressed and becomes almost triangular in cross-section near its base. The convex ventrolabial surface has a corrugated appearance due to four to five shallow grooves that run from incisor tip to base. The lingual side is a relatively flat surface bearing a distinct ridge running from incisor tip to base. The ventral edge of the laterally compressed incisor forms a ridge separating the lingual from the ventrolabial surface. The dorsal side of the incisor is a dorsally concave, narrow, triangular surface with its apex at the incisor tip. One leg of the triangle is formed by a ridge emanating on the lingual side of the incisor tip; this ridge separates the dorsal from the lingual surface of the incisor. The other leg is a prominent ridge, running from incisor tip to base, which separates the dorsal from the ventrolabial surface. This ridge bears five small cuspules aligned directly behind the incisor tip, which extends about halfway along the ridge. The incisor is completely enamel-covered; near the tooth base on the dorsolingual surface the preserved color is lighter, which suggests that the enamel may be thinner here. CCM 68–104 preserves P_3 and its morphology is similar to that described by Clemens (1964, p. 87). The P_4 of UCMP 107405 has eight serrations, the last two of which are closely approximated; there are eight labial and lingual ridges. The more posterior labial ridges have been muted by wear. P_4 of CCM 68–104 is damaged but probably had ten serrations; there are nine labial and eight lingual ridges. M_1 of UCMP 107405 has a cusp formula of 5:4; CCM 68–104 has a formula of 7:4. The first external cusp in both M_1's is smaller than other cusps in the external row. In UCMP 107405 it is conical and in CCM 68–104 it is subcrescentic. In UCMP 107405 the ultimate posteroexternal cusp (fifth cusp) is partially subdivided by a groove on its lingual surface. In CCM 68–104 there is a fully developed sixth external cusp plus a smaller seventh cusp on the low posterior ridge that connects the external and internal cusp rows. The ultimate cusp (fourth cusp) of the internal row in CCM 68–104 is partially subdivided by a groove on its labial surface. Only UCMP 107405 preserves the M_2. Although the bone is somewhat distorted it appears that this tooth is not fully erupted. It has a cusp formula of 4:2. The last external cusp is not fully separated from the preceding cusp. The P_4 to M_1 length ratios for UCMP 107405, CCM 68–104, and YPM 10613 (right mandible with P_{3-4}, M_1, Marsh, 1892, pl. V, fig. 7, Lance Formation, Niobrara County, Wyoming) are, respectively, 0.79, 0.68, and ~0.86. This compares to the values of 0.83 for UCMP 46875 and 0.70 for AMNH 57400 which Clemens (1964, p. 88) calculated for these two specimens from the Lance local fauna. Based on these five specimens, the probable tooth eruption sequence for the adult dentition in *Meniscoessus robustus* is $M_1 \rightarrow P_4 \rightarrow P_3 \rightarrow M_2 \rightarrow I$ (or $I \rightarrow M_2$).

It is possible that the jaw deepens with age. If available jaws are arranged according to the tooth eruption sequence (CCM 68–104 → UCMP 107405 → UCMP 46427 → YPM 10613), a slight but perceptible increase in jaw depth can be seen (Table 12). This is highly speculative, as these few jaws come from different, often greatly separated localities which possibly differ in age. Sexual dimorphism also cannot be discounted.

Skull and palate. A skull fragment (LACM 24524) with associated teeth was found by H. J. Garbani in the Hell Creek Formation, Garfield County, in 1962 (Pl. 1). This is the first substantial portion of a multituberculate skull reported from the Cretaceous of North America. Only the anterior half of the skull is preserved and this is severely broken and distorted. The following description and restoration are tempered by the poor quality of preservation.

TABLE 12. Associated lower teeth of *Meniscoessus robustus*.

			P_4				M_1		M_2		*Jaw Depth Below*
	L/1	*L*	*H*	*W*	*L1/L*	*H/L*	*L*	*W*	*L*	*W*	P_4
UCMP 107405	3.25	6.18	4.17	3.49	0.53	0.67	7.86	3.99	6.76	4.96	11.0
CCM 68-104	—	6.33	—	3.49	—	—	9.30	4.32	—	—	11.5
[1]YPM 10613	3.05	6.47	3.99	2.96	0.47	0.62	a7.56	a4.11	—	—	12.8
[2]UCMP 46427	—	—	—	—	—	—	8.06	4.11	8.39	5.25	a12.0
[2]UCMP 46875	3.02	6.49	3.98	3.14	0.47	0.61	7.97	3.88	—	—	—

a Approximate measurement.
[1]Marsh, 1892
[2]Clemens, 1964, remeasured.

The skull preserves the left second upper incisor, P^{1-2}, right second upper incisor, and alveoli for P^{1-2}. In addition, the following isolated teeth were found in association with the skull: left P^4, worn fragment of M^1 in bone, worn M^2 in bone; right ?M^1 roots in bone, worn M^2 in bone. The palatal exposure of the premaxilla on the right side is relatively undistorted. Large ovoid incisive foramina are present. The location of sutural contacts between the palatal portions of the premaxilla and maxilla is not clear. A fragment of a left maxilla (LACM 55753, Fig. 26) preserves a portion of the premaxilla-maxilla suture just anterior to P^1. This is the basis for the restoration of this suture in Fig. 27. The only suture on the skull that can be recognized with any certainty is the premaxilla-maxilla suture on the right side of the skull. The fronto-maxillary, naso-maxillary, and naso-premaxillary sutures are partially preserved, but due to distortion their positions in Fig. 27 are only approximate. On LACM 55753, medial to P^{2-3}, the palatal portion of the maxillary thins and forms what appears to be the finished edge of a palatal vacuity (Fig. 26a). UCMP 48155, from loc. V5711 (Clemens, 1964), is a maxillary fragment preserving M^1 and alveoli for P^{2-4}. The bone medial to the alveoli and M^1 noticeably thins and the anteromedial edge appears to be unbroken, possibly representing the lateral edge of a palatal vacuity. In the skull, LACM 24524, palatal vacuities cannot be discerned (Pl. 1). However, the bone in this region is so crushed and distorted that the presence or absence of these openings cannot be determined with any certainty.

Clemens (1973, p. 85, fig. 30d–e) described a portion of the maxilla of *Meniscoessus robustus* preserving the anterior portion of the zygomatic arch and two infraorbital foramina. LACM 55753 is in agreement with this specimen in regard to the position of origin of the zygomatic arch on the maxilla.

The anterior edge of the upper second incisor is level with the anterior border of the incisive foramen. This tooth appears to be only slightly medial to the premolar-molar tooth row. Most of the crown of the second upper incisor has been obliterated by wear. The wear surface is approximately ovoid, and because of greater wear on the lingual side the wear surface slants lingually. Unworn I^2's have not been reported in association with other

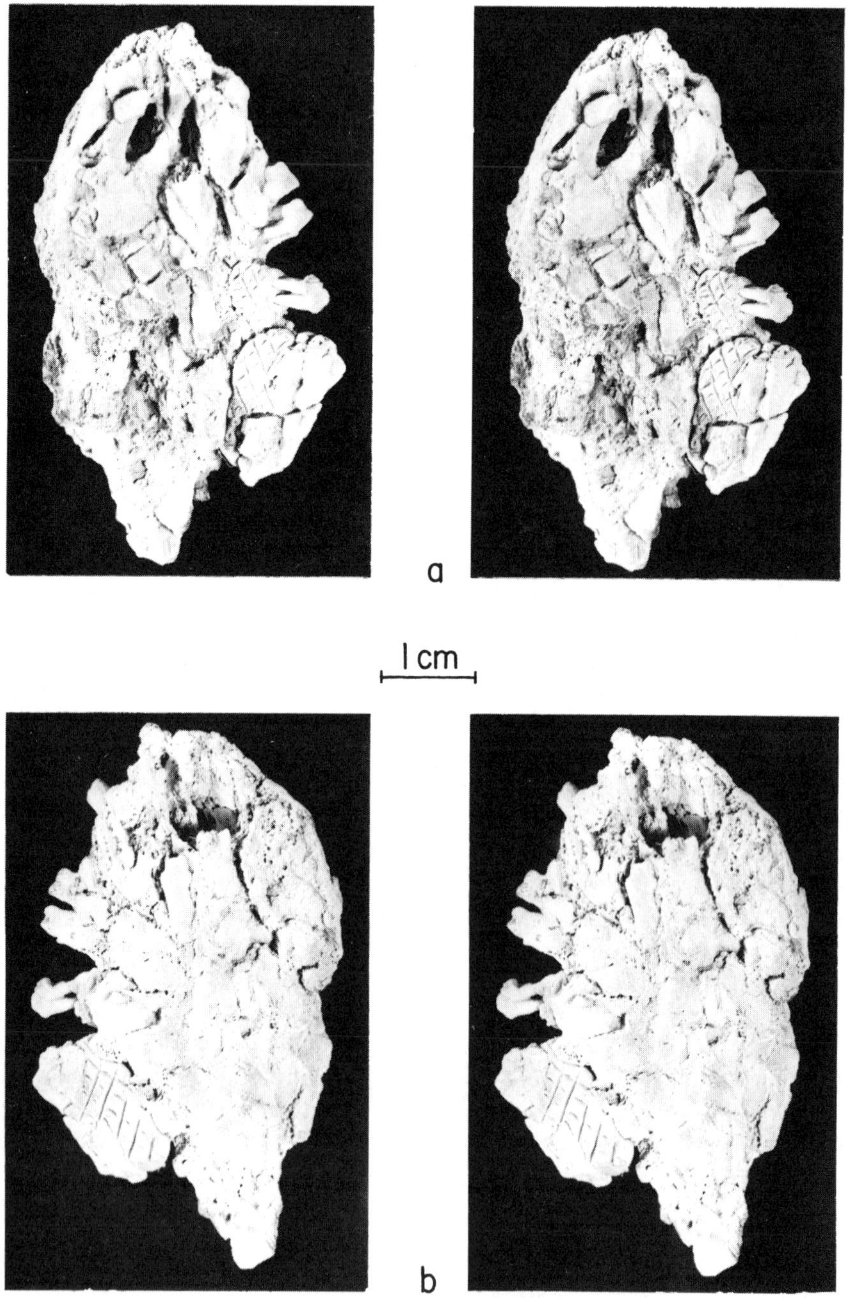

Plate 1. *Meniscoessus robustus* (Marsh). Stereophotographs of crushed skull fragment with left I^2, P^{1-2}, P^4, M^{1-2} fragments, right I^2, alveoli for P^{1-2} (right M^{1-2} fragments not shown), LACM 24524, LACM loc. 3075 (= UCMP loc. V72206): (a) slightly right-lateral of ventral; (b) dorsal and left-lateral. Specimen is coated with ammonium chloride; hatched area is clay holding left P^4 and M^{1-2} fragments.

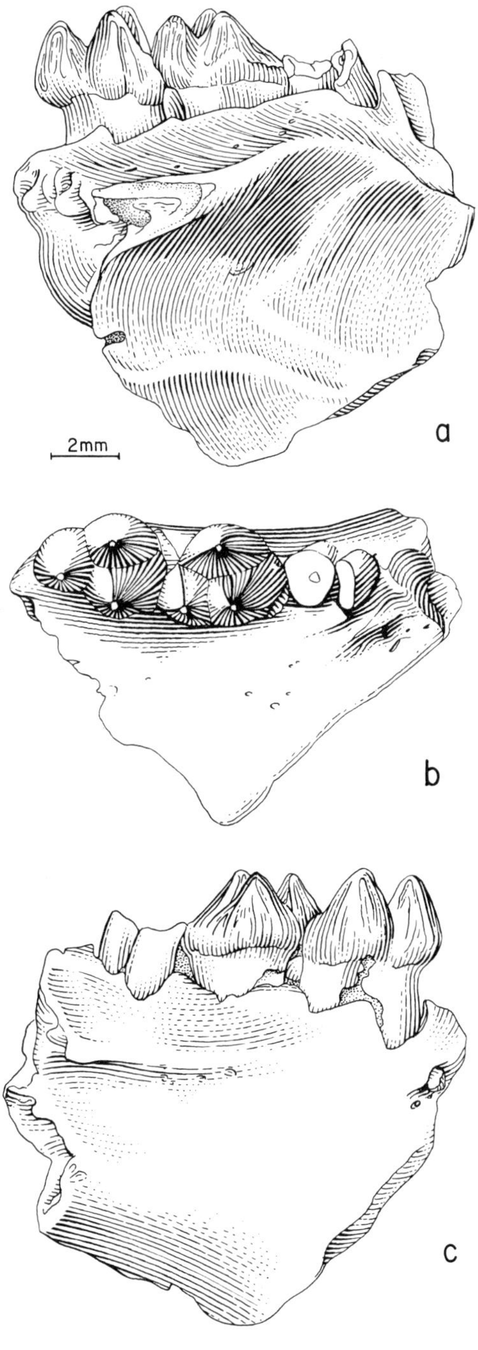

FIG. 26. *Meniscoessus robustus* (Marsh). Left maxilla fragment with P^{1-2}, P^3 roots; LACM 55753, LACM loc. 3485: (a) lingual view (note suture anterodorsal to P^1 and finished edge dorsolingual to P^2 and P^3 roots; see text for explanation); (b) occlusal view; (c) labial view. (a, c, P. L.; b, A. L.)

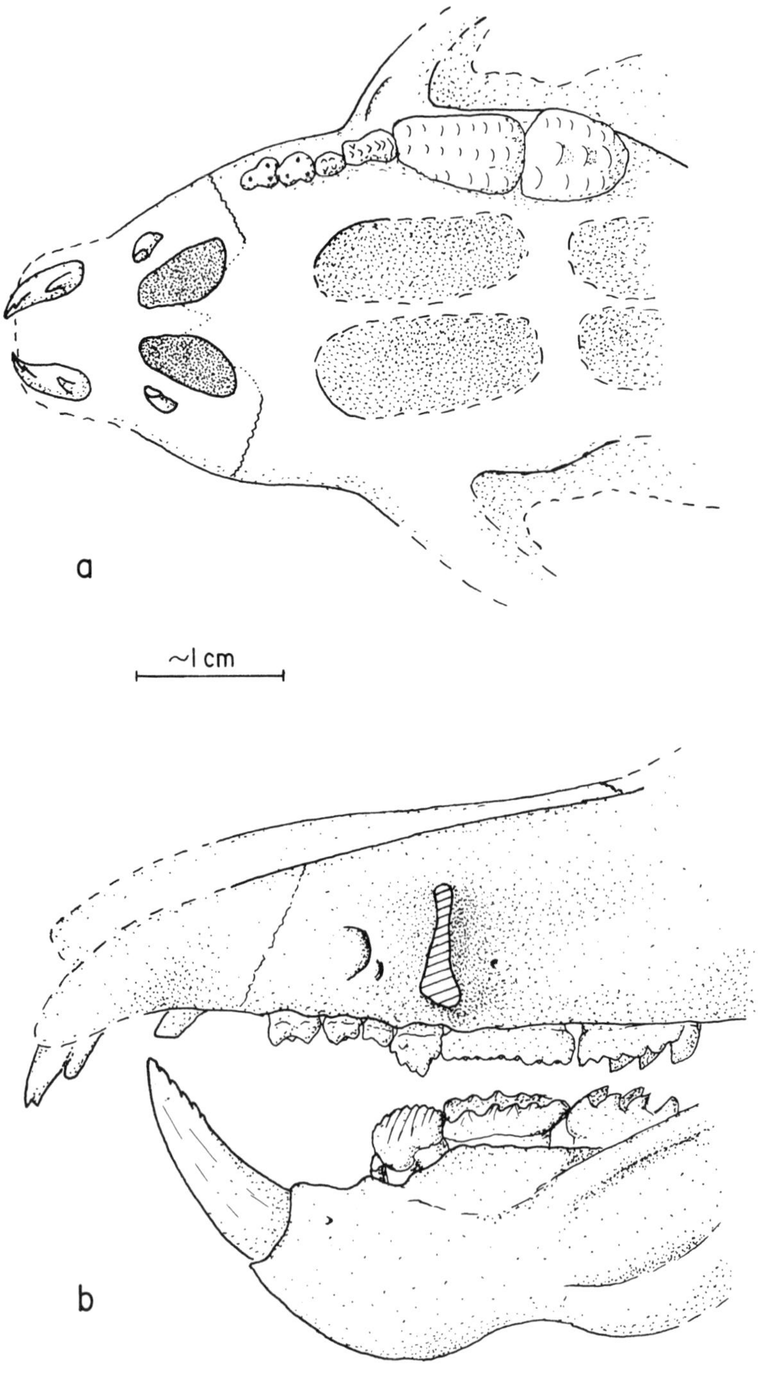

FIG. 27. *Meniscoessus robustus* (Marsh). Reconstruction of anterior skull: (a) palatal view; (b) lateral view with dentary.

teeth; however, bicuspid teeth like the one shown by Osborn (1893, pl. VII, fig. 9) could be unworn examples of this tooth.

It should be noted that Osborn (1893) labeled the first and second upper incisors of *Meniscoessus* as I^2 and I^3, respectively. It is not clear why he did this, but recent data tend to support his numbering sequence of the upper incisors. Hahn (1971) has shown that four genera of plagiaulacoids have three upper incisors, which suggests that in earlier evolutionary stages multituberculates had three. I^1, the smallest, was probably lost, leaving I^{2-3}.

No upper first incisors have been found in association with more posterior teeth. Osborn (1893, pl. VII, fig. 9) considered the first upper incisor of *Meniscoessus* (which he labeled I^2) to be the teeth named *Halodon sculptus* and *H. serratus* by Marsh (1889a, p. 87 and pl. III, figs. 1–6). Subsequent authors (Simpson, 1927a, 1929a; Clemens, 1964) have agreed with this assignment. Upper first incisors are shown in Fig. 28a & b; Clemens, 1964, fig. 45; and Osborn, 1893, pl. VII, fig. 9. The body of the tooth has two small cusps at its apex (possibly three in AMNH 57873: Clemens, 1964, fig. 45). There is also a lower but prominent posterolabial cusp separated from the main body of the incisor by a deep labial groove. This groove is absent on the lingual side. I agree with Osborn's assessment (1893, p. 318) that this cusp "apparently serves as a check for the lower teeth [incisor]." This is partially borne out by a worn first upper incisor, UCMP 116976, from loc. V73097. Where the two small apical cusps were located it possesses a small wear facet which continues toward the base of the tooth along its posterolingual margin and terminates on the apex of the smaller posterolabial cusp. This cusp is noticeably shortened by wear against the lower incisor.

The upper P^4 found with the skull, but not in place, fits the morphology described by Clemens (1964, p. 90) for this tooth. This P^4 has a cusp formula 1:4:?. There is a minor

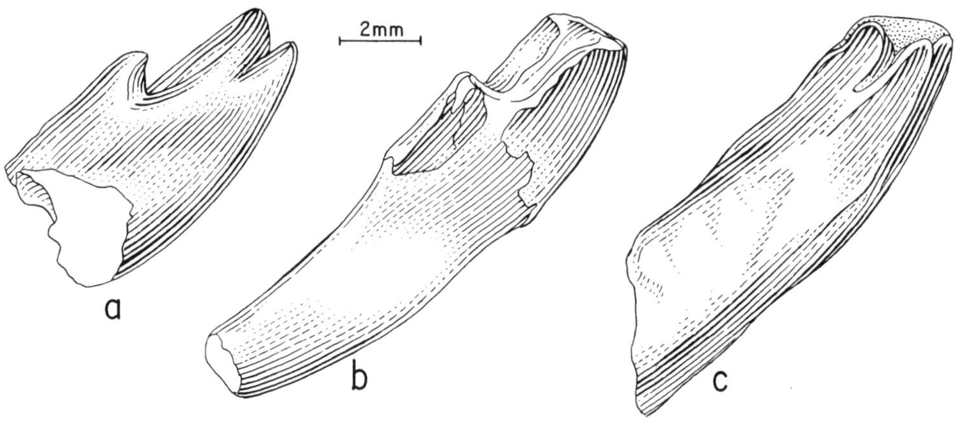

FIG. 28. *Meniscoessus robustus* (Marsh). Posterolingual view of isolated incisors: (a) unworn right I^1, UCMP 108971, loc. V73078; (b) worn right I^1, UCMP 116976, loc. V73097; (c) left I_1 showing apical wear, UCMP 116977, loc. V75178. (P. L.)

swelling on its posterolabial border. It cannot be determined whether any basal cusps were present on the posterior or posterolingual margins of the tooth, as a large wear facet covers most of the posterior and lingual surfaces of the tooth.

Among the molars only the upper M^2's associated with the skull are relatively complete. These teeth, like the others associated with the skull, are a grayish white (most of the chemically unweathered teeth from the Hell Creek Formation are dark brown or black). Most of the molar cusps are obliterated, but based on the appearance of the wear facets on these teeth, it seems most likely that this obliteration is due to dental attrition, not chemical alterations. The animal was probably an old individual.

Measurements of the following isolated teeth are given in Table 13.

M_1. Two of three M_1's (from three different localities) have a cusp formula 5:4, and the third has a cusp formula 6:4.

M_2. The cusp formula is 4:2 on five of the isolated M_2's. One M_2, UCMP 116962, loc. V72213, has a small fifth external cusp.

P^4. In addition to the P^4 associated with the skull fragment (LACM 24524), seven isolated P^4's are listed in Table 12. However, two distinct morphological groups can be recognized. Four of the P^4's—UCMP 108980, 116964, 120389, and 120390—resemble the P^4's that Clemens (1964) referred to *Meniscoessus robustus*. Each of the four teeth has one anteroexternal cusp, and UCMP 116964 has a smaller cuspule posterior to this cusp. UCMP 116964 and UCMP 120390 each have four cusps in the medial row and UCMP 108980 and UCMP 120389 each have three. Each of these four P^4's has one posteromedial basal cusp and one posterolingual basal cusp, connected by a narrow cingulum. The other three P^4's—UCMP 116965, 120394, and 120395—are unusual for *M. robustus*. The anteroexternal margin on each of the three P^4's is swollen labially, but no cusps are present in this region. None of the three teeth possesses a posterior basal cusp, although on each, muted posteromedial and posterolingual ridges descend from the back of the last medial cusp, forming a very small basal cingulum. Each of the three P^4's has four medial cusps.

During most of this study it was believed that the P^4 morphology displayed by the second group simply represented an atypical condition for *Meniscoessus*. However, in 1978, a left maxilla of *Essonodon browni* (UCMP 120397) preserving a worn, broken P^4 was recovered from Clam Hollow (V76138). Although it was not originally part of this study, W. A. Clemens has generously allowed me to describe the specimen. The P^4 is so badly worn and broken that a complete comparison with the preceding specimens is not possible. Enough of this P^4 has been preserved to suggest that the second morphological group of P^4's described above may belong to *Essonodon*. Discussion of these similarities is deferred to the description and discussion of *Essonodon*. It will suffice for the time being to note that there is too little evidence to refer these rather atypical P^4's confidently to *Essonodon*. Therefore, they have been tentatively retained within *Meniscoessus*. One of these atypical P^4's is shown in Fig. 29 for comparison with more typical *Meniscoessus* P^4's (Clemens, 1964, fig. 46) and with the fragmentary P^4 of *Essonodon* (Fig. 32).

M^1. Five M^1's were recovered from the Hell Creek Formation, Garfield County, four of which are complete enough to determine the cusp formula 5–6:7:6–7. The four M^1's have seven medial cusps each, while the modal value for the internal row is six (two of three M^1's) and for the external cusp row is six (three of four M^1's). Three other M^1's can be referred to *Meniscoessus robustus*. An M^1 fragment from B. Brown's Crooked Creek locality, AMNH 14407, was tentatively referred to *M. borealis* by Simpson (1927a, p. 4). Although this tooth is large—probably 6 mm wide when complete—it can be referred to

M. robustus. The other two M^1's are complete and have a cusp formula 6:7:6. The first, USNM 17084, a right M^1, is (along with eleven other mammalian specimens) from Powder River County, Montana, and probably comes from R. P. Bryson's fv–15 locality (Bryson, 1951, p. 36, table 3). It was identified by C. L. Gazin as *Meniscoessus* cf. *robustus* in Bryson, 1951. Bryson (1951, p. 45) reports locality fv–15 as occurring in the "Hell Creek Formation(?)," 105 ft. 8 in. below the overlying contact with the "Lower member of Fort Union formation." The second complete M^1, AMNH 21930 (plus an ?I^1), is reported as coming from the "Lance Formation, Hell Creek Beds; 35 mi. S. Lismas, Montana; 1908 Expedition," probably in what is now Garfield County. Lismas is now under the waters of Fort Peck Reservoir.

M^2. Including the two worn M^2's associated with the skull LACM 24524, ten M^2's can be referred to *Meniscoessus robustus*. The cusp formula for the less worn teeth is

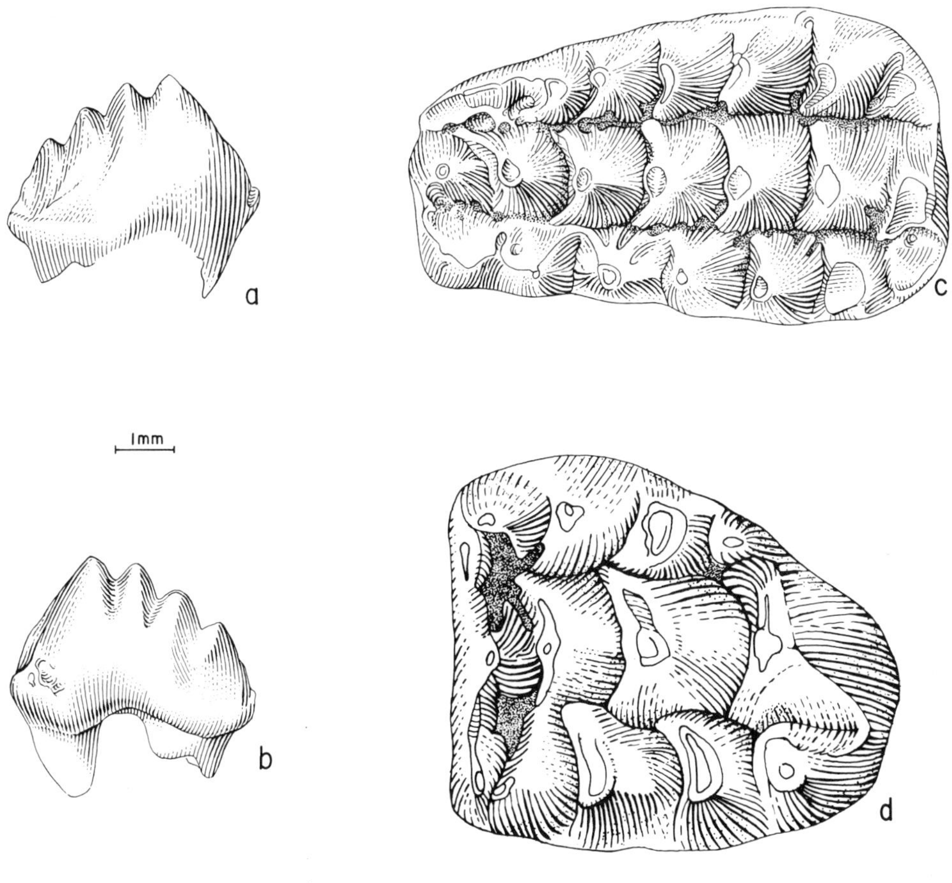

FIG. 29. *Meniscoessus robustus* (Marsh). Isolated teeth. Right P^4 showing atypical morphology (see text), UCMP 116965, loc. V72213: (a) labial view; (b) lingual view. Right M^1, UCMP 116966, loc. V73103: (c) occlusal view. Left M^2, UCMP 116971, loc. V73089: (d) occlusal view. (a, b, c, P. L.)

3–5:3–4:4–5. The number of external cusps can be determined on only five of the M^2's, and three or four external cusps are equally common. Most of the M^2's have four medial and internal cusps. Other M^2's which can be referred to *M. robustus* are AMNH 14411 (type of *M. borealis* Simpson, 1927a), from B. Brown's Crooked Creek locality, and USNM 215182, from locality fv–15 of Bryson (1951).

TABLE 13. Isolated teeth (except LACM 24524 from skull) of *Meniscoessus robustus.*

Specimen Number	*Locality*	*Tooth Site*	*L*	*W*
LACM 24524	LACM 3075	P^4	4.08	2.76
*UCMP 116965	V72213		4.16	2.54
UCMP 116964	V72213		4.27	3.02
UCMP 108980	V73076		3.97	2.97
*UCMP 120389	V77130		4.10	3.05
*UCMP 120390	V77130		3.60	2.65
UCMP 120394	V77130		4.40	2.50
UCMP 120395	V77130		4.00	2.50
UCMP 116958	V72207	M_1	9.40	4.40
UCMP 116959	V72213		6.99	3.48
UCMP 116960	V76145		a7.29	4.01
LACM 55728	LACM 3072	M_2	7.27	4.73
LACM 55731	LACM 3081		7.09	4.62
LACM 55734	LACM 3095		7.90	5.41
UCMP 116961	V75178		6.78	—
UCMP 116962	V72213		7.40	4.76
UCMP 116963	V73103		—	a5.01
LACM 55730	LACM 3072	M^1	—	5.33
UCMP 116966	V73103		9.14	5.22
UCMP 116967	V72213		a8.57	—
UCMP 116968	V73087		a8.90	4.77
UCMP 116969	V73089		8.75	5.25
AMNH 21930	"LISMAS"		9.54	5.64
USNM 17084	fv–15		9.88	5.71
LACM 24524 left	LACM 3075	M^2	a7.26	a5.95
LACM 24524 right	LACM 3075		a7.21	a6.70
LACM 55729	LACM 3072		a6.94	a5.79
UCMP 108963	V73083		6.34	5.98
UCMP 116970	V72213		7.18	6.46
UCMP 116973	V72200		7.75	—
UCMP 116971	V73089		7.49	6.62
UCMP 116972	V72207		6.94	6.74
UCMP 116974	V72207		6.98	—
UCMP 116975	V72207		7.07	6.22

a Approximate measurement.
* Only tentatively referred to *Meniscoessus* (see text).

Discussion. Beginning with Simpson's discussions of the systematics of *Meniscoessus* (1927a, 1929a), three species of this genus have generally been recognized as occurring in the latest Cretaceous of North America. These species are: *M. conquistus* Cope, 1882c, which was restricted by Clemens (1964, p. 83) to the lectotype, AMNH 3011, a broken M^2; *M. robustus* (Marsh) 1889a, which is known from a relatively large sample of teeth and jaws (Clemens, 1964), the type being an M_2, YPM 11234; and *M. borealis* Simpson, 1927a, described on a few teeth, the type being based on an M^2, AMNH 14411. Subsequent to the work of Clemens (1964), specimens belonging to this genus have been reported from a variety of Edmontonian and Lancian localities. *M. conquistus* was reported by Sloan and Russell (1974) from the St. Mary River Formation, Alberta, and by R. W. Wilson (pers. comm. and data presented in Sloan and Russell, 1974) from the Hell Creek Formation, South Dakota. *M. robustus* has been noted by Clemens (1973) from the Hell Creek Formation, Carter County, Montana, and by Waage (1968) from the Fox Hills Formation, South Dakota. *M. borealis* has been reported as occurring in the upper Hell Creek Formation of Montana by Sloan (unpubl. data reported in Sloan and Russell, 1974) and by Sloan and Van Valen (1965).

Three additional species of *Meniscoessus* have been described from the Aquilan through Judithian: *M. ferox* Fox, 1971 (Upper Milk River Formation); *M. intermedius* Fox, 1976b (lowermost Oldman and possibly Foremost formations); and *M. major* (Russell) 1937 in Sahni, 1972 (Judith River Formation). Although undoubtedly important for considering the evolution of the genus, these species do not enter into the following discussion of the synonomy of Lancian species.

The new specimens of *Meniscoessus* described in the preceding pages and a reevaluation of older material indicate that *M. borealis* is best regarded as a junior synonym of *M. robustus. M. conquistus* may be the valid senior synonym of all Lancian specimens referable to this genus; however, the fragmentary nature of the type specimen of *M. conquistus* and the uncertainties surrounding the location of the type locality and its stratigraphic position do not favor the use of this name as a senior synonym. Research currently being conducted by R. W. Wilson (pers. comm.) on specimens of *Meniscoessus* from the Late Cretaceous of South Dakota may shed light on this problem. I agree with Clemens (1964) that *M. conquistus* should be restricted to the lectotype, AMNH 3011, at least for the present.

The largest samples of M^2's of this genus come from UCMP locs. V5620 and V5711 in the Lance Formation, Wyoming (Clemens, 1964). Clemens (1964, pp. 93–94) noted that the type of *Meniscoessus borealis*, although high in measured parameters of size, cusp formula, and length-width ratio, was within the observed range of specimens of *M. robustus* from V5711. An additional M^2 (AMNH 57445) from loc. V5711 discovered by AMNH parties subsequent to the research of Clemens exceeds the type of *M. borealis* in every dimension. This tooth has a length of 8.26, width of 6.95, cusp formula 6:4:4 or 5, and a length/width ratio of 1.19. The type of *M. borealis*, when remeasured, was found to have a length of 8.18, width of 6.91, cusp formula 5:4:4, and length/width ratio of 1.18. Fig. 30 is a bivariate plot of length vs. width of upper M^2's of *Meniscoessus*. This graph includes the specimens mentioned above, plus the UCMP samples from the Lance Formation UCMP locs. V5620 and V5711, the type of *M. conquistus* (AMNH 3011), two specimens from the Hell Creek Formation of South Dakota (mentioned in Sloan and Russell, 1974, p. 6), AMNH 57442 from loc. V5711, and specimens from various UCMP localities in the Hell Creek Formation, Garfield County, Montana. The type of *M. borealis* and several specimens from loc. V5711 are somewhat isolated near the upper size range, and the type of *M. conquistus* and a specimen from the Hell Creek Formation of Garfield County are at the lower end of the graph. However, there is no clear indication of bi- or trimodality in this sample. Even with this mixed sample from localities separated by hundreds of miles and possibly some differences in age, the CV for length is only 6.02 ($\bar{x}=7.38$; $s=0.444$; $N=41$) and for width is 5.16 ($\bar{x}=6.39$; $s=0.330$; $N=32$). Only the CV for length is slightly greater than the "good average values" of 5 to 6 given by Simpson, Roe, and Lewontin (1960) for linear dimensions of homogeneous mammalian samples. The slightly high CV for length is best regarded as an artifact of wide geographic separation and possible time differences within one species rather than as an indication of the presence of two or three species.

All the M^2's of *Meniscoessus robustus* available in the UCMP collections from locs. V5620 and V5711 were remeasured (length and width). In addition, two M^2's in the AMNH collections (AMNH 57445 and AMNH 57442), from loc. V5711 were included with this sample. For length the sample mean = 7.48, s = 0.410, CV = 5.49, and N = 27. For width the sample mean = 6.45, s = 0.289, CV = 4.48, and N = 20. T-tests were made

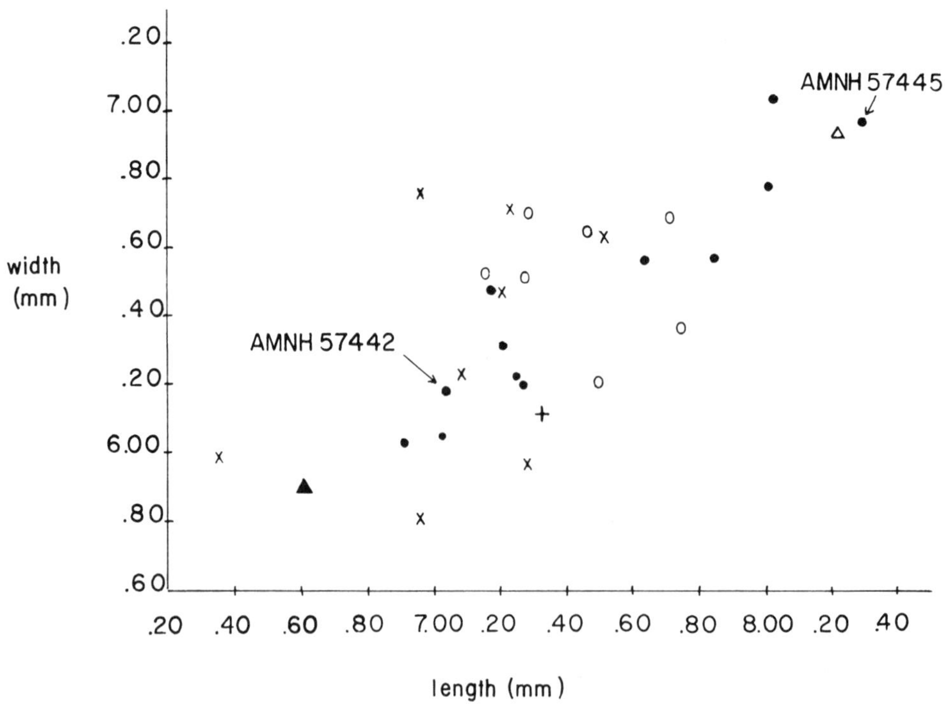

FIG. 30. Bivariate plot (length vs. width) of M^2's of *Meniscoessus*. x = Hell Creek Formation, Montana (various localities); ○ = UCMP loc. V5620, and • = UCMP loc. V5711, both Lance Formation, Wyoming (UCMP specimens unless labeled otherwise); △ = *M. "borealis,"* type, AMNH 14411; ▲ = *M. conquistus*, type, AMNH 3011; + = two specimens from the Hell Creek Formation, South Dakota, measurements from Sloan and Russell, 1974.

comparing the length and width of the type of *M. borealis* ($L = 8.18$; $W = 6.91$) to the above Lance sample. It was found that $0.10 < P < 0.20$ that the type of *M. borealis* could come from the same statistical population as the Lance sample. Based on size, *M. borealis* could belong to the same species as the Lance sample (i.e., *M. robustus*). Unless two species (*M. robustus* and *M. ?borealis*) are present in this Lance sample—and this is not supported by the low CV's or bimodality in the sample—there is no statistical basis for separating the type of *M. borealis* from *M. robustus*. The type of *M. conquistus* was remeasured and the length is estimated to be 6.6 (due to breakage, this is only approximate) and the width is 5.89. A similar t-test was done comparing the type of *M. conquistus* and the above Lance sample. The probability that the type of *M. conquistus* and the Lance sample were drawn from the same statistical population is >0.02, but <0.05 for length, and is >0.05, but <0.10 for width. These lower probabilities suggest that *M. conquistus* may belong to a separate species. However, since only the type specimen is clearly referable to this species, its validity must await confirmation by future discoveries near the type area in South Dakota.

Sloan and Russell (1974) referred a worn M^2 from the St. Mary River Formation, Alberta, to *Meniscoessus conquistus* because of its small size. Although I do not agree with some of their statistical tests and some of the conclusions drawn from them, their material indicates that a smaller species of *Meniscoessus* may have been present in the Edmontonian or early Lancian.

Family ?CIMOLOMYIDAE

Essonodon Simpson, 1927a

Essonodon browni Simpson, 1927a

(Figs. 31–34 & Table 14)

Essonodon browni Simpson, 1927a, p. 2

Type. AMNH 14410, an isolated right M_2 (Simpson, 1927a, fig. 1).

Type locality. Near the head of the East Fork Crooked Creek, Hell Creek Formation, Montana.

Referred specimens. UCMP 108967, left maxillary fragment with M^2 and part of alveoli for M^1, and UCMP 108965, left M_2, from UCMP loc. V73083. UCMP 109025 and UCMP 116979, both M^1 fragments from UCMP loc. V73087. UCMP 116980, left M^1 fragment, and UCMP 116981, left M^2 fragment, from UCMP loc. V73076. UCMP 120397, left maxillary fragment with alveolus and root for P^3, broken and worn P^4, and M^{1-2}, from UCMP loc. V76138. UCMP 120393, right M^1; UCMP 120396, right M_1 fragment; and UCMP 120391, left M_1 in dentary fragment, from UCMP loc. V77130. MCZ 19982, right M_2 fragment; CM 37160, left M^2; and various other isolated teeth from Bug Creek Anthills. All the preceding are from the Hell Creek Formation, Montana. YPM 34527, isolated right M_1, from Lance Formation, Wyoming, exact locality unknown. USNM 215120 and USNM 244198, both left M^2's, from Lance Formation, Wyoming.

Localities. See Referred specimens.

Distribution. Hell Creek Formation, Montana, and Lance Formation, Wyoming (both Lancian).

Revised diagnosis. Incisors and premolars unknown except for single, broken P^4; length ratios of P^4 to M^1, M^2 to M^1, and M_2 to M_1 very low; known portion of P^4 crown very tall, probably with extremely steep anterior edge; P^4 possibly lacking anteroexternal cusps; molar cusps complicated by numerous accessory ridges; crowns of M^1_1 very low and broad; internal cusp row on M^1 greater than half M^1 length; M_2's and M^2's wider than long; M_1 with small, posterolabial cuspules; M^2 with only two distinct cusp rows; M^1_1 with numerous accessory roots.

Comments. The present study was essentially completed in 1977. In the following year, new and important specimens of *Essonodon* were recovered from the Hell Creek Formation, and W. A. Clemens has kindly permitted me to describe those specimens: a left maxillary fragment from loc. V76138, and three isolated molars from loc. V77130 (see Referred specimens). The latter is a new and potentially important Hell Creek faunal-facies locality that is only now beginning to be sampled. A number of other taxa have been recovered at loc. V77130, but in addition to the isolated molars of *Essonodon* the only specimens from loc. V77130 described in this study are four P^4's mentioned previously under *Meniscoessus* and discussed more fully below.

Description. Simpson (1927a) named this genus and species on an isolated right M_2. The discovery of additional specimens referable to this taxon (Clemens, 1964; Sloan and Van Valen, 1965; and here) supports Simpson's recognition of this rather aberrant multituberculate.

For completeness, all new specimens of this species examined by the author are included in the Referred specimens. R. Sloan is currently studying the multituberculates from Bug Creek Anthills; specimens from that locality are not included in the description.

M_1. YPM 34527 (from the Lance Formation) and UCMP 120391, loc. V77130, are the only complete M_1's referable to *Essonodon browni.* A second, posterior half of an M_1, UCMP 120396, was recovered from loc. V77130. The crown morphology of the two complete teeth is essentially identical except that in occlusal view UCMP 120391 is somewhat more squared in outline and is approximately 10 percent larger than YPM 34527. The posterolabial portion of UCMP 120391 is so heavily worn that an accurate count cannot be made of the cuspules labial to the external cusp row. The following description is applicable to both M_1's, except where noted otherwise. The cusp formula is 3–4:7:6. The cusp height is lower than in comparably sized M_1's of *Meniscoessus.* This plus two additional features gives the M_1 its low, broad appearance. First, there is a narrow ridge bearing three or four small cuspules labial to the last five cusps in the external cusp row. Second, the broader lingual sides of the internal cusps are inclined at an angle noticeably less than the labial sides of the same cusps. This shifts the apices of the internal cusps nearer to the midline of the tooth. The second and third external cusps are the largest and tallest cusps in this row (taller than their internal counterparts). The first external cusp is slightly smaller, and those behind the third cusp gradually decrease in size and height. The external cusps are subcrescentic with apices directed posteriorly in YPM 34527, but are more pyramidal in UCMP 120391. The larger external cusps bear two to three faint ridges, radiating from the apices, on their curved anterior side. The external cusps also have one or two lingual ridges which join with their counterparts from the internal cusp row in the midline of M_1. This gives the appearance of a latticework of ridges joining the cusps across the longitudinal valley between the external and internal cusp rows. The six internal cusps gradually increase in size and height posterior to the fourth and fifth cusps. The sixth cusp is slightly smaller. The morphology of the internal cusps of *Essonodon* is unique among multituberculates. The labial side of each cusp is steep and bears two to three ridges, and these usually join the complex latticework of ridges in the longitudinal valley. The lingual side of each internal cusp is at a lower angle than the labial side. The smooth, relatively

expansive lingual sides of the internal cusps form ridges that are broadly confluent with the lingual side of M_1. The anterior margin of the crown is slightly indented and slightly overhangs the anterior root, helping to form a shallow depression for the P_4. Only the proximal portion of the roots is present in YPM 34527, and all but the damaged anterior roots and part of the posterior roots in UCMP 120391 are embedded in bone. However, the basic forms and variations of the roots can be described by combining information from the two specimens. The proximal portions of both the anterior and posterior roots are rectangular in cross-section and are similar in size. More apically (distally) both roots divide (based on UCMP 120391). At least on the damaged anterior root of UCMP 120391 it can be seen that a pulp cavity is present, starting from the base of the crown, in each of the two divisions. YPM 34527 bears two lingual and at least five labial accessory roots.

A fragment of the mandible preserved with UCMP 120391 sheds little light on the morphology of this bone. However, a few features are noteworthy. As would be expected, based on the large size of the M_1, the dentary is moderately deep. The maximum depth of the preserved portion is about 7 mm, and clearly more bone was originally present. At about the midpoint on the external side below the M_1, the bone is slightly thickened. This is the point from which the thickened anterior margin of the ascending ramus arose. A smooth, slightly concave surface on the ventral surface of the fragment is probably the roof of the mandibular canal and possibly the incisor alveolus.

M_2. The M_2 holotype, AMNH 14410, and the M_2 referred to *Essonodon* by Clemens (1964), YPM 14907, were remeasured and are included in Table 14. These measurements are in accord with those given by Simpson (1927a) and Clemens (1964), except for the length of YPM 14907 given by Clemens. By these new measurements, the length/width ratio for AMNH 14410 is 0.68; for YPM 14907 it is 0.74. These are close to the values of 0.67 and 0.79 given by Clemens (1964). Another M_2, UCMP 108965, referable to *Essonodon* has a much higher length/width ratio, 0.95. Except for the higher length/width ratio there are only minor morphological differences between UCMP 108965 and the other M_2's. The third external (ridgelike) cusp in UCMP 108965 is smaller than its counterparts in the other two M_2's. In the latter M_2's the third external cusp is oriented linguolabially, while in UCMP 108965 it runs posterolabially to anterolingually.

Maxilla and upper dentition. The maxilla recovered in 1978 (UCMP 120397) preserving a fragmentary P^4 and M^{1-2} represents the only known associated teeth of *Essonodon browni*. It also shows that isolated M^1's and M^2's earlier thought to belong to *Essonodon* were correctly associated and the questionable allocation of this genus to the Cimolomyidae has been strengthened (see Discussion). The bone fragment is composed entirely of the maxillary bone. The external surface preserves the ventral portion of the zygomatic process. The most posterior margin of the zygomatic process lies immediately dorsal to the midpoint of P^4. A shallow groove bordering the dorsal surface of the broken, ventral portion of the zygomatic process is clearly a continuation of the infraorbital foramen. The dorsal aspect of the maxillary fragment forms a jagged, broken surface except for a small, dorsally concave groove directly above the midpoint of M^1. This is most probably the ventral edge of the sphenopalatine foramen. Dorsal to the posterior half of M^1 and M^2 the external side of the maxilla forms a distinct maxillary tuberosity, the upper surface of the tuberosity constituting the most ventral aspect of the orbital fossa. A roughly ovoid opening on the dorsal surface of the tuberosity above the contact of M^1 and M^2 appears to be a break in the maxilla, but it could be the broken margins of an alveolar canal.

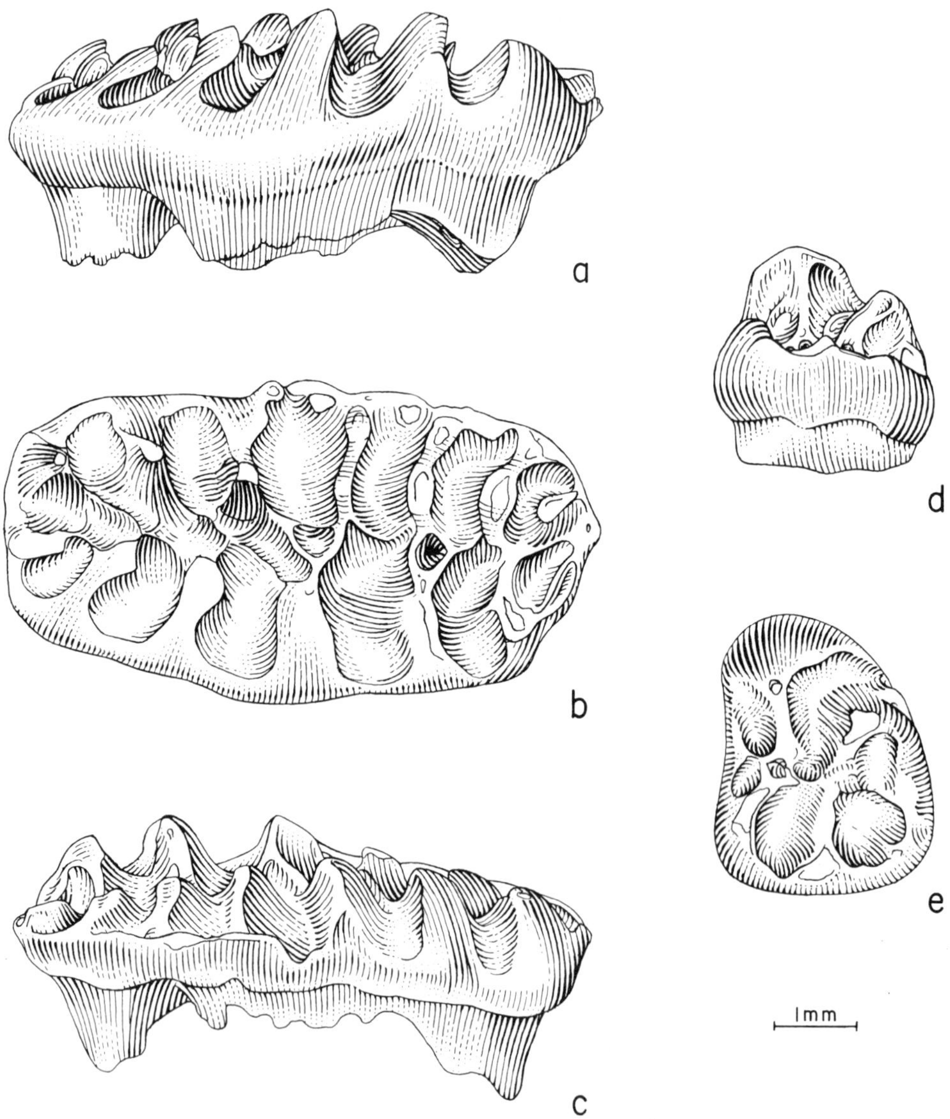

FIG. 31. *Essonodon browni* Simpson. Isolated right M_1, YPM 34527, Lance Formation, Wyoming: (a) lingual view; (b) occlusal view; (c) labial view. Isolated left M_2, UCMP 108965, loc. V73083: (d) labial view; (e) occlusal view. (P. L.)

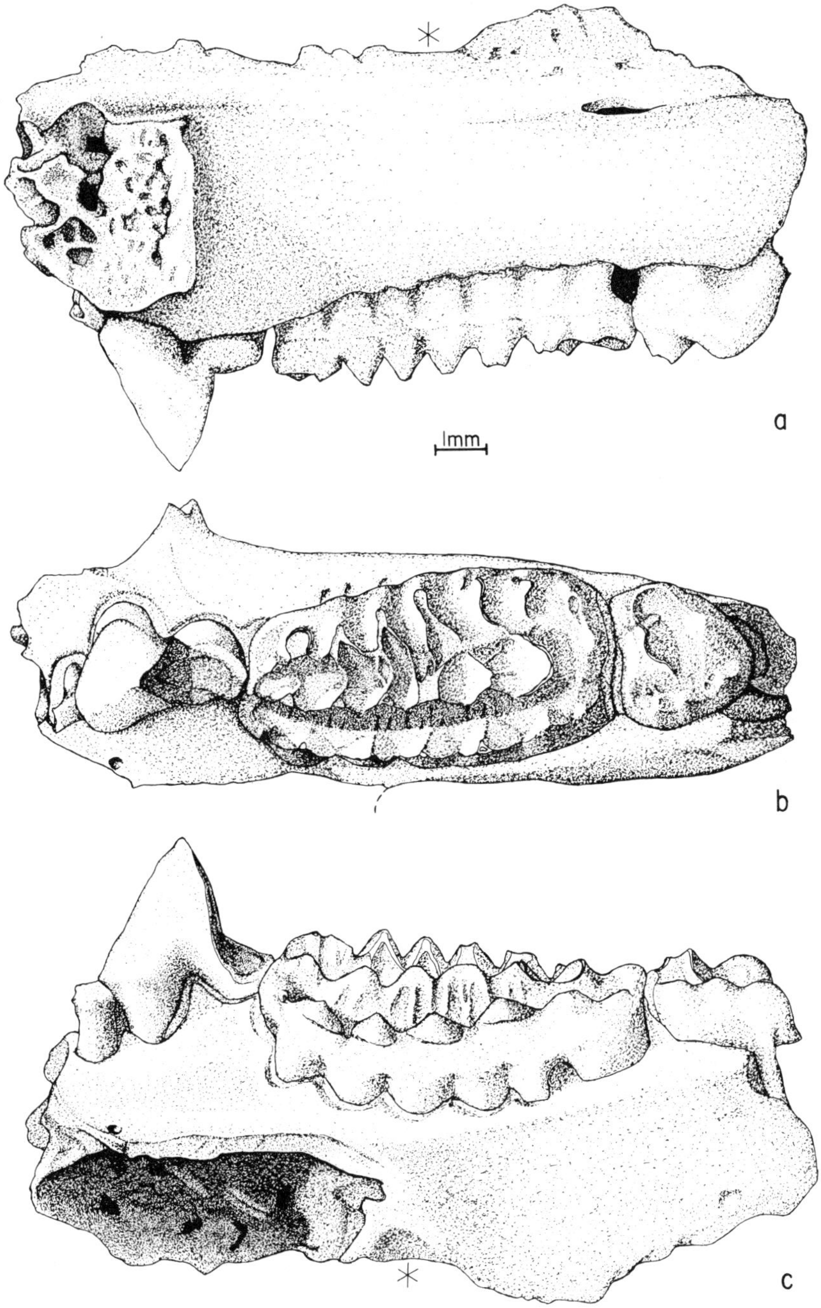

FIG. 32. *Essonodon browni* Simpson. Left maxillary fragment with anterior alveolus and fragment of P^3, broken and worn P^4, and M^{1-2}, UCMP 120397, loc. V76138: (a) labial view; (b) occlusal view; (c) lingual view. Asterisks in (a) and (c) mark position of sphenopalatine foramen. Dotted line in (b) marks anterior border of presumed palatal vacuity.

The internal surface of the maxillary fragment presents some interesting problems of interpretation. The broken edge of the palatine process of the maxilla clearly extends from the anterior edge of the bone fragment to approximately the anterior one-third of M^1. Posterior to this until the broken, posterior margin of the maxilla is reached at the level of M^2 there is no suggestion of a palatine process. One interpretation is that on its lateral side the secondary palate extended posteriorly only to about the level of the anterior one-third of M^1. A second possibility is that a palatal vacuity was present medial to the posterior two-thirds of M^1 and M^2. Unfortunately, neither of these interpretations is strongly supported by published accounts of the palatal morphology of multituberculates. In all published accounts of multituberculate palates, the secondary palate extends posteriorly at least to the level of the posterior margin of M^1. Thus, if the first interpretation were to be accepted *Essonodon* would have the shortest secondary palate (relative to its tooth row) of any multituberculate. In reference to the second interpretation, there are no published accounts of multituberculate palates in which a palatal vacuity is present lingual to the posterior two-thirds of M^1 and M^2. In all other instances palatal vacuities, if present, are distinctly anterior in position. A third interpretation is that the fragment is too damaged for the correct palatal morphology to be discerned. This explanation seems highly unlikely and is rejected. Granting that there is little supporting evidence, it seems most likely that the second explanation is correct (i.e., a large, very posteriorly positioned palatal vacuity was present—see also Discussion).

The teeth preserved in UCMP 120397, except for portions of the M^1 and M^2, are devoid of their enamel covering. Only the posterior root and a small fragment of the crown of P^3 are preserved. A small portion of the alveolus for the anterior root of P^3 is present. Like the posterior alveolus, it was probably ovoid in shape, but it was probably slightly larger in diameter. The P^4 is closely appressed to the P^3 and the crown of the P^4 overhangs that of the P^3. Only the anterior half of the P^4 crown is present and this is completely devoid of enamel, making any detailed description of cusp morphology impossible. The crown of the P^4 is very distinctive in one respect: the preserved portion of the P^4 is approximately 3 mm high and fully one-half of this height is above the level of the molar cusps. If the anterior half of the P^4 does not represent the greatest height of the crown the ratio is even greater. The anterior edge of the P^4 is slightly wavy which suggests that three or four cusps may have been present in this portion of the tooth. The anteroexternal section of the crown is expanded but there is no suggestion of cusps in this region. The anterior half of the crown, including the expanded anteroexternal section, is supported by a root that is ovoid in cross-section. This root is almost twice the size of the posterior root. The posterior root morphology was observed in several atypical P^4's tentatively assigned to *Meniscoessus*. (Possible implications of this observation are discussed later.)

The M^1 associated with the UCMP 120397 is somewhat better preserved than the P^4. A complete but slightly more damaged M^1, UCMP 120393, was recovered from loc. V77130. Both of these M^1's were discovered subsequent to the completion of the present study. Two M^1 fragments, UCMP 109025 and UCMP 116979 (both preserving the posterior end), recovered from loc. V73087 had previously been referred to *Essonodon*. The recovery of UCMP 120397 confirms this identification. From what can be seen, the crown morphologies of all four M^1's are very similar. A complete cusp formula can be determined only on UCMP 120397; it is 8:7:7. All three cusp rows extend the full length of M^1. The curious "torsion" of the cusp rows around the anteroposterior axis of M^1 is also seen, to a lesser degree, in some other Late Cretaceous multituberculates. In the more anterior portion of M^1 there is a distinct stairstep arrangement from the internal to the external cusp row. In this region the apices of the cusps in the internal row are approximately level with bases of the cusps in the medial row. A similar relationship is present between the medial and the external row of cusps. More posteriorly, the apices of cusps in the medial row become almost level with their counterparts in the external row. Although the apices of the cusps in the internal row do increase in height relative to their counterparts in the medial row going posteriorly on M^1, they never approach being level with the apices of the medial cusps. Because of their low position, it is very unlikely that any of the more anterior internal cusps on M^1 occluded with their counterparts in the lower M^1 until both teeth were heavily worn. This is partially confirmed by occluding the isolated, unassociated M^1's and M_1's (a right and a left for both M^1 and M_1). It is possible that the P_4 (this tooth is unknown) occluded in the anterior portion of the longitudinal valley between the internal and medial cusp rows of M_1. The seven more posterior cusps in the external row are approximately equal in size and are conical in shape. Each of the seven

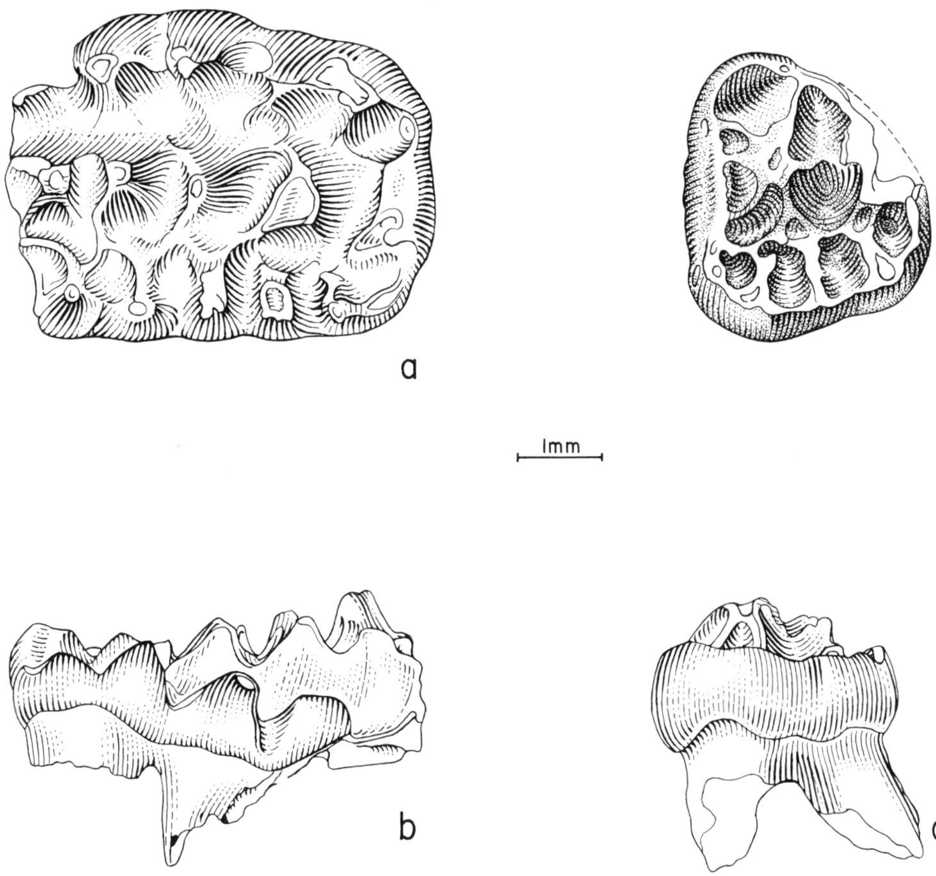

FIG. 33. *Essonodon browni* Simpson. Isolated right M^1 fragment, UCMP 116979, loc. V73087: (a) occlusal view; (b) lingual view. Isolated left M^2, USNM 215120, Lance Formation, Wyoming: (c) occlusal view; (d) lingual view. (P. L.)

cusps bears two lingual ridges radiating from the apex. These ridges join one or two counterparts from the medial cusp row, forming a latticework of ridges in the longitudinal valley analogous to the morphology described above for M_1. The anteromost cusp in the external row is distinctly smaller than the other seven and it bears only one ridge joining with the anterior face of the first cusp in the medial row. The first five medial cusps are pyramidal or quadrate, while the last two are more nearly subcrescentic. The medial cusps increase in size, but only slightly in height, going posteriorly. The medial cusps bear one to three labial ridges joining counterparts from the external cusp row. The medial cusps are united to one another by ridges at their antero- and posterolingual corners. The more central medial cusps bear up to three lingual ribs or ridges, and the more posterior two or three cusps may also bear a single posterior ridge. The internal cusps are subequal in size, except for the very small anteromost cusp (and the posteromost cusp on UCMP 120397, which is heavily abraded). The internal cusps are conical to pyramidal and are united to each other by anterior and posterior ridges. All but the first internal cusp also bear a labial ridge. A low posterior ridge joins the last cusps in each of the three cusp rows. There is a large posterior root that is subrectangular in cross-section near its base, but the root divides more apically (distally) to form a larger lingual and smaller labial root. A smaller, anterolabial root (observable only on UCMP 120393) is also subrectangular in cross-section. There are also five lingual and five labial accessory roots.

The M^2 preserved in the maxillary fragment, UCMP 120397, is heavily abraded and lacks most of its enamel, but it confirms the referral of better-preserved, unassociated M^2's to *Essonodon*. The M^2 with UCMP 120397, one other M^2, UCMP 108967, from the Hell Creek Formation, and two M^2's, USNM 215120 and USNM 244198, from the Lance Formation, Wyoming, have a cusp formula 0–R:2:3–4. UCMP 108967 is nearly complete except for the antero- and posterolingual corners, and it is preserved in a piece of maxilla which also preserves four of the more posterior alveoli for M^1. There are no external cusps on UCMP 108967; however, the ridge that nearly encompasses the perimeter of the crown is slightly extended labially in the anterolabial corner of the M^2. This portion of the ridge is more pronounced in the other three M^2's but bears no cusps there, either. The anterior medial cusp is conical and in three of the M^2's,—UCMP 108967, UCMP 120397, and USNM 215120—is widened transversely by labial and lingual ridges. The cusp also bears three anterior ridges and one to two posterior ridges. The large, posterior medial cusp is subcrescentic. This cusp bears two small anterolingual ridges joining the latticework between the internal and medial cusp rows. One of the M^2's, UCMP 108967, has four internal cusps, and USNM 215120 may have the same number. USNM 244198 has three internal cusps, as may UCMP 120397. The internal cusps are transverse ridges running from the lingual side to the midline of the M^2. Smaller ridges project labially from the internal cusps, joining the latticework of small ridges in the longitudinal valley separating the medial and internal cusp rows. Although all four M^2's are wider than they are long, the differences between width and length are not a great as in the M_2's. This is seen in the higher length/width ratios: 0.89 for USNM 215120; 0.94 for UCMP 108967; and 0.95 for USNM 244198 and UCMP 120397. The M^2's have a large, transversely widened, anteroposteriorly narrowed anterior root and one or two smaller posterior roots that are ovoid in cross-section.

Although the P^4 preserved in UCMP 120397 is lacking its enamel, the measurement of its length given in Table 14 is probably close to the true length. Based on this, the P^4 to M^1 length ratio for UCMP 120397 is 0.44. The M^2 to M^1 length ratio for the same specimen is 0.36.

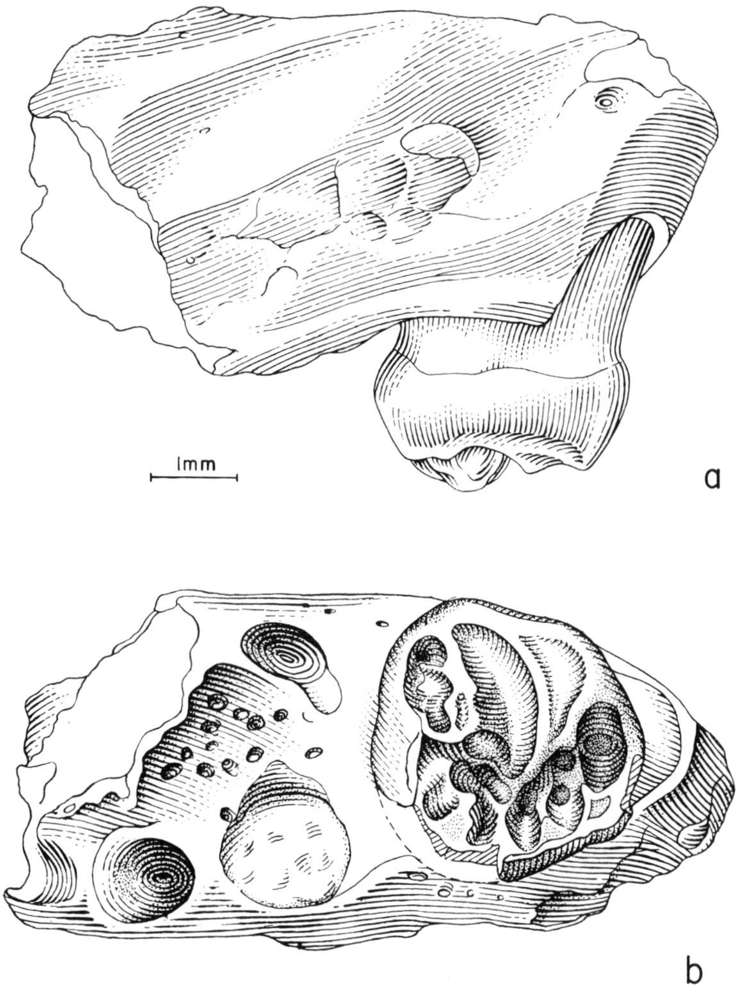

FIG. 34. *Essonodon browni* Simpson. Left maxillary fragment with M^2 and posterior alveoli for M^1, UCMP 108967, loc. V73083: (a) labial view; (b) occlusal view. (P. L.)

TABLE 14. Measurements of teeth of *Essonodon browni* from various localities.

Specimen Number	Locality	Tooth Site	L	W
YPM 34527	Lance Fm. (Wyoming)	M_1	7.12	3.83
UCMP 120391	V77130	M_1	a7.90	4.15
[1]AMNH 14410 (type)	Crooked Ck.	M_2	2.45	3.61
[2]YPM 14907	Lance Fm. (Wyoming)	M_2	2.39	3.23
UCMP 108965	V73083	M_2	2.58	2.71
[3]UCMP 120397	V76138	P^4	a3.30	a2.10
[3]UCMP 120397	V76138	M^1	7.45	3.85
UCMP 120393	V77130	M^1	a7.60	3.60
UCMP 116979	V73087	M^1	—	3.86
UCMP 109025	V73087	M^1	—	3.84
[3]UCMP 120397	V76138	M^2	2.65	2.80
UCMP 108967	V73083	M^2	3.09	3.30
UCMP 116981	V73076	M^2	—	3.05
USNM 215120	Lance Fm. (Wyoming)	M^2	2.95	3.30
USNM 244198	Lance Fm. (Wyoming)	M^2	2.70	2.83

a Approximate measurement.
[1]Simpson, 1927a, remeasured.
[2]Clemens, 1964, remeasured.
[3]Associated teeth in UCMP 120397.

Discussion. When Simpson (1927a) named *Essonodon browni* he referred it to the Ptilodontidae. This assignation has continued until the present (Simpson, 1929a, 1945; Clemens, 1964, 1973; Sloan and Van Valen, 1965). In the late 1920s all Cretaceous and Tertiary multituberculates were allocated to two families, the Ptilodontidae and Taeniolabididae (e.g., Simpson, 1929a, pp. 11–12). At that time the fossil record probably warranted recognition of only two families for this group of multituberculates and the referral of *Essonodon* to the ptilodontids was justifiable. Based on our current understanding of multituberculate diversity and phyletic relationships, however, placement of *Essonodon* in the Ptilodontidae is no longer advisable.

Essonodon is one of the most unusual multituberculates. With the possible exception of the lack of well-developed crescentic cusps, all characters of *Essonodon* are best considered as derived in comparison to the common(?) taeniolabidoid-ptilodontoid ancestor. These derived characters include: (1) relatively low-crowned molars; (2) some molar cusps developed into ridges and numerous accessory ridges forming a latticework between cusp rows; (3) low length to width ratio of M_2; (4) secondary(?) reduction of internal cusp row on M^2; (5) very low M_2 to M_1 and M^2 to M^1 length ratios; (6) increase in number of molar cusps on M^1_1 (and M^2_2?); (7) accessory posterolabial cuspule row on M_1, (8) well-developed internal cusp row on M^1; (9) very low P^4 to M^1 length ratio; and (10) multiple accessory roots on M^1_1 and occasionally on M^2. Of these the first four character states are best regarded as autapomorphies. Although some ptilodontoids and taeniolabidoids do show low M_2 to M_1 and M^2 to M^1 length ratios, none is as low as in *Essonodon* (M_2 to M_1 length ratio 0.30 to 0.36, based on isolated teeth; M^2 to M^1 length ratio 0.36, based on UCMP 120397). Thus, these two correlated character states (character 5) are probably also autapomorphic for *Essonodon*. The remaining five character states are shared with some taeniolabidoids, ptilodontoids, and cimolomyids. The resemblances of most, if not all, of these last five character states among the above three groups are probably the result of parallelisms and are considered below in the cladistic analysis of the multituberculate dentition. Without knowledge of the incisors and P_4 and a better knowledge of the P^4 it is impossible to assign *Essonodon* with any certainty to any of the preceding three taxa. However, with closer scrutiny of some of the above character states and some other details of dental morphology not listed, the affinities of *Essonodon* can be narrowed to two families or possibly a single family.

At least *Cimolodon, Cimolomys, Meniscoessus* (Clemens, 1964), *Catopsalis,* and *Taeniolabis* (Clemens, pers. comm.) have multiple roots like those seen in *Essonodon*. Although only the latter three genera have multiple accessory roots, Clemens (pers. comm.) has suggested that the development of accessory roots may be related to an increase in occlusal area and thus have questionable phyletic significance.

Although the cusp morphology is almost completely obliterated on the P^4 preserved in UCMP 120397, the P^4 to M^1 length ratio could be calculated. The very low value of this ratio (0.44) is very distinctive and is found in only a few other multituberculates, particularly when the general outline of P^4 and the general morphology of the M^1 are also considered. The only other multituberculates that fit this overall pattern are the cimolomyid *Meniscoessus* and the taeniolabidids *Taeniolabis* and *Catopsalis*.

One feature of *Essonodon* that may distinguish it from P^4's of the above genera is the extreme height of its crown. Even though the posterior half of the P^4 crown is missing and may have had additional height, the remaining anterior portion towers over the apices of the molar cusps. This feature is somewhat reminiscent of P^4's of advanced ptilodontoids. If the worn anteroexternal margin of the P^4 reflects the unworn morphology of this region, the tooth lacked anteroexternal cusps. Some advanced taenolabidids also lack anteroexternal cusps, but the P^4 in these forms is also more greatly reduced overall, reflecting a reduction in P^4_4 shear. Three P^4's described earlier and tentatively referred to *Meniscoessus* bear some resemblance to the P^4 of *Essonodon* (compare Figs. 29 and 32). Unlike more typical P^4's referred to *Meniscoessus*, these P^4's lack anteroexternal cusps. There also appear to be similarities in the root size and configuration between the better-preserved of these three atypical P^4's and the P^4 of *Essonodon*. As was noted above, the anterior root of the P^4 of *Essonodon* is ovoid in cross-section and is almost twice as large as the posterior root, which is round in cross-section. The anterior root supports the anterior and anteroexternal portions of the crown. This description applies equally well to the root morphology of the better-preserved atypical P^4's of *Meniscoessus*. Although large samples of typical *Meniscoessus* P^4's were not available when this comparison was made, the typical P^4's examined all had anterior and posterior roots that were of subequal size and were subcircular to ovoid in cross-section. Based on this and on the lack of anteroexternal cusps, the atypical P^4's may actually belong to *Essonodon*; however, two aspects of morphology argue against this referral, at least for the time being. First, none of the three P^4's shows as high a crown as the P^4 of *Essonodon*. Second, all three P^4's are markedly larger than the P^4 preserved in UCMP 120397.

The three M_1's or M_1 fragments of *Essonodon* all possess an accessory posterolabial cuspule row. M_1's of *Meniscoessus* frequently (less frequently in *Cimolomys*) possess this accessory row of cuspules. Among taeniolabidids a similar condition is found in at least some M_1's of *Taeniolabis*. As could be true for accessory roots, the presence of an accessory posterolabial cuspule row may reflect large size rather than a close phyletic relationship.

Another potential region for comparison is the secondary bony palate, especially in regard to the presence or absence of palatal vacuities. The presence or absence of palatal vacuities, where evidence is available, appears to follow familial groupings of multituberculates except among paulchoffatiids, but above the family level these two character states are not consistent (Clemens and Kielan-Jaworowska, 1979). Based upon the previous discussion of the P^4–M^1 relationships *Essonodon* appears to be either a taeniolabidid or a cimolomyid; therefore the discussion of palatal vacuities will be restricted to these two families. If the recent literature is to be accepted (e.g., Clemens and Kielan-Jaworowska, 1979), taeniolabidids lack palatal vacuities; however, this point is not at all clear in the various accounts, and before comparisons with *Essonodon* can be made it is necessary to comment upon this.

From my examination of the literature, it appears that *Taeniolabis* (e.g., Matthew, 1937; Granger and Simpson, 1929), *Prionessus* (Granger and Simpson, 1929), and *Kamptobaatar* (Kielan-Jaworowska, 1970, 1971) do lack palatal vacuities. The palatal region of *Sphenopsalis* is unknown. The literature (and unpublished figures) is ambiguous in regard to palatal vacuities in Asian species of *Catopsalis*. Descriptions and figures of the palate of North American *Catopsalis* have yet to be published. When Simpson (1925) named and described *Djadochtatherium matthewi* (now *Catopsalis matthewi*, Kielan-Jaworowska and Sloan, 1979), he indicated that the anterior end of a small palatal vacuity was present opposite P^3. He repeated this observation later (1937b). When first described the genus was not referred to any family, but by 1945 Simpson regarded it as a ptilodontid of uncertain subfamilial status. It should be noted that eucosmodontids (*sensu stricto*) were at this time considered a subfamily of the Ptilodontidae. Sloan and Van Valen (1965) elevated the Eucosmodontinae to familial level and transferred it to their suborder Taeniolabidoidea. They transferred *"Djadochtatherium"* to the Eucosmodontidae and noted that it "has a *Stygimys*-like palate" (presumably including the fact that both taxa have palatal vacuities). Kielan-Jaworowska (1970, p. 45) accepted this classification and noted that a new Asian genus of this family, *Kryptobaatar*, "resembles also *Djadochtatherium* Simpson and *Stygimys* Sloan and Van Valen in having a similar pattern of palatal vacuities, this, however, must be regarded as tentative, as in *Djadochtatherium* and *Stygimys* the most posterior part of [the] palate is unknown." In the same paper she added that the taeniolabidid *Catopsalis* lacks palatal vacuities, according to a personal communication from R. E. Sloan. Later (1974a), she transferred *"Djadochtatherium"* to the Taeniolabididae. In the same paper she described a new species, *Djadochtatherium catopsaloides* (now *Catopsalis catopsaloides*). The reconstruction of the palate of this new species is illustrated in her text (fig. 6a) and the palates of two of the three known skulls are shown (plates XVIII, 1d, type specimen Z. Pal. No. MgM-I/78; and XXI, 1b, Z. Pal. No. MgM-I/79). It is clear from these illustrations that *C. catopsaloides* does not have palatal vacuities whose anterior margin is opposite P^3, as was recognized by Simpson (1925) in *C. matthewi*. The palatal region of *C. catopsaloides* was not described in detail, however, and it was not indicated whether Simpson was incorrect regarding the presence of palatal vacuities in *C. matthewi*. The only possible reference to this is found on p. 38, where she noted twice that taeniolabidids lack palatal vacuities. Beyond this point, if it is assumed that the palatal reconstruction of *C. catopsaloides* in her text fig. 6a is complete, then this species has a palatal configuration unlike other multituberculates but possibly like *Essonodon*. The presumed posterior margin of the secondary palate extends posteriorly only to the anterior one-third of M^1; this was one interpretation offered earlier for the palate of *Essonodon*. Again presuming that text fig. 6a is meant to represent a restored palate, there are problems with this reconstruction of *C. catopsaloides*. In the stereophotographs of the palates of the two skulls (plates XVIII, 1d, and XXI, 1b) the sutures are very obvious, yet no complete palatine bones appear to be preserved. However, on the right side of the skull, lingual to the posterior half of M^1 and M^2, on specimen Z. Pal. No. MgM–I/79 (plate XXI, 1b), two fragments of bone are present which I would interpret as pieces of the right palatine. If this interpretation is correct, the secondary palate of *C. catopsaloides* was much more extensive and was similar to that seen in other multituberculates. The position of these fragments also suggests that palatal vacuities were absent, but this is by no means certain. This interpretation of the palate of *C. catopsaloides* is supported by a more recent reconstruction found in fig. 1a of Kielan-Jaworowska and Sloan, 1979. Based on this rather lengthy digression, I conclude that currently recognized taeniolabidids, with the possible exception of *C. matthewi*, lack palatal vacuities.

The evidence concerning palatal vacuities in cimolomyids is even more scanty and problematic. Of the two genera, *Cimolomys* and *Meniscoessus,* referred to this poorly known family, only *Meniscoessus* is represented by skull fragments. As was discussed above, the evidence is somewhat contradictory, but *Meniscoessus* was reconstructed as having palatal vacuities (see Fig. 27). However, if both *Meniscoessus* and *Essonodon* did possess palatal vacuities, at least the anterior margins would have been different in the two genera. In *Meniscoessus* they would have reached the level of P_2, but in *Essonodon* would have reached only to the anterior one-third of M^1.

In attempting to ascertain the correct phylogenetic relationships of *Essonodon* some weight must be given to whether the character states discussed above are primitive or derived. The presence of accessory molar (or premolar) roots and the presence of a posterolabial accessory cuspule row on M_1 are assuredly derived for multituberculates in general. However, the presence of these structures may well be size-related and thus their utility above the generic level must for the time being be suspect. The very low P^4 to M^1 length ratio seen in the cimolomyid *Meniscoessus* and many taeniolabidids is also clearly derived relative to multituberculates in general. As will be shown later, this low ratio developed separately in the two families. A similarly low ratio in *Essonodon* could represent a third lineage, but I prefer to invoke parsimony and suggest that the low ratio is due to a close phyletic relationship with one of the above families. An assessment of the polarity of the presence or absence of palatal vacuities is much less clear-cut. Although palatal vacuities could have, and may well have, developed in parallel in different multituberculate lineages it seems more reasonable to suppose that their presence is primitive for the great majority of the order. Thus, based on the above discussion of palatal vacuities, *Essonodon* can tentatively be excluded from the Taeniolabididae but not from the Cimolomyidae. Utilizing the admittedly scant evidence, *Essonodon* is questionably referred to the Cimolomyidae.

The development of a latticework of ridges in the longitudinal valleys separating the cusp rows and the ridgelike development of some of the cusps (e.g., the internal row of M_1) suggest that *Essonodon*'s shearing function of the molars was increased over that of most other multituberculates. The earliest stages of wear on the latticework of ridges, the height of the cusps, and the orientation of the cusps and ridges indicate that jaw movement was still fore-aft, as in other multituberculates.

Suborder and Family, *incertae sedis*

Cimexomys Sloan and Van Valen, 1965

Cimexomys Sloan and Van Valen, 1965, p. 221

Genotypic species. Cimexomys minor Sloan and Van Valen, 1965.

Referred species. Cimexomys sp. unnamed Sloan and Van Valen, 1965; *Cimexomys antiquus* Fox, 1971; *Cimexomys hausoi* n. sp.

Distribution. Upper Milk River Formation, Alberta (Aquilan); Hell Creek Formation, Montana, and Lance Formation, Wyoming (both Lancian); lowermost Tullock Formation (pre-mantuan); Polecat Bench Formation, Wyoming (mantuan).

Revised diagnosis. P_4 with fewer serrations (8–10) than in comparably sized species of *Mesodma*, arcuate in outline, last two or three serrations formed into distinct cusps without internal or external ridges; M_1 cusp formula (5–7:4–5) greater than *Paracimexomys* and less than or equal to *Mesodma*; height of P^4 lower than in comparably sized species of *Mesodma*, P^4 with five to six medial cusps and always two to three posterior basal cusps; M^1 cusp formula (4–5:5–6:R–4) greater than *Paracimexomys* but less than *Mesodma*, M^1 not divergent anteriorly, internal cusp row less than 50 percent of M^1 length; molar cusps tending to be subcrescentic to crescentic; molars not waisted.

Comments. Cimexomys has been restricted to include only the type species *C. minor*, plus *C. antiquus* (Fox, 1971) from the Aquilan, an unnamed species of *Cimexomys* (Sloan and Van Valen, 1965) from the Lancian and Puercan, and *C. hausoi* n. sp. from the pre-mantuan. A new genus, *Paracimexomys*, is named to include several species formerly included in *Cimexomys*. The differences between *Cimexomys* and *Paracimexomys* are discussed under the latter genus.

Cimexomys minor Sloan and Van Valen, 1965

(Fig. 35 & Table 15)

Cimexomys minor Sloan and Van Valen, 1965, p. 221

Type. SPSM 62-2115, left mandible with P_{3-4} and alveoli for I, M_1, and M_2 (Sloan and Van Valen, 1965, fig. 2).

Type locality. Bug Creek Anthills, Hell Creek Formation, Montana.

Referred specimens. 3 P_4's; 2 P^4's; 4 M^1's from UCMP loc. V74111.

Locality. UCMP loc. V74111.

Distribution. Hell Creek Formation, Montana, and Lance Formation, Wyoming (both Lancian); lowermost Tullock Formation (pre-mantuan).

Diagnosis. Sloan and Van Valen, 1965, p. 221.

Description. Only additions to Sloan and Van Valen (1965) are given.

P_4. Only one of the P_4's is undamaged and preserves nine serrations and six internal and external ridges. The other two P_4's probably had eight or nine serrations. Based on size and morphology these teeth are indistinguishable from P_4's of *Cimexomys minor* from Bug Creek Anthills (Sloan and Van Valen, 1965; Sloan, unpublished data).

TABLE 15. Measurements of teeth of *Cimexomys minor*, UCMP loc. V74111.

		L1	*L*	*H*	*W*	*L1/L*	*H/L*
P_4	$\bar{x}$	1.39	2.66	1.55	0.99	0.52	0.59
	OR	a1.38 −1.40	a2.61 −2.74	1.44 −1.66	0.91 −1.06	0.51 −0.53	0.53 −0.64
	N	2	3	2	3	2	2

		L	*W*
P^4	$\bar{x}$	2.31	1.24
	OR	2.30 −2.32	1.22 −1.26
	N	3	3

		L	*W*
M^1	$\bar{x}$	2.47	1.44
	OR	a2.25 −2.61	1.33 −1.50
	N	4	4

a Approximate measurement.

P^4. Sloan and Van Valen (1965) did not describe P^4's of *Cimexomys minor*. Two isolated P^4's from loc. V74111 are referred to this species based on frequency of occurrence, size, and morphology (Fig. 35a–c); they resemble isolated P^4's in the UCMP Bug Creek Anthills sample, which probably also belong to this species. Both P^4's have a cusp formula 3:5:2. The three anteroexternal cusps are conical and increase in size posteriorly. The five medial cusps increase in size, but not in individual height (notch to apex) going posteriorly. The total crown height increases posteriorly, but not as much as in *Mesodma*. The larger and more posterior medial cusps have weak striations on the lingual surface. There are two posterior basal cusps. One is almost in line with the medial cusp row and a distinct crest runs from it to the apex of the ultimate medial cusp. In lateral view the outline of the crest is slightly convex, giving the ultimate medial cusp a rounded, less conical profile. The second posterior basal cusp is lingual to the midline and has a less pronounced crest merging with the last medial cusp approximately halfway up its lingual side. A small depression is formed between the two posterior basal cusps and their respective crests.

M^1. The four M^1's of *Cimexomys minor* have a cusp formula 4:6–?7:R–4. The morphology and size of the M^1's agree with the descriptions by Clemens (1964 and 1973) and Sloan (unpubl. data).

M_1, *M_2*, and *M^2*. Although *Cimexomys minor* is relatively rare at loc. V74111, isolated M_1's and M^2_2's are probably included in the sample. If present, they are indistinguishable from M_1's and M^2_2's of *Mesodma*.

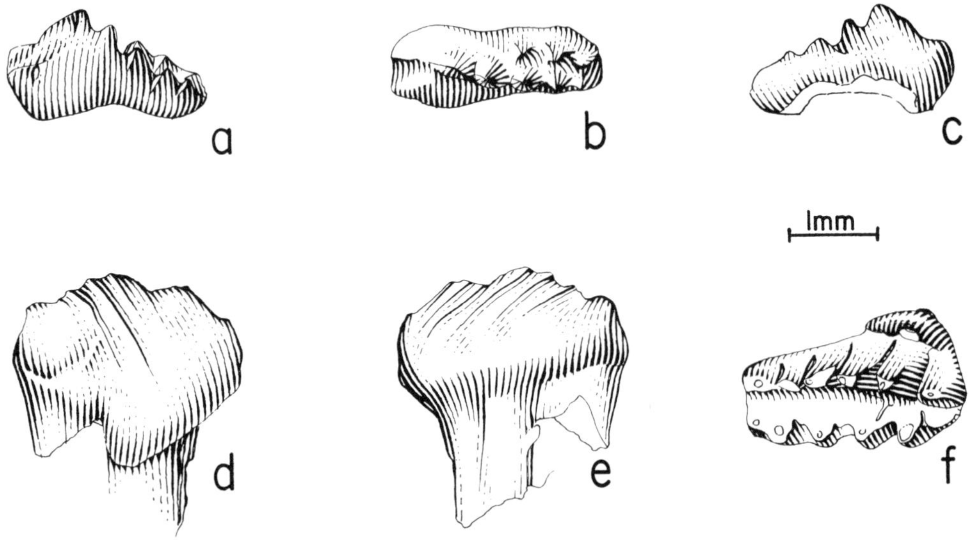

FIG. 35. *Cimexomys minor* Sloan and Van Valen. Isolated teeth, loc. V74111. Right P^4, UCMP 116986: (a) lingual view; (b) occlusal view; (c) labial view. Right P_4 (damaged), UCMP 116983: (d) labial view; (e) lingual view. Right M^1, UCMP 116989: (f) occlusal view. (A. L.)

Discussion. The only other previously published species here referred to *Cimexomys* is *C. antiquus*. Fox (1971) described this species on the basis of two P^4's, four M^1's, and two M_1's. At that time P^4's of *C. minor* had not been described. The description here of P^4's of *C. minor* now makes this comparison possible. The sample size of P^4's for both species is small (N = 2 for both species). Like the M^1_1's, the P^4's of *C. antiquus* are smaller than their counterparts in *C. minor*. The overall morphology of the P^4's of these two species is very similar, corroborating Fox's referral of these two P^4's to *C. antiquus*. The few differences that do exist could be due to the small sample size as well as to differences in age. *C. antiquus* has two anteroexternal cusps and *C. minor* has three. Fox reported that *C. antiquus* lacks striations on the sides of the cusps. The single relatively unworn P^4 of *C. minor* from loc. V74111 has weak striations on the lingual side of its more posterior cusps. Based on Fox's illustrations (1971, fig. 3) the height of the P^4 crown may be slightly higher than in *C. minor*. Although Fox does not include them in his P^4 cusp formula, it is apparent from his illustrations (fig. 3) that *C. antiquus* has two posterior basal cusps. This agrees with similarly placed posterior basal cusps in *C. minor*. If the teeth of *C. minor* and *C. antiquus* had been found at one locality and not been separated by almost ten million years, it might be difficult to differentiate between them.

Cimexomys hausoi, n. sp.

(Figs. 36–38 & Table 16)

Etymology. Named for the Hauso family, out of gratitude for their hospitality in allowing us to collect on their land.

Type. UCMP 117000, right mandible with I root, P_{3-4}, M_{1-2}.

Type locality. UCMP loc. V74111, lowermost Tullock Formation, Montana.

Referred specimens. 3 P_4's; 6 M_1's; 5 M_2's; left mandible fragment with M_2; 4 P^4's; 5 M^1's; 7 M^2's from UCMP locality V74111.

Locality. UCMP loc. V74111.

Distribution. Lowermost Tullock Formation, Montana (pre-mantuan).

Diagnosis. Larger than *Cimexomys antiquus* or *C. minor*, but (?)similar in size to unnamed species from Harbicht Hill and Mantua lentil; number of molar cusps and P_4 serrations greater than in *C. minor* or *C. antiquus.* Cusp formulae: P_4, 9–10 serrations; M_1, 6–7:4; M_2, 4:2–3; P^4, 3–4:5–6:2–3; M^1, 4–5:5–6:R–?5; M^2, ?3–5:3:R–3.

Description. In 1965, when Sloan and Van Valen named the new genus and species *Cimexomys minor*, they mentioned the presence of a larger species of this genus at Harbicht Hill and Mantua lentil (p. 222). This larger species has not been named. That material could not be found; however, Dr. Sloan examined the specimens upon which *C. hausoi* is based and felt they were not referable to the unnamed species of *Cimexomys.*

Mandible. Two mandible fragments belonging to *Cimexomys hausoi* were recovered from loc. V74111. The holotype, UCMP 117000, is a right mandible which has been slightly damaged by corrosion, probably by acid from plant roots. The jaw preserves the root for the lower incisor, P_{3-4}, and M_{1-2} (Fig. 36). Not enough of the incisor is preserved to tell whether the enamel was restricted to a ventral band or if it surrounded the incisor. However, the latter is assumed to have been the case. In cross-section below the P_4 the incisor root is deeper than it is wide. The width/depth ratio is approximately 0.67. The width/depth ratio for the incisor alveolus of *C. minor* is also 0.67, based on UCMP 92666 in the UCMP Bug Creek Anthills sample. The jaw of *C. hausoi* is moderately deep and robust. The lateral profile of the ventral margin of the jaw forms a straight line roughly paralleling the alveolar margin. In lingual view the deepest portion of the jaws is directly below the P_4–M_1 contact. The depth decreases gradually both anterior and posterior from this point. The bone is worn, but the anteroventral margin of the ascending ramus appears to merge with the corpus near the posterior margin of M_1. There is a longitudinal depression separating the labial border of the M_2 from the ascending ramus. The anterior surface of the pterygoid fossa forms a slightly concave, posteriorly facing, vertical surface running from just below the posterior edge of M_2 to the broken ventral margin of the corpus. This area is almost at right angles to the lingual surface of the mandible. These two areas join at a distinct vertical crest running from just below the posterolingual corner of M_2 to the broken ventral margin of the corpus. The mandibular(?) foramen is situated 2 mm below the M_2 in the pterygoid fossa, just labial to this vertical crest.

P_3. Only the P_3 preserved in the holotype (UCMP 117000) is referable to *Cimexomys hausoi*. The single root is exposed on the labial surface of the jaw (Fig. 36a). From the root tip to the crown apex the P_3 is approximately 2.7 mm long. The crown height accounts for only about 0.6 mm of this length. The root is convex labially, conforming to the curvature of the mandible. The tip of the root slightly overlaps the labial side of the incisor root. The

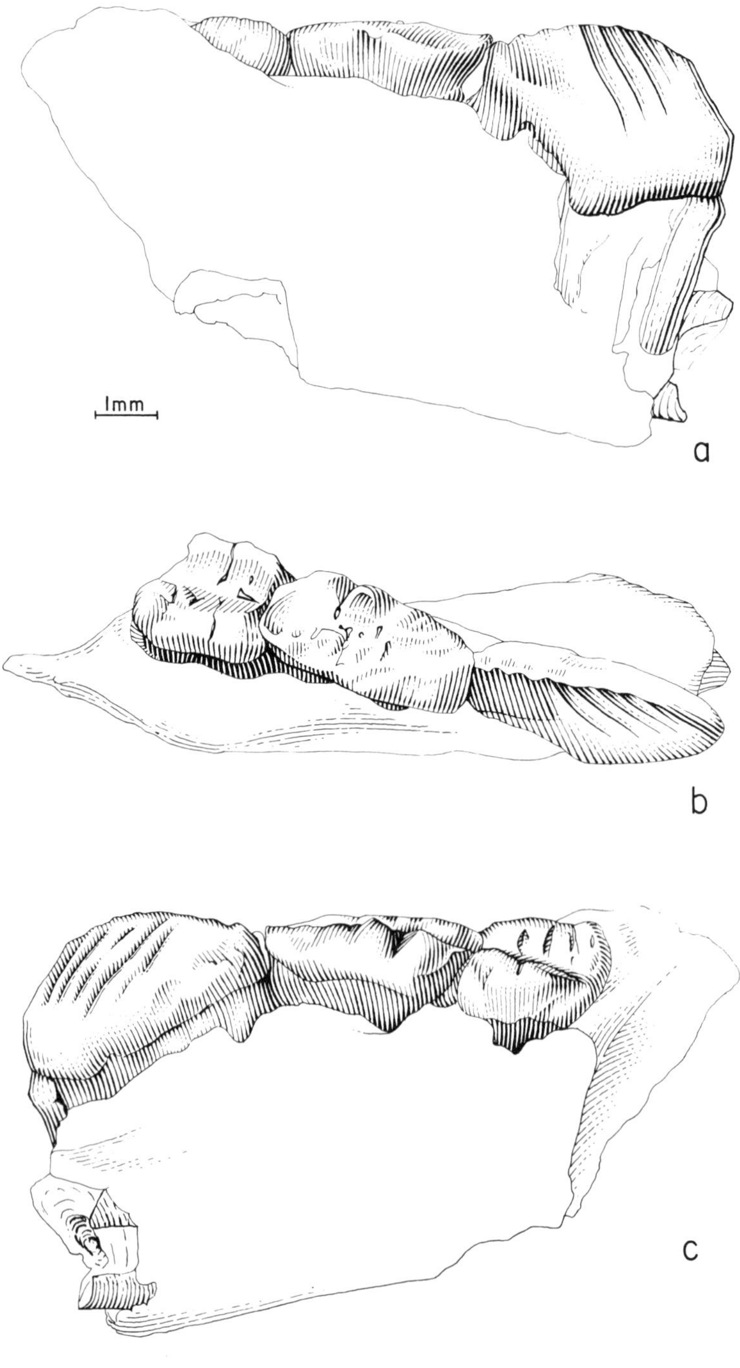

FIG. 36. *Cimexomys hausoi*, n. sp., type. Slightly corroded right dentary with incisor root, P_{3-4}, M_{1-2}, UCMP 117000, loc. V74111: (a) labial view; (b) occlusal view; (c) lingual view. (A. L.)

crown of the P_3 is not well preserved, and most of its dorsal surface is overhung by the P_4. The P_3 is approximately 0.6 mm long (anteroposteriorly). The crown is globose and the exposed portion of the crown has two(?) very small cuspules on its apex, although more could be hidden by the overhang of P_4.

P_4. Including the worn P_4 in the holotype (UCMP 117000), four P_4's are assignable to *Cimexomys hausoi*. All are worn, making the determination of serration count difficult. The P_4 in the holotype has nine or ten serrations and five or six external and internal ridges. The three isolated P_4's probably have nine serrations and six external ridges. These P_4's may have five or six internal ridges. The last two to three serrations are formed into distinct conical cusps. The penultimate and ultimate serrations do not usually have internal or external ridges. As in *C. minor*, the P_4 greatly overhangs the P_3. The overhang is about one-third of the length from the posterior margin of the anterior P_4 root to the anterior of the blade. Due to damage or wear on all three P_4's, the morphology of the ridges at the anterior margin of the P_4 cannot be determined. More posteriorly the external and internal ridges are better preserved and parallel each other throughout their extent. The ridges are oriented at a lower angle (to the horizontal) than in *C. minor*. Although poorly preserved, the anterior margin of the P_4 appears to be relatively short dorsoventrally and slightly convex anteriorly. The serrate edge forms a smooth arc with its highest point slightly posterior to the midpoint of the tooth. This is shown by the L1/L ratio, which averages 0.54 for three of the teeth (OR = 0.52–0.57). A small ridge is directed ventrally from the last serration. A short distance below this serration the ridge swings anterolabially and expands to form a narrow posterolabial basal shelf. A small circular wear facet develops just dorsal to this shelf. With further wear the facet expands dorsally, and eventually the labial surfaces of the more posterior serrations are beveled. There are two roots: the posterior root is circular in cross-section; the anterior root is long and thin in cross-section and has a distinct groove on its anterior surface for P_3. An inter-radicular crest connects the two roots.

M_1. All six M_1's have four internal cusps. On some of the larger teeth a small ridge runs anteriorly from the first internal cusp. The internal cusps are almost pyramidal in shape, with their apices directed posteriorly. Well-developed notches separate the cusps. The more posterior internal cusps have shallow, irregularly shaped grooves on their labial surfaces which are obliterated in early stages of wear. The lower, external cusp row has six or seven cusps. The more anterior external cusps are small and closely spaced. With slight dental wear it is difficult to determine the true number of cusps in this region. None of the M_1's has completely unworn external cusps. The least-worn external cusps are vertical and chisel-shaped, with the labial and lingual surfaces forming the chisel faces. The external cusps are closely spaced. In early stages of wear the cusps form an almost continuous chisel-shaped ridge, interrupted only by deep, vertically sided notches. With progressive wear this ridge becomes worn down. The last external cusp is longer than the preceding ones and is incompletely divided, by a shallow groove on its labial side, into two cusps. The M_1's have two almost equal-sized roots.

M_2. Six of the seven M_2's have two internal cusps. The seventh has three internal cusps. The larger internal cusps are subcrescentic and have irregular grooves on their labial surfaces. The external cusp row has four very closely spaced and incompletely separated cusps. The labial grooves separating the cusps are shallow and develop only between the more anterior cusps. The lingual grooves are better developed but become progressively shallower between the more posterior cusps. Wear facets first develop in the longitudinal

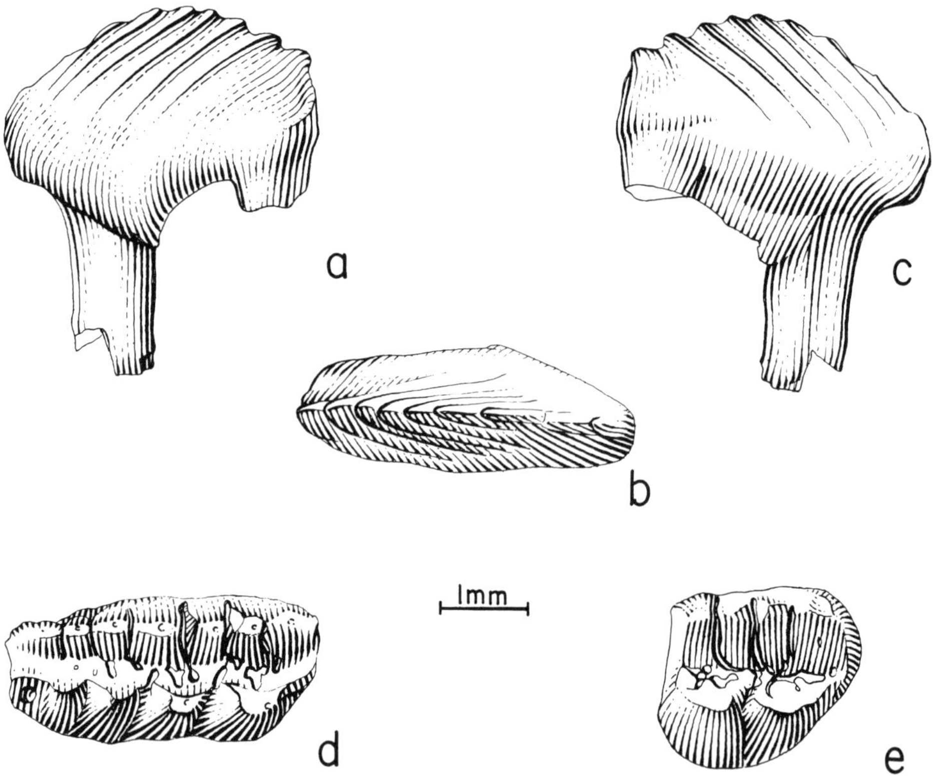

FIG. 37. *Cimexomys hausoi*, n. sp. Isolated teeth, loc. V74111. Left P_4, UCMP 116994: (a) labial view; (b) occlusal view; (c) lingual view. Right M_1, UCMP 116997: (d) occlusal view. Right M_2, UCMP 117003: (e) occlusal view. (A. L.)

valley separating the cusp rows. The next stage of wear involves the labial surface of the external cusp row. Finally, the apices of the cusps are worn down, first in the external cusp row, then in the internal row.

P^4. The four P^4's have a cusp formula 3–4:5–6:2–3. The anteroexternal cusps are conical and increase in size posteriorly. In the P^4's with four anteroexternal cusps (N = 2), the first cusp is very small, and in the P^4's with only three cusps (N = 2), a small ridge takes the place of this very small anterior cusp. The number of cusps in the medial row is evenly divided between five and six. When six cusps are present the first is very small. The cusps increase in size but not in individual height (notch to apex) going posteriorly. The total crown height gradually increases posteriorly but is never as high as in *Mesodma*. The medial cusps are conical and on unworn P^4's they have vertical striations on their labial and lingual surfaces. Only on the labial side of the last medial cusp are the striations absent. Two of the four P^4's have three basal cusps; the other two have two cusps. In all four teeth one of the basal cusps is lingual to the medial cusp row. This cusp has a low, rounded

ridge going about halfway up the lingual side of the last medial cusp. The other one or two basal cusps are approximately in line with the medial row. The last medial cusp is elongated posteriorly, forming a broad ridge which continues down to the one or two posteromedial basal cusps. A deep notch separates the last medial cusp and the first posteromedial cusp. When only one posteromedial cusp is present it is almost as large and as high as the third or fourth medial cusp. A small, usually crenulated depression is present between the posteromedial and posterolingual basal cusps. None of the P^4's shows much dental wear. When present, wear is restricted to two facets: the larger facet is along the posterolingual side of the medial cusp row; the smaller wear facet develops on the lingual face of the one or two posteromedial basal cusps. The anterior and posterior roots are ovoid in cross-section and are approximately the same size. There may be a very short inter-radicular crest joining the two roots.

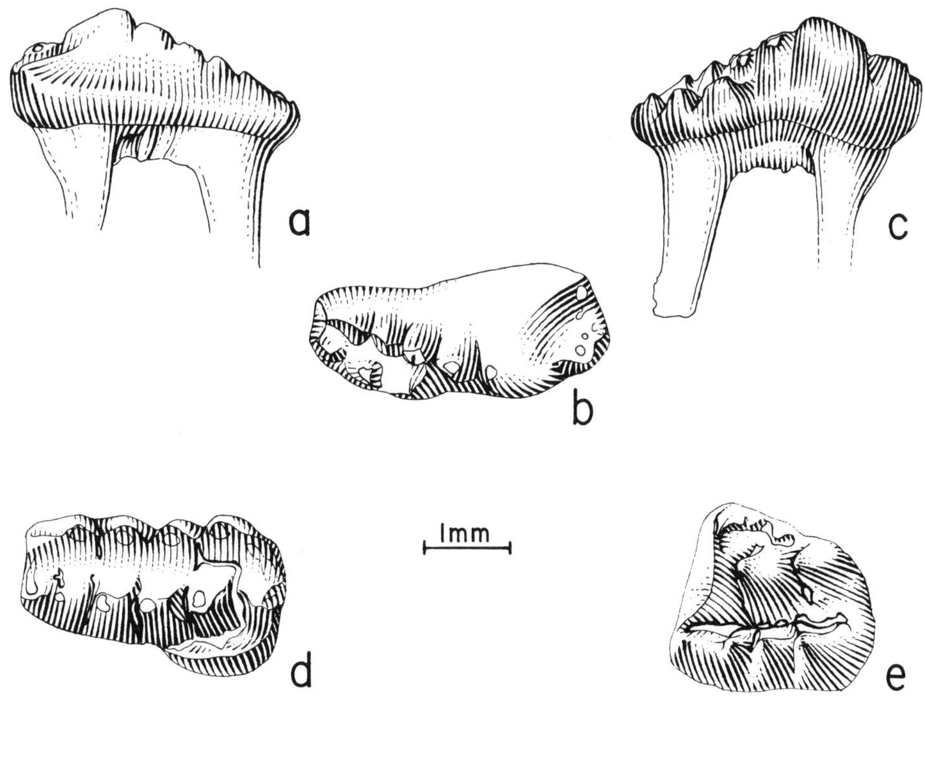

FIG. 38. *Cimexomys hausoi*, n. sp. Isolated teeth, loc. V74111. Right P^4, UCMP 117009: (a) lingual view; (b) occlusal view; (c) labial view. Left M^1, UCMP 117013: (d) occlusal view. Left M^2, UCMP 117018: (e) occlusal view. (A. L.)

M^1. The five complete M^1's have a cusp formula 4–5:5–6:R–?5. The internal cusp row is less than 50 percent of the M^1 length and is best described as a ridge. Only in unworn teeth is there a hint of small cuspules along this ridge. Of the five M^1's, three have six medial cusps and two have five cusps. The height but not the size of the medial cusps increases slightly from front to back. The medial cusps are closely spaced and are separated by moderately deep, vertically sided notches. The cusps become progressively more subcrescentic going posteriorly. Four of the five M^1's have five external cusps. The slightly elongate, conical cusps in the external row are about equal in size, except for the first, which is smaller. In earlier stages wear is restricted to the longitudinal valley separating the external and medial cusp rows, and along the labial surface of the internal ridge. Finally, the apices of the cusps are worn down. The two roots are rectangular in cross-section and are equal in size. The anterior root is elongate in the anteroposterior direction and the posterior root is elongate in the labiolingual direction.

M^2. Cusp formula is ?3–5:3:R–3. The three(?) to five (mode=4) internal cusps are small and of equal size. The apices of the cusps slant anteriorly. The four complete M^2's

TABLE 16. Measurements of teeth of *Cimexomys hausoi*, n. sp., UCMP loc. V74111.

Measurements of the Type, UCMP 117000

P_4						M_1		M_2		Length ratio P_4/M_1	Length ratio M_2/M_1
L1	*L*	*H*	*W*	*L1/L*	*H/L*	*L*	*W*	*L*	*W*		
2.19	4.24	a2.01	1.51	0.52	a0.47	3.43	1.58	2.32	1.86	1.24	0.68

Measurements of Isolated Teeth, Including Type, UCMP 117000

		L	*L1*	*H*	*W*	*L1/L*	*H/L*
P_4	$\bar{x}$	4.07	2.18	2.11	1.50	0.54	0.52
	OR	3.72–4.26	1.93–2.43	a2.01–2.27	1.41–1.57	0.52–0.57	a0.47–0.55
	N	3	3	3	3	3	3

		$\bar{x}$	*OR*	*N*
P^4	L	3.26	3.05–3.44	4
	W	1.68	1.65–1.73	4
M_1	L	3.42	a3.35–3.58	6
	W	1.56	1.45–a1.62	6
M_2	L	2.24	a2.04–a2.40	6
	W	1.87	1.76–1.95	6
M^1	L	3.35	3.13–3.62	5
	W	1.91	1.76–2.04	6
M^2	L	2.36	a2.23–2.52	5
	W	2.17	2.10–2.25	6

a Approximate measurement.

have three medial cusps. The three cusps are subcrescentic to crescentic and of equal size. The medial cusps are larger than the internal cusps, and the notches separating the former are deeper. The anterior face of the first medial cusp forms the anterior margin of the M^2. In an unworn condition the external cusp row appears to have up to three small cuspules. Most of the wear is restricted to the longitudinal valleys separating the cusp rows and to apices of the medial cusps. The two roots are rectangular in cross-section and the anterior root is the larger of the two.

Discussion. Most of the differences between *Cimexomys minor* and *C. hausoi* are attributable to size.

Based on the holotype (UCMP 117000) the P_4 to M_1 length ratio is 1.24. The isolated P_4's and M_1's have an OR of 1.04–1.31 (mean = 1.19) for this ratio. No comparable measurements have been published for *C. minor*.

Suborder and Family, *incertae sedis*

Paracimexomys, n. gen.

Etymology. Para, Gr., near; alludes to its resemblance to *Cimexomys* Sloan and Van Valen, 1965.

Genotypic species. Cimexomys priscus Lillegraven, 1969, p. 24.

Referred species. Cimexomys magister Fox, 1971, p. 922; probably *Cimexomys judithae* Sahni, 1972, p. 365; and possibly *Cimexomys magnus* Sahni, 1972, p. 366.

Distribution. Upper Milk River Formation, Alberta (Aquilan); Judith River Formation, Montana, and Oldman Formation, Alberta (Judithian); possibly Fruitland and Kirtland formations, New Mexico (Edmontonian); Scollard Formation, Alberta, and Hell Creek Formation, Montana (Lancian).

Diagnosis. M^1_1 cusp formula lower than in *Cimexomys* (M^1, 4–5:5:1–2; M_1, 4–5:3–4); external and medial cusp rows of M^1 diverge anteriorly; molar cusps not crescentic but may slant anteriorly; M^1_1 (and possibly M^2_2) with waist in occlusal view.

Discussion. Cimexomys minor was diagnosed on material from the Bug Creek Anthills and regarded as a small neoplagiaulacid (= ectypodid of Sloan and Van Valen) by Sloan and Van Valen (1965, p. 221). It possesses several characters separating it from *Mesodma formosa*. It possesses fewer serrations on P_4, fewer cusps on M^1_1, and a shorter (<50%) internal cusp on row M^1 than *Mesodma* (and most other neoplagiaulacids). P^4's recovered from loc. V74111 referable to *C. minor* indicate that these teeth also differ from *Mesodma* in several features, most importantly in having fewer and lower cusps in the medial cusp row. Subsequently, Lillegraven (1969), Fox (1971), and Sahni (1972) named five new species which they referred to *Cimexomys*. M^1_1's and M^2_2's (but no P^4_4's) of "*Cimexomys*" *priscus* Lillegraven, 1969, have been found at UCMP loc. V73087, extending the range of this species south into Montana. From UCMP loc. V74111 teeth have been recovered that are referable to *C. minor* and a new species, *C. hausoi*. Through study of these specimens and descriptions of Lillegraven (1969), Fox (1971), and Sahni (1972), it is possible to formalize Lillegraven's suggestion (1969, p. 25) that "*C.*" *priscus* "may indeed ultimately be referred to a distinct genus." It should be stressed that most of these species (except *C. minor*, *C. hausoi*, and *Paracimexomys priscus*) are known only from isolated teeth and referral of these teeth to a particular species is tenuous, and in at least one case I believe it is incorrect.

The diagnosis above includes only characters that separate *Paracimexomys* from *Cimexomys*. As *P. priscus* is the only species of this genus I have studied at first-hand, it has served to diagnose the genus; referral of the other species is based on the literature and on personal communication from R. Fox. The specific problems in referring other species to *Paracimexomys* are as follows: (1) Sahni (1972, p. 365 and fig. 10N) referred an M_1 (AMNH 77103) to "*Cimexomys*" *judithae*, which has a cusp formula 6:4. The M^1 (AMNH 77110) he referred to this species has a cusp formula 5:5:1. It resembles the M^1 of *P. priscus* and can be quite confidently placed in *Paracimexomys*. However, the morphology of the M_1 (AMNH 77103) is quite different from that seen in the M_1 of *P. priscus*. Because of the discrepancies in the morphology between the M^1 and M_1, and the higher cusp formula of the M_1, referral of the latter tooth to *P. judithae* is questionable. (2) The M^1 (AMNH 77120) Sahni (1972) designated as the holotype of "*Cimexomys*" *magnus* shares a lower cusp formula (5:5:1) and a slight(?) waisting with *P. priscus*. However, the external and medial cusp rows do not appear to diverge anteriorly (Sahni, 1972, fig. 10U) as in *Paracimexomys*. The referral of "*C.*" *magnus* to *Paracimexomys* is based on very tenuous characters and is only tentatively placed in the latter genus.

Paracimexomys, like *Cimexomys* and the cimolomyids, cannot be placed in any currently recognized suborder of multituberculates. These taxa, expecially *Cimexomys* and *Paracimexomys*, have a number of similarities. As discussed later, most of these similarities are regarded as primitive for the hypothetical taeniolabidoid-ptilodontoid ancestor and thus cannot be used to show close phyletic relationships among the above three groups or to one of the suborders.

Paracimexomys priscus (Lillegraven, 1969)

(Fig. 39 & Table 17)

Cimexomys priscus Lillegraven, 1969, p. 24

Type. UA 3231, isolated right M^1 (Lillegraven, 1969, fig. 1a–c).

Type locality. Loc. KUA-1, Scollard Formation, Alberta.

Referred specimens. 4 M_1's; 3 M_2's; 5 M^1's; 3 M^2's from UCMP loc. V73087.

Locality. UCMP loc. V73087.

Distribution. Scollard Formation, Alberta, and Hell Creek Formation, Montana (both Lancian).

Revised diagnosis. Compared to other species of the genus: P_4 with fewest number of serrations(?); M^1_1 with lowest cusp formula (M^1, 4:4:2; M_1, 4:3); M^1_1 and M^2_2 with greatest amount of waisting.

Description. Lillegraven's descriptions (1969, pp. 24–26) are adequate and only additions or differences will be given here.

P_4. No P_4's in the collections from the Hell Creek Formation match the P_4 described by Lillegraven (p. 25).

M_1. Of the four M_1's referable to *Paracimexomys priscus*, three are relatively unworn and show a cusp formula 4:3. The last external cusp is morphologically distinct from the preceding cusps. The cusp is spatula-shaped, with the anterior edge forming a high ridge to the third external cusp and the posterior edge forming a low, notched ridge with the last internal cusp. The extreme of this development is seen in UCMP 117026 (Fig. 39c). Although not as extreme as in UCMP 117026, Lillegraven's illustrations (1969, fig. 10–1) indicate that this cusp is similarly developed in the M_1's from his sample. All the M_1's

from loc. V73087 are smaller than the comparable measurements of M_1's given by Lillegraven (see Table 17).

M_2. The three M_2's from loc. V73087 agree with the description given by Lillegraven, except that the labial surface of the third cusp of the external row is not noticeably inflated on any of them. All three teeth definitely have two internal cusps. Because of moderate to heavy wear the determination of the external cusp formula is uncertain. All three M_2's probably had three external cusps, but, based on the number of pulp cavities, one tooth may have had four external cusps. The M_2's from loc. V73087 are all smaller than the one M_2 measured by Lillegraven.

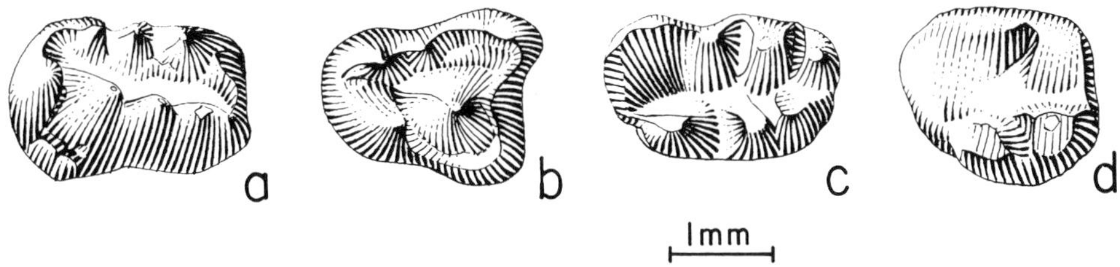

FIG. 39. *Paracimexomys priscus* (Lillegraven), n. gen. Isolated teeth, loc. V73087. Right M^1, UCMP 117033: (a) occlusal view. Left M^2, UCMP 117038: (b) occlusal view. Left M_1, UCMP 117026: (c) occlusal view. Left M_2, UCMP 117029: (d) occlusal view. (A. L.)

Lillegraven had only one M_2 of *Paracimexomys priscus* and it was found in association with an M_1. He noted that the M_2 was wider and nearly as long as the M_1. The M_2's and M_1's from loc. V73087 are isolated and smaller than their Alberta counterparts. However, these teeth agree with the M_1–M_2 size relationships noted by Lillegraven.

M^1. Five M^1's from loc. V73087 are referable to *Paracimexomys priscus*. Of these five teeth, four are complete or nearly so. All four complete M^1's have four external and medial cusps. One of the M^1's has a cuspule anterior to the first medial cusp. On two of the less worn M^1's and on the M^1 fragment the internal cusp row is sufficiently complete to determine that two small cuspules are present.

M^2. Lillegraven (1969) questionably referred only one M^2 fragment, which he did not illustrate, to *Paracimexomys priscus*. Based on frequency of occurrence, size, and morphology, three complete M^2's from loc. V73087 are referable to this species. Two of the M^2's are unworn and can be used to determine the morphology and cusp formula. Both teeth have three cusps in the medial row. The medial cusps increase in size and height posteriorly. The first two cusps are conical and incompletely separated by a shallow transverse notch. The smaller, first cusp is part of a low continuous ridge that joins the external and internal cusp rows. The third medial cusp is separated from the preceding two cusps by a large transverse notch. The third cusp is the tallest and largest in the medial row and on M^2. The anterior face of the cusp is nearly vertical and its apex slightly overhangs the anterior face. Most of the external and internal cusp rows form a ridge which is nearly

continuous around the perimeter of the M^2. Because of this ridge the determination of cusp formulae for the internal and external cusp rows is highly subjective. The internal cusp row is best described as a continuous ridge extending labially from the first medial cusp and then turning posteriorly to join the large, third medial cusp. The highest point on this ridge is at the anteroexternal corner of the tooth. The ridge is slightly swollen at this point and posteriorly the ridge is swollen, but to a lesser degree, at one or two other places. On the slightly worn M^2 (UCMP 117040) the ridge is notched where it joins the large, third medial cusp. The anterior two-thirds of the external cusp row forms a ridge (quite high in UCMP 117040) that is separated by a transverse notch from a posterior, conically shaped external cusp. This posterior external cusp is also separated from the large, ultimate medial cusp by a larger, longitudinal valley. In the unworn M^2 (UCMP 117038) this is the only point around the perimeter of the tooth at which the ridge is discontinous. The anterior two-thirds of the external cusp row (ridge) is slightly swollen in two or three places along its length, most noticeably at its posterior margin. As in the medial cusp row of M^1, the medial cusp row of the M^2 in occlusal view trends toward the lingual side of the tooth going anteriorly

TABLE 17. Measurements of teeth of *Paracimexomys priscus*, UCMP loc. V73087.

		$\bar{x}$	*OR*	*N*
M_1	L	2.23	a2.06–2.31	4
	W	1.37	1.30–1.41	4
M_2	L	2.07	1.99–2.23	3
	W	1.68	1.61–1.80	3
M^1	L	2.41	2.21–2.58	4
	W	1.45	1.34–1.60	4
M^2	L	2.14	2.12–2.18	3
	W	1.75	1.69–1.83	3

a Approximate measurement.

Discussion. One of the diagnostic characters of *Paracimexomys* (except possibly for ?*P. magnus*) is the anterior divergence of the medial and external cusp rows of M^1. This divergence is real but is only 4–8°. The divergence is accentuated by two other features. First, even with the presence of a small internal cusp row the anterior and posterior widths of M^1 are nearly equal, which gives the tooth a rectangular outline in occlusal view. The medial cusp row starts at the anterointernal corner of this rectangle and slants diagonally, terminating in the midline of the tooth at its posterior end. The external cusp row also follows this trend but to a lesser degree. Second, as noted by Lillegraven (1969), the longitudinal valley separating the external and medial cusp rows goes from a V-shape posteriorly to a broad U-shape anteriorly.

In the M^2's referred to *Paracimexomys priscus*, the medial cusp row also trends toward the lingual side anteriorly on the tooth. Possibly because it is a shorter tooth, this trend is not as obvious on the M^2's.

Comments on the Phylogenetic Relationships of Multituberculates

In classroom lectures it is often said that multituberculates are the longest-lived order (or subclass) of mammals. Although this statement is in part an artifact of classification, the multituberculates do share an evolutionary history separate from other mammals at least from the mid-Jurassic and more likely from the latest Triassic (if haramiyids are ancestral or closely related to multituberculates). The record of multituberculates is far from complete, with major gaps throughout much of the Mesozoic; it becomes relatively continuous only in the Late Cretaceous through early Tertiary (Kielan-Jaworowska, 1974b). The known distribution of this group is Holarctic, but with few Mesozoic and early Tertiary mammal records from Gondwana, the known distribution of multituberculates may be only an artifact of collecting.

In studying Lancian and pre-mantuan multituberculates the paucity of information concerning the long pre-Lancian evolutionary history of this group must be considered, because of the great possibility for parallelism. This is especially true for the Mesozoic multituberculates from North America, where most of the phylogenetic information comes largely from one small part of the animal, i.e., the dentition.

Work by Hahn (1969, 1971) suggests that a careful study needs to be made of the upper dentition of plagiaulacoids, ptilodontoids, and taeniolabidoids to determine the homologies between the upper premolars of these groups; such a study is beyond the scope of this paper, however.

The homologies of the lower dentition among these three suborders appear to be, at least superficially, more straightforward. In the lower jaw the plagiaulacoids possess an incisor, three to four premolars, and two molars (Hahn, 1971); the ptilodontoids and taeniolabidoids reduce the number of premolars to one or two. [It should be noted that R. Sloan (pers. comm., and noted in Schiebout, 1974, and Holtzman and Wolberg, 1977) feels that the P_4 of at least some multituberculates is actually a molar; that is, it has no deciduous precursor.] Because the homologies of the lower teeth are more certain among these suborders, the following cladistic analysis and phylogenetic inferences will draw most heavily from this part of the dentition. However, the naming of the four lower premolars is somewhat problematic. The earliest and presumably more primitive plagiaulacoids have at most four lower premolars, while some have five upper premolars and one, *Paulchoffatia delgadoi*, has six (Hahn, 1969, 1971). It seems probable that at one point multituberculates or their immediate ancestors had more than four lower premolars to match the number in the upper dentition. However, care should be exercised in assuming for multituberculates a therian type of dental evolution and replacement (which generally supposes a one-to-one match of upper and lower postcanine teeth in the primitive therian). Therefore, although it is somewhat incongruous, the dental terminology of $P_{1\text{-}4}$ has been used here.

Hahn (1971) has suggested that primitively the lower premolars in such paulchoffatiids as *Paulchoffatia* and *Kuehneodon* were used for "grinding" rather than "cutting." However, because of the differences in I^3 and M_2 he believes that the later development of "cutting" premolars was achieved separately in the paulchoffatiids and plagiaulacids.

Van Valen (1976) has noted that the M_2 of *Paulchoffatia* and the other paulchoffatiids resembles isolated teeth of the Rhaetic haramiyids. He regards this resemblance as the primitive state, which would not preclude the family from the ancestry of other multituberculates. Hahn (1971) had concluded that an I^3 with several cusps, as in paulchoffatiids, was a specialized condition. Van Valen (1976) indicates that the situation regarding the I^3 is less clear-cut, but that multicuspid teeth (incisors?) have been described by Peyer (1956) and Parrington (1947). In the first description (Peyer) the tooth was identified as a probable haramyid incisor. Van Valen notes: "It is at least plausible that the paulchoffatiid condition (i.e., referring to the I^3) is primitive"; I agree. Another point

TABLE 18. Character states for a number of characters considered important for recognizing major groups of multituberculates. The character states for the Plagiaulacidae are regarded as primitive compared to the character states for the groups to the right.

	Paulchoffatiidae	*Plagiaulacidae*	*Paracimexomys*	*Cimexomys*
1 Number of lower premolars	3–4	3–4	?2	2
2 Number of P_4 serrations	4–7	6–8	8–10	8–10
3 Posterolabial basal ridge on P_4	yes	yes	yes	yes, but small
4 Serrate edge of P_4 in relation to M_{1-2}	level to ?below	level to ?below	??level or below	level
5 P_4 to M_1 length ratio	al.8	1.31–1.83	al.46–al.67	1.04–1.31
6 Cusp formula: M_1	3–4:2–3	2–3:3	4–5:3–4	5–7:4–5
7 M_2	0:1	3:3	3–4:2	3–4:2–3
8 M^1	4–5:3:0	4^+:4^+:0	4–5:4–5:R–2	4–5:5–6:R–4
9 M^2	?0:2–3^+:3–6	0:2:3	1:3:3–4	1–5:3:1–3
10 Molar cusp shape	conical[1]	conical	conical-subcrescentic	conical-crescentic
11 Length of internal cusp row compared to total M^1 length	none	none	< 50%	< 50%
12 Multiple accessory roots on M_1^1	no	no	no	no
13 Posterobasal cusps on P^4	—	—	yes	yes
14 Enamel restricted to ventrolabial surface of lower incisor	no	no	no	no
15 M_2 to M_1 length ratio	al.00	1.00–1.18	0.78–0.92	0.68–1.08
16 Medial and external cusp rows on M^1 diverge	no	no	yes	no

a Approximate measurement.

[1]Conical includes pyramidal and quadrate, and for this chart equals not-subcrescentic and not-crescentic.

	Cimolomyidae	*Ptilodontoidea*	*Taeniolabidoidea*
1 Number of lower premolars	(?1)–2	1–2	1–2
2 Number of P_4 serrations	8–11	10–16	3–15
3 Posterolabial basal ridge on P_4	yes	variable	variable
4 Serrate edge of P_4 in relation to M_{1-2}	level to ?below	level to above	level to below
5 P_4 to M_1 length ratio	0.70–1.19	1.35–2.30	0.50–1.90
6 Cusp formula: M_1	5–8:4–7	5–11:4–7	4–9:3–8
7 M_2	3–5:2–3	3–7:2–3	2–6:2–6
8 M^1	5–8:6–10:4–7	6–10:7–13:4–10	4–7:4–8:R–8
9 M^2	0-5:2–4:3–7	1–5:2–4:2–6	1–2:2–5:2–6
10 Molar cusp shape	conical-crescentic to ridgelike	conical-crescentic	conical-subcrescentic
11 Length of internal cusp row compared to total M^1 length	$>50\%$	$>50\%$	$<50\%$ to $>50\%$
12 Multiple accessory roots on M_1^1	yes	no	variable
13 Posterobasal cusps on P^4	variable	variable	variable
14 Enamel restricted to ventrolabial surface of lower incisor	no	no	yes
15 M_2 to M_1 length ratio	?0.30–1.24[2]	0.50–0.71	0.47–0.79
16 Medial and external cusp rows on M^1 diverge	no	no	no

[2]The evidence is ambiguous, but some cimolomyids may have secondarily increased the size ratio of M_2–M_1.

raised by Van Valen that could be used to support Hahn's thesis (but was not mentioned by Hahn) involves the way in which M^2_2 occlude. The full discussion can be found in Van Valen (1976, p. 198), but basically Van Valen (utilizing Hahn's comments) notes that in *Paulchoffatia* the M^2 is directly behind M^1 and not lingual as in other multituberculates. Unless a dental shift was involved, *Paulchoffatia* is derived in this feature. However, Van Valen goes on to show that the M^2 in *Paulchoffatia* could be lingual as in other multituberculates, based on alveoli shown in pl. 8 and fig. 9 of Hahn, 1969. Clearly, additional work needs to be done to establish the phyletic relationships of the paulchoffatiids and plagiaulacids (and haramiyids).

Nevertheless, because of the similarities in the lower serrate blades of plagiaulacids and later multituberculates, I regard the plagiaulacids as the most plausible sister group of later forms (see Hahn, 1971, fig. 85).

However, are the taeniolabidoids and ptilodontoids a monophyletic group, or do they have separate plagiaulacid ancestry? To my knowledge, except for a general reduction in the number of premolars and an increase in cusp and serration count the taeniolabidoids and ptilodontoids do not possess any obvious shared derived characters. Thus, the cladistic and phylogentic alignment of these two suborders can only be regarded as speculative.

In studying the Lancian and pre-mantuan multituberculates from the Hell Creek Formation of Garfield County, Montana, problems arose in referring *Cimexomys*, *Paracimexomys*, and the cimolomyids to the recognized suborders Ptilodontoidea and Taeniolabidoidea. This prompted the cladistic analysis summarized in Table 18 and Fig. 40 and discussed briefly below.

Table 18 is a summary of data compiled from the material described in the previous pages and from Clemens (1964, 1973), Fox (1971, 1976b), Granger and Simpson (1929), Hahn (1969, 1971), Jepsen (1940), Kielan-Jaworowska (1970, 1971, 1974a), Lillegraven (1969), Matthew (1937), Sahni (1972), Simpson (1928, 1929a), Sloan (unpubl. data), and Sloan and Van Valen (1965). Because of the unwieldy amount of data the genera and species have been allocated to their respective suborders or families, where possible, on the basis of shared derived characters. Characters 1 through 16 are mostly those features which can be used to recognize higher taxonomic groupings. The problematic taxa—*Cimexomys*, *Paracimexomys*, and the cimolomyids—have been considered in reference to these characters. The accuracy of the data in Table 18 generally increases from left to right.

Especially for *Paracimexomys* and *Cimolomys* the two ratios (length of P_4 to M_1 and length of M_2 to M_1) are based almost entirely on isolated teeth. This problem has been compensated for by using the extremes of length at each tooth site to calculate the ratios. Therefore, the observed ranges of the ratios are, if anything, greater than would be found by using associated teeth.

The two most important derived characters that can be used to recognize the suborders Taeniolabidoidea and Ptilodontoidea are, respectively, the restriction of the enamel to the ventrolabial surface of the incisor (14′) and the development of the large P_4 blade (5A′). (The prime and letter indicate a shared derived character state.) It would be better to have more characters upon which to base these two suborders, but because of the fragmentary nature of the Mesozoic record of multituberculates in North America this is not possible. In future studies, and with the discovery of additional fossil material, it is hoped that these subordinal characters can be more rigorously tested.

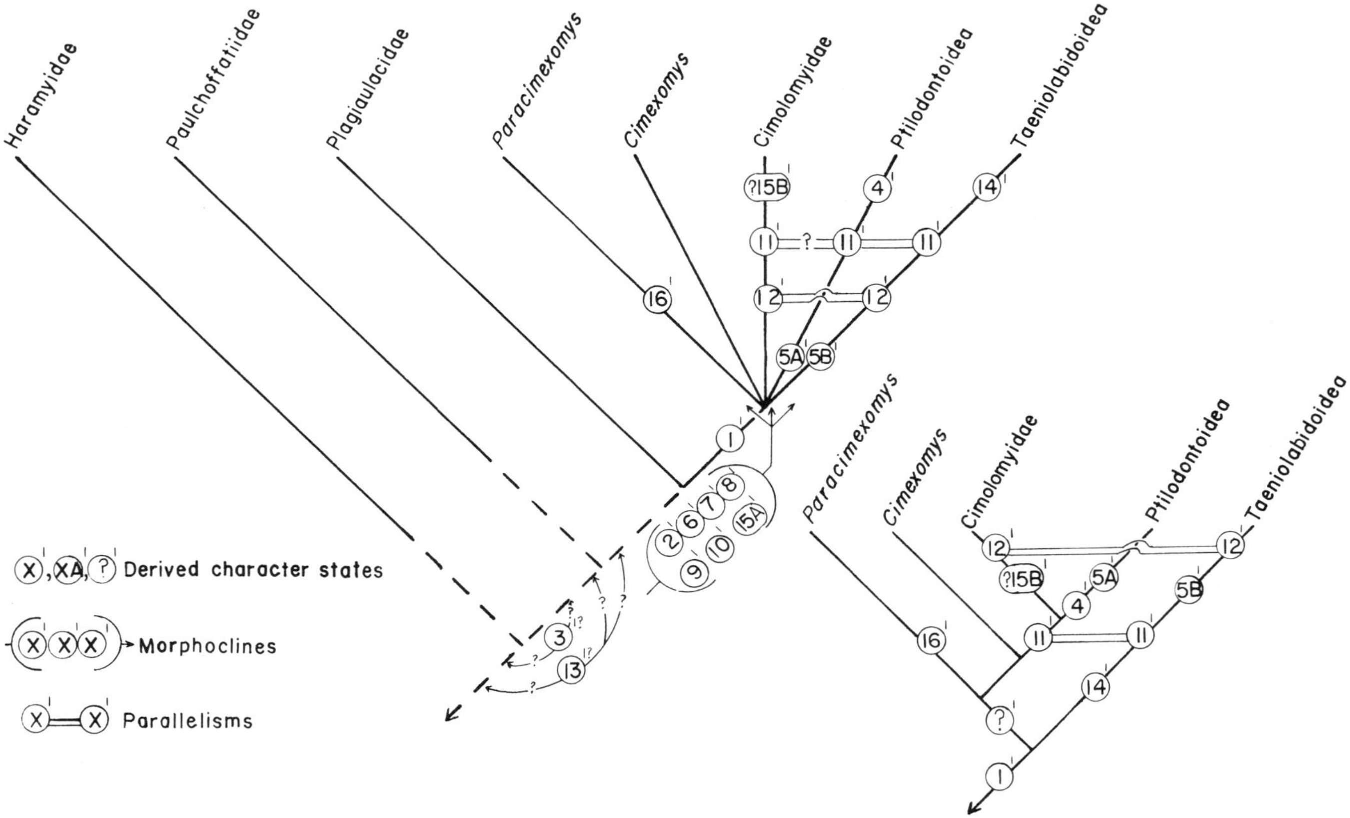

FIG. 40. The two most probable cladistic relationships of later multituberculates. (See text and Table 18 for further explanation.)

A number of the characters represent morphoclines basically changing from left to right. These characters are: decrease in the number of lower premolars (1 ′); increase in number of P_4 serrations (2 ′); increase in number of molar cusps (6 ′–9 ′); change from conical (including pyramidal and quadrate) to crescentic cusps (10 ′); increase in length of internal cusp row on M^1 (11 ′); and decrease in M_2 to M_1 length ratio (15A ′). The most notable exceptions to these trends are the diminution in size and number of serrations on the P_4 of taeniolabidids (5B ′) and the increase in overall size of the M_2's in some later cimolomyids such as *Meniscoessus robustus* and *Cimolomys gracilis* (?15B ′). Both of these exceptions are regarded as derived characters unique to the respective families. Two other characters included in Table 18 are best regarded as primitive characters because of their commonality in the order. The first is the presence of a posterolabial basal ridge of P_4(3). The size and presence or absence of cusps on the ridge vary widely throughout the order, but because of the uniformity of the ridge throughout the order it is probably a primitive multituberculate character. Using Hahn's observation that the P_4 shifts from "grinding" to "cutting" in the plagiaulacoids it can be speculated that in the earliest multituberculate, or its immediate ancestor, the posterolabial basal ridge may have been an integral part of a "grinding" surface. The second primitive multituberculate dental feature is the presence of posterobasal cusps on P"4" (13). This may also reflect a "grinding" ancestry for this tooth, but because of the problems of homologizing the P^4 of later multituberculates with the two posterior upper premolars in plagiaulacoids, the time of origin of the posterobasal cusps is uncertain.

Paracimexomys, *Cimexomys*, and the cimolomyids cannot be included in either the ptilodontoids or taeniolabidoids without stretching or modifying the concepts of these suborders as used here. Excluding for a moment the autapomorphies of these taxa, a morphocline in many characters of the lower dentition can be recognized in the sequence plagiaulacids→*Paracimexomys*→*Cimexomys*→cimolomyids→ptilodontoids. The P_4 to M_1 length ratio decreases in this cline and then dramatically increases in the ptilodontoids. At the beginning of this study it was thought that the lower P_4 to M_1 length ratios of the three intermediate taxa in this cline might be the result of common ancestry. Although this still may be the case, a more satisfactory explanation suggests that the low ratios are primitive in relation to the ptilodontoids. Fig. 41 diagramatically shows this favored explanation. In the plagiaulacids M_1 and M_2 are similar in size and have a low cusp count, and P_4 is relatively long in comparison to M_1. In the sequence *Paracimexomys*→*Cimexomys*→cimolomyids the cusp count and the size of M_1 increase relative to M_2, and although P_4 develops a few more serrations it does not increase as markedly in size. At the ptilodontoid grade M_1 and M^1 continue to increase in size and cusp count, but P_4 has become a more important part of the dentition, possibly as the result of a dietary shift. Although the large P_4 blade was already present in the Late Cretaceous ptilodontoids, *Mesodma* and *Cimolodon*, the extremes in development of P_4, such as in *Ptilodus*, did not occur until the probable immigration of taeniolabidids in the Late Cretaceous. This very tenuously suggests character displacement and partitioning of the small, rodentlike herbivore niche by the taeniolabidoids and ptilodontoids near the beginning of the Tertiary in North America.

The two cladograms judged most likely (based on Table 18) are shown in Fig. 40. There is little basis for favoring one alternative over the other. Biogeographic considerations favor the aligning of *Paracimexomys*, *Cimexomys*, and the cimolomyids with the ptilodontoids. However, there are few, if any, shared derived characters (indicated by (? ′))

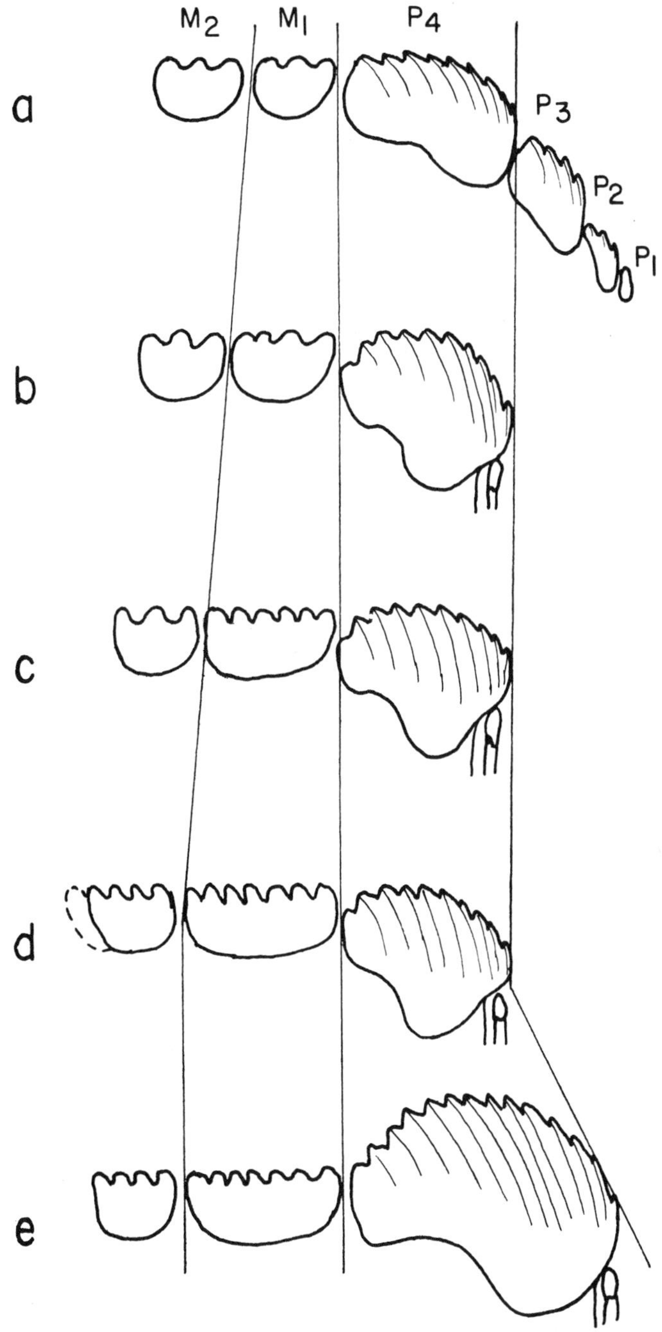

FIG. 41. Cartoon showing stages in development of ptilodontoid lower dentition. No phyletic relationship is inferred by this diagram (see text for discussion): (a) plagiaulacid grade; (b) *Paracimexomys* grade; (c) *Cimexomys* grade; (d) cimolomyid grade; (e) ptilodontoid grade.

uniting these groups. *Paracimexomys*, *Cimexomys*, the cimolomyids, and the ptilodontoids may have jointly begun to increase the number of molar cusps; and the latter two groups may have jointly increased the length of the internal cusp row on M^1.

Unfortunately, the fossil record cannot be used to support or falsify either of these relationships. Fox (1971) has shown that these groups are recognizable back at least to the Aquilan in North America, suggesting that if these taxa share a common ancestry (separate from the taeniolabidoids) separation was a pre-Aquilan event. The M^1's of the Aquilan ptilodontoids (Fox, 1971) show that the internal cusp row had already reached 50 percent of the total M^1 length, although in general they do not appear to be as well developed as in their Lancian counterparts. An M^1 of a cimolomyid has not been described from the Aquilan, but Fox (1976b) named and figured *Meniscoessus intermedius* from the slightly younger Oldman Formation to which can probably be referred a complete M^1. As in the Aquilan ptilodontoids, the internal cusp row is greater than 50 percent of the total M^1 length, but it is less than in the Lancian species of *Meniscoessus*.

Unlike some proponents of cladistics, I regard the finding of new, and in this case earlier, fossils important for determining the most likely relationship among these four North American groups.

Several additional comments and possible exceptions to the proposed cladograms are noteworthy. Some of the exceptions suggest that either changes in the cladograms or in the conclusions of other authors may be necessary. Fox (1971) described two different morphs which he assigned to ?*Cimolomys* sp. A and ?*Cimolomys* sp. B. ?*C.* sp. A is represented by incomplete P_4's, M_1, and P^4. A P_4 of ?*C.* sp. A, UA 5384, which Fox figured (fig. 69), is almost 6 mm long, although incomplete. The largest M_1 noted by Fox is 4.1 mm long. The samples are too small to make any meaningful statements; however, if the P_4 fragment and M_1 are correctly associated this makes the P_4–M_1 length ratio greater than 1.46. This ratio is larger than in any other cimolomyid which suggests: an incorrect association of the P_4 and M_1; an incorrect identification as ?*Cimolomys*; a specialized species of ?*Cimolomys*; or that the lower P_4–M_1 length ratios in Judithian and Lancian cimolomyids are not primitive for the family, as the cladistic analysis suggests. As an aside I should add that I think the two P^4's Fox referred to ?*C.* sp. B could also belong to *Paracimexomys magister* (Fox, 1971). Fox (1971) noted resemblances between ?*C.* sp. A from the Aquilan and *C. trochuus* (Lillegraven, 1969) from the Lancian. The latter species is known from a P_4 in a jaw fragment and two M^1's. The most interesting feature of this species is the lack of a P_3. If this feature is not anomalous, then this is a unique cimolomyid. This would not bar *C. trochuus* from the Cimolomyidae but would be another example of loss of P_3, which has occurred at least three other times within the Multituberculata. One point of correction concerns Lillegraven's statement (1969, p. 31): "The presence or absence of the P_3 in *C. gracilis* is unknown." Although I could find no comment by Clemens (1964, 1973), on examining the isolated P_4's of *C. gracilis* from the Lance Formation, Wyoming, I found an anterobasal cup on these blades, indicating that a P_3 had been present.

In view of the preceding cladistic analysis a final comment is necessary concerning *Gobibaatar parvus* from the Djadokhta Formation of Mongolia. In naming this species Kielan-Jaworowska (1970) referred it to the Neoplagiaulacidae (= Ectypodontidae) and the Ptilodontoidea. She felt that it most closely resembled *Cimexomys* (and now *Paracimexomys*) and that *Gobibaatar* might be the earliest of a lineage that had migrated to North America and given rise to *Cimexomys*. Most recently (1975b and pers. comm.) she feels that the *Gobibaatar*–*Cimexomys* lineage may represent a new family of ptilodontoids. I agree that there are similarities between *Gobibaatar* and *Cimexomys* (and *Paracimexomys*); however, I would exclude *Gobibaatar* from the Ptilodontoidea for the same reasons that I exclude *Cimexomys* (and *Paracimexomys*). *Cimexomys* does not have any autapomorphies that would prevent it from having been the structural ancestor of cimolomyids and ptilodontoids. Based on Kielan-Jaworowska's figures and description (1970), *Gobibaatar parvus* is as primitive as or more primitive than *Cimexomys*, but it does not have the peculiarities distinctive to *Paracimexomys*. Therefore, I do not regard the similarities between *Gobibaatar* and *Cimexomys* as indicative of a close phyletic relationship, but, rather, as the retention of primitive characters in both genera.

Subclass THERIA
Infraclass METATHERIA
Order MARSUPIALIA

Unlike the multituberculates from Garfield County, the marsupials do not show any overlap of taxa at the generic level between Lancian and pre-mantuan localities. The Hell Creek faunal-facies (Lancian) includes a total of twelve or thirteen species representing three families. In contrast, the pre-mantuan has only one species of marsupial, *Peradectes* cf. *P. pusillus*.

Sloan and Van Valen (1965) reported a total of six species of marsupials from their two stratigraphically lowest Bug Creek faunal-facies localities, Bug Creek Anthills and Bug Creek West. No marsupials were reported as occurring at Harbicht Hill, the stratigraphically highest locality. Subsequent collecting by the UCMP at Harbicht Hill has turned up at least one species of small marsupial, probably a didelphine or pediomyid.

The differences in abundance, diversity, and species representation of marsupials between the Bug Creek faunal-facies of McCone County and the pre-mantuan of Garfield County are related to differences in age. These biostratigraphic aspects are discussed in the final section of this study.

The discovery of the Flat Creek local fauna in Garfield County has almost doubled the number of marsupials known to have been present in the Hell Creek faunal-facies during the Lancian. The eleven marsupials that Clemens (1973, p. 96) reported as occurring in the Lance Formation, Wyoming, are represented in the Hell Creek faunal-facies by the same or closely related species. The only difference is that the occurrence of a new species of *Glasbius* in the Hell Creek Formation that replaces *G. intricatus* known from the Lance Formation.

In the Hell Creek faunal-facies marsupials are more abundant than proteutherians (palaeoryctids, plus *Gypsonictops*). This is comparable to the findings of Clemens (1966, 1973) for the Lance local fauna and contrasts with the Trochu local fauna, where Lillegraven (1969, p. 50) found that the proteutherians "challenge the abundance, taxonomically and numerically, of the metatherians." Unfortunately, the sample sizes of marsupials (and eutherians) are quite small, even in the richest Flat Creek local fauna. This has made it difficult to allocate some specimens to specific taxa. In several instances the referral of isolated teeth to a particular species is at best tentative.

Clemens (1966, 1973) and Lillegraven (1969) have adequately described and figured Lancian marsupials. Therefore, in the following systematic description of the Hell Creek Formation marsupials only new data are given.

The classificatory scheme for the marsupials follows Clemens and Marshall (1976).

Superfamily DIDELPHOIDEA (Gray, 1821) Osborn, 1910
Family DIDELPHIDAE Gray, 1821
Subfamily DIDELPHINAE (Gray, 1821) Simpson, 1927b
Alphadon Simpson, 1927b

Comments. Clemens (1966) recognized three species of *Alphadon* in the Lance local fauna—*A. marshi*, *A. lulli*, and *A. rhaister*. Lillegraven (1969) found all but *A. lulli* to be present in the Trochu local fauna. He also demonstrated the presence of another species of *Alphadon*, *A. wilsoni*, in both the Trochu and Lance local faunas. All four of these species appear to be represented in the Hell Creek faunal-facies, although two of the species, *A. lulli* and *A. rhaister*, are rare.

Alphadon marshi Simpson, 1927b
(Tables 19, 20)

Alphadon marshi Simpson, 1927b, p. 125

Type. YPM 13659, right M^3 (Simpson, 1929a, fig. 45F).

Type locality. Lance Formation, Niobrara County, Wyoming.

Referred specimens. 4 M^1's; 5 M^2's; 3 M^3's; 3 M^4's; 4 M_1's; 5 M_2's; 5 M_3's; 3 M_4's from UCMP loc. V73087. Eight isolated teeth from other localities.

Localities. UCMP locs. V72208, V72212, V72213, V73077, V73078, V73083, V73085, and V73087.

Distribution. Lance Formation, Wyoming; Scollard Formation, Alberta; Hell Creek Formation, Montana (all Lancian); possibly Fruitland and Kirtland formations, New Mexico (Edmontonian).

Revised diagnosis. Clemens, 1966, p. 5, and Lillegraven, 1969, p. 33.

Description. *Upper molars.* The upper molars of *Alphadon marshi* from loc. V73087 can most easily be distinguished from *A. wilsoni* on the basis of size. As can be seen by a comparison of Tables 19 and 21, there is no overlap in the observed ranges of upper molars of the two species. Lillegraven (1969, fig. 17) indicates that there is some overlap in size between his upper molar samples of the two species. His sample sizes were considerably larger than those from loc. V73087, so the lack of size overlap in samples from loc. V73087 could very well be due to the extremely small sample sizes.

Other than proportional differences, Lillegraven (1969) found only one consistent morphological feature that differed in the upper molars of *Alphadon marshi* and *A. wilsoni* from the Trochu local fauna. He found that in M^3 of *A. wilsoni* stylar cusp C is larger than D; the reverse was found to be true for M^3's of *A. marshi*. Clemens (1973) reported that in the Lance local fauna stylar cusp C is either of equal size or larger than cusp D in *A. marshi*. The sample of M^3's from loc. V73087 for both of these species is very small (N = 3 for both species), but they appear to follow the trend Lillegraven observed for the specimens from Alberta.

Lillegraven (1969) also noted proportional differences in the upper molars of *Alphadon marshi* and *A. wilsoni*. He observed that upper molars of *A. wilsoni* are proportionally longer (relative to anterior and posterior widths) than those of *A. marshi*. This relationship holds true for the upper molars of these two species from the Flat Creek local fauna (loc. V73087).

Lower molars. All but one specimen from the Hell Creek Formation, Garfield County,

TABLE 19. Measurements of isolated upper molars of *Alphadon marshi* from UCMP loc. V73087.

		L	*W–A*	*W–P*	*L/W–A*	*L/W–P*	*W–A/W–P*
	N	4	4	4	4	4	4
M^1	OR	2.13 –2.27	2.24 –2.27	2.42 –2.61	0.94 –1.01	0.87 –0.91	0.86 –0.91
	$\bar{x}$	2.23	2.26	2.52	0.99	0.89	0.90
	N	4	5	5	4	4	5
M^2	OR	2.28 –2.45	2.43 –2.62	2.60 –2.80	0.90 –0.98	0.86 –0.88	0.91 –0.96
	$\bar{x}$	2.37	2.53	2.71	0.94	0.88	0.93
	N	2	1	2	1	1	1
M^3	OR	2.32 –2.39	2.70	2.60 –2.89	0.89	0.83	0.94
	$\bar{x}$	2.36		2.75			
	N	3	3	3	3	3	3
M^4	OR	1.74 –1.76	2.37 –2.66	1.66 –1.91	0.66 –0.73	0.92 –1.05	1.32 –1.49
	$\bar{x}$	1.75	2.50	1.79	0.70	0.98	1.40

referable to *Alphadon* are isolated teeth. The exception is a dentary fragment of *A. wilsoni* preserving two molars. The allocation of isolated upper molars to the various species of *Alphadon* can be done with some confidence; however, this is not the case with isolated lower molars. Only loc. V73087 (Flat Creek local fauna) produced a small sample of upper and lower molars of *A. marshi* and *A. wilsoni*.

M_1's and M_4's were relatively easy to recognize based on the criteria noted by Clemens (1966) and Lillegraven (1969). On M_1 the paraconid and metaconid are more separated than comparable cusps on M_{2-4}. The talonid on M_4 is relatively much narrower than on M_{1-3}. The M_1's and M_4's were clearly divisible into two size groupings. Based on the frequency of upper molars, the M_1's and M_4's were allocated to either the larger *Alphadon marshi* or the smaller *A. wilsoni*.

It was somewhat difficult to differentiate between M_2's and M_3's. As was shown by Lillegraven (1969, fig. 16), the trigonid of M_2's of *Alphadon marshi* and *A. wilsoni* tends to be narrower than the talonid, while the reverse is true for the M_3's. Based on this criterion two groups were recognizable, presumably representing M_2's and M_3's. Each of these groups was further subdivided into two groups based on absolute size. The larger M_2's and M_3's are referred to *A. marshi*, while the smaller are referred to *A. wilsoni*. This procedure is not wholly satisfactory and there are probably several misallocations to species and molar site. It is also very likely that a few of the M_2's and M_3's (and possibly M^1's and M^4's) really belong to *A. lulli*. This species is recognized at loc. V73087 based on three isolated upper molars.

Morphologically, the lower molars of *Alphadon marshi* and *A. wilsoni* from loc. V73087 do not differ from those described by Clemens (1966, 1973) and Lillegraven (1969). Lillegraven (1969) noted that the lengths of the lower molars, proportional to the trigonid and talonid widths, are relatively longer in *A. wilsoni* than in *A. marshi*. This relationship is also true for *A. wilsoni* and *A. marshi* from the Flat Creek local fauna (loc. V73087).

TABLE 20. Measurements of isolated lower molars of *Alphadon marshi* from UCMP loc. V73087.

		L	*W-TR*	*W-TA*	*L/W-TR*	*L/W-TA*	*W-TR/W-TA*
M_1	N	4	4	4	4	4	4
	OR	2.09–2.25	1.05–1.35	1.22–1.42	1.67–1.99	1.58–1.71	0.86–0.09
	$\bar{x}$	2.21	1.23	1.34	1.81	1.65	0.92
M_2	N	5	5	5	5	5	5
	OR	2.20–2.41	1.26–1.40	1.32–1.41	1.66–1.86	1.62–1.78	0.91–0.99
	$\bar{x}$	2.33	1.30	1.37	1.79	1.70	0.95
M_3	N	5	5	5	5	5	5
	OR	2.18–2.37	1.23–1.41	1.16–1.26	1.65–1.77	1.75–1.96	0.98–1.13
	$\bar{x}$	2.26	1.29	1.21	1.71	1.86	1.07
M_4	N	3	3	3	3	3	3
	OR	2.05–2.44	1.10–1.22	1.02–1.10	1.68–2.12	2.01–2.21	1.01–1.20
	$\bar{x}$	2.27	1.18	1.07	1.94	2.12	1.10

Alphadon wilsoni Lillegraven, 1969
(Tables 21 & 22)

Alphadon wilsoni Lillegraven, 1969, p. 39

Type. UA 3681, fragmentary right maxilla with M^{2-3} (Lillegraven, 1969, fig. 19–4).

Type locality. Loc. KUA-1, Scollard Formation, Alberta.

Referred specimens. Right dentary with M_{1-2}, UCMP 117694; 2 M^1's; 2 M^2's; 3 M^3's; 3 M_1's; 6 M_2's; 5 M_3's; 1 M_4.

Locality. UCMP loc. V73087.

Distribution. Scollard Formation, Alberta; Lance Formation, Wyoming; Hell Creek Formation, Montana (all Lancian).

Diagnosis. Lillegraven, 1969, p. 39.

Description. This species has been adequately described by Lillegraven (1969). The specimens of *Alphadon wilsoni* from loc. V73087 do not differ in any substantial way from those described by Lillegraven. Comparison of the measurements of the upper and lower molars given in Tables 21 and 22 with Lillegraven's figs. 16 and 17 shows that for most measurements the loc. V73087 sample falls within one standard deviation of the Alberta sample. The small sample from loc. V73087 does not warrant a more rigorous statistical comparison to the larger Alberta sample.

As noted in the previous description of *Alphadon marshi*, both the *A. wilsoni* and *A. marshi* specimens from loc. V73087 appear to follow the differences in M^3 stylar cusp morphology observed by Lillegraven (1969) for these species.

The basis upon which lower molars of *Alphadon wilsoni* have been separated from *A. marshi* was given in the description of the latter species. The sample of *A. wilsoni* from loc. V73087 includes the only associated teeth of this genus recovered from the Hell Creek Formation, Garfield County. The specimen, UCMP 117694, is a fragmentary right dentary preserving M_{1-2}

TABLE 21. Measurements of isolated upper molars of *Alphadon wilsoni* from UCMP loc. V73087.

		L	*W–A*	*W–P*	*L/W–A*	*L/W–P*	*W–A/W–P*
	N	2	2	2	2	2	2
M^1	OR	1.92 –2.07	1.77 –1.92	2.02 –2.24	1.08	0.92 –0.95	0.86 –0.88
	$\bar{x}$	2.00	1.85	2.13	1.08	0.94	0.87
	N	2	2	2	2	2	2
M^2	OR	2.01 –2.20	2.02 –2.22	2.33 –2.45	0.99 –1.00	0.82 –0.94	0.82 –0.95
	$\bar{x}$	2.11	2.12	2.39	1.00	0.88	0.89
	N	2	2	2	2	2	2
M^3	OR	2.00 –2.08	2.22 –2.38	2.34 –2.39	0.84 –0.94	0.85 –0.87	0.93 –1.02
	$\bar{x}$	2.04	2.30	2.37	0.89	0.86	0.98

TABLE 22. Measurements of isolated lower molars of *Alphadon wilsoni* from UCMP loc. V73087.

		L	*W–TR*	*W–TA*	*L/W–TR*	*L/W–TA*	*W–TR/W–TA*
	N	3	3	3	3	3	3
M_1	OR	1.69 –1.83	0.78 –0.96	0.86 –1.10	1.91 –2.25	1.66 –1.97	0.87 –0.91
	$\bar{x}$	1.77	0.85	0.96	2.11	1.86	0.88
	N	7	7	7	7	7	7
M_2	OR	1.88 –2.22	1.01 –1.23	1.07 –1.26	1.74 –1.93	1.54 –1.81	0.85 –0.99
	$\bar{x}$	2.04	1.11	1.19	1.84	1.70	0.92
	N	5	5	5	5	5	5
M_3	OR	2.12 –2.26	1.08 –1.26	1.15 –1.23	1.73 –1.96	1.75 –1.95	0.94 –1.04
	$\bar{x}$	2.20	1.21	1.20	1.84	1.84	1.00
	N	1	1	1	1	1	1
M_4	OR	1.80	0.99	0.73	1.82	2.47	1.36
	$\bar{x}$	—	—	—	—	—	—

Alphadon lulli Clemens, 1966

(Table 23)

Alphadon lulli Clemens, 1966, p. 8

Type. UCMP 47047, fragmentary left maxilla with M^{1-2}, alveoli of M^3, parts of alveoli of M^4 and P^3 (Clemens, 1966, fig. 7).

Type locality. UCMP loc. V5620, Lance Formation, Wyoming.

Referred specimens. 3 M^3's.

Locality. UCMP loc. V73087.

Distribution. Lance Formation, Wyoming, and Hell Creek Formation, Montana (both Lancian).

Diagnosis. Clemens, 1966, p. 9.

Description. Three isolated M^3's from loc. V73087 do not differ in any major feature from M^3's of *Alphadon lulli* described by Clemens (1966). In distinguishing *A. lulli* from *A. marshi* Clemens (1966) noted that in the former species stylar cusp C is small or absent, the paracone is distinctly higher but noticeably more slender than the metacone, and the valley between the paracone and metacone tends to be relatively narrower. The first two of these characters are also present in the M^3's from loc. V73087. The third character, the distance between paracone and metacone, is more difficult to assess because of the very small sample sizes. Stylar cusp C is present on all three M^3's but varies in its development. On UCMP 109031 it is barely distinguishable, while on UCMP 117695 and 117696 it is a small but distinct cusp. UCMP 117695 is lacking the metastylar region, but on the other two molars it can be determined that cusp D is subequal to or larger than cusp C. Cusp B is the largest stylar cusp on all three molars. The only difference between the M^3's from loc. V73087 and from the Lance local fauna is the former molars' slightly larger size. The M^3's of *A. lulli* from loc. V73087 are intermediate in size between those of *A. wilsoni* and *A. marshi* from the same locality.

TABLE 23. Measurements of isolated M^3's of *Alphadon lulli* from UCMP loc. V73087.

UCMP Number		*L*	*W–A*	*W–P*	*L/W–A*	*L/W–P*	*W–A/W–P*
109031		2.18	2.60	2.61	0.84	0.84	1.00
117695		—	2.59	—	—	—	—
117696		2.16	2.55	2.61	0.85	0.83	0.98
	$\bar{x}$ =	2.17	2.58	2.61	0.85	0.84	0.99

Alphadon rhaister Clemens, 1966

Alphadon rhaister Clemens, 1966, p. 11

Type. UCMP 50292, fragmentary left maxilla with M^{1-3}.

Type locality. UCMP loc. V5815, Lance Formation, Wyoming.

Referred specimens. Worn M_x, UMCP 117697, loc. V73087. Two trigonids, UCMP 110624 and 110633, loc. V72208. Tentative referral of an M^4, UCMP 110623, loc. V72208.

Localities. See Referred specimens.

Distribution. Lance Formation, Wyoming; Scollard Formation, Alberta; Hell Creek Formation, Montana (all Lancian); possibly St. Mary River Formation, Alberta (Edmontonian).

Diagnosis. Clemens, 1966, p. 11.

Description. The complete but worn lower molar, UCMP 117697, from loc. V73087 and the trigonid fragments from loc. V72208 can be quite confidently referred to *Alphadon rhaister*. UCMP 117697 is very worn and the roots are lacking; however, when complete the talonid was probably narrower than the trigonid (L = 3.64; W–TR = approx. 2.09; W–TA = approx. 1.80). This suggests that UCMP 117697 is an M_4.

The M^4, UCMP 110623, loc. V72208, is questionably referred to *Alphadon rhaister*. It is smaller (L = 2.28; W–A = 3.88; W–P = 2.52) than the single M^4 given by Clemens (1966, table 3) and differs morphologically. The stylar shelf is worn, but the morphology can be inferred by the portion remaining. The parastylar lobe is immense relative to the primary trigon. Both stylar cusps A and B were probably tall when unworn. A strong paracrista joins the paracone and stylar cusp B. The metastylar lobe is small. A small cusp is present slightly posterior to the usual position of stylar cusp C. There may have been another small cusp just anterior to the first. Cusps D and E were probably absent. The metacone is broken (or worn) away, but when present was probably smaller and lower than the paracone, judging from the diameter of its base. A distinct paraconule is present. There is no evidence for a metaconule, but a very small one may have been present. Other than its small size, UCMP 110623 differs from the M^4's of *A. rhaister* in possessing a very small cusp C. In the M^4's of *A. rhaister* described by Clemens (1966, pp. 12–13) stylar cusp C is large.

Peradectes Matthew and Granger, 1921

Comments. W. A. Clemens (ms. and pers. comm.) is currently revising North American Puercan didelphids and has generously allowed me to refer to his unpublished study. In referring to his work it is anticipated that his study will be published before this paper. For a more complete discussion of Paleocene didelphines, the reader is directed to Clemens' work. The extension of *Peradectes* back into the Late Cretaceous, as cited by Clemens (1979), was based on my earlier incorrect placement of locs. V74110 and V74111 in the Hell Creek Formation and my belief that these localities were latest Cretaceous in age. However, the genus still may be present in the latest Cretaceous if a very worn dentary in the UCMP collections from Harbicht Hill belongs to this genus.

Peradectes cf. *P. pusillus* (Matthew and Granger, 1921) Clemens, ms.

(Figs. 42–44 & Tables 24 & 25)

Thylacodon pusillus Matthew and Granger, 1921, p. 2

Type. AMNH 16414, dentary fragment with $M_{2\text{-}3}$, M_1 talonid (Matthew, 1937, fig. 82, and Clemens, ms., fig. 1).

Type locality. ". . . upper level of the Puerco formation near Ojo Alamo. San Juan Basin, New Mexico . . ." (Matthew and Granger, 1921, p. 2).

Referred specimens. 1 DP^3 fragment; 12 M^1's; 10 M^2's; 5 M^3's; 2 M^4's; 5 upper molar fragments; 1 DP_3; 11 M_1's; 9 M_2's; 7 M_3's; 7 M_4's, all from UCMP loc. V74111. Three isolated molars from UCMP loc. V74110.

Localities. UCMP locs. V74110 and V74111.

Distribution. Nacimiento Formation, New Mexico; Tullock Formation, Montana (both Puercan); Mantua lentil, Polecat Bench Formation, Wyoming (mantuan); lowermost Tullock Formation, Montana (pre-mantuan).

Revised diagnosis. Clemens (ms.).

Comments. Much of the material examined by W. A. Clemens in his study of this species was not on hand when the present analysis was done. Most of the comparisons of the Garfield County material were made with Clemens' manuscript illustrations and data and previously published descriptions. Therefore these specimens are only tentatively referred to *Peradectes pusillus* (indicated by "cf."). Clemens will include this material in his study and may be able to show more positively the affinities of these specimens.

Description. *Upper molars.* None of the upper (or lower) molars of *Peradectes* cf. *P. pusillus* from loc. V74111 (Hell's Hollow local fauna) were found in association. However, it was quite easy to recognize four groupings corresponding to the four upper molar sites. The task was made easier because a reasonably large sample (N = 28) of complete upper molars was available. The four groupings were assigned to a specific molar site

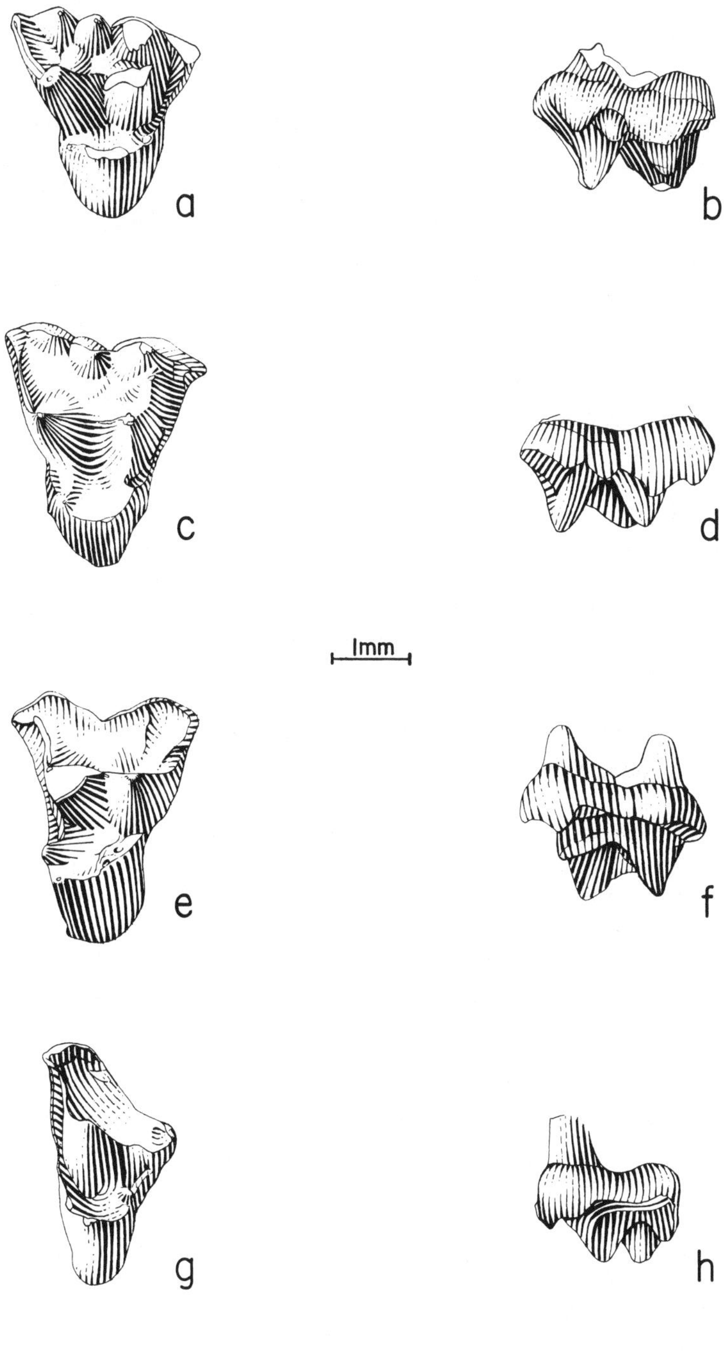

FIG. 42. *Peradectes* cf. *P. pusillus* (Matthew and Granger). Isolated upper molars, loc. V74111. Right M^1, UCMP 117744: (a) & (b). Right M^2, UCMP 117764: (c) & (d). Left M^3, UCMP 117770: (e) & (f). Left M^4, UCMP 117771: (g) & (h). Occlusal view: (a), (c), (e), & (g); labial view: (b), (d), (f), & (h). (A.L.)

through comparison with *Alphadon marshi* and *A. wilsoni* (see Lillegraven, 1969, fig. 17). As in these two species of *Alphadon*, the length of the upper molar decreases relative to the anterior width going from M^1 to M^4. Concomitantly, the anterior width increases relative to the posterior width going from M^1 to M^4. When these two ratios were used as the axes of a bivariate plot, four groupings of upper molars were distinguishable corresponding to M^1 through M^4. One of these shape differences can be seen in Fig. 42, where the stylar shelf labial to the paracone is increasingly wider in M^1 through M^4.

Clemens (ms.) was able to distinguish *Peradectes pusillus* from other species of *Peradectes* based on its larger size and relatively larger and more distinct paraconules and metaconules. The molars from loc. V74111 are comparable in size to those of *P. pusillus*. This was the major criterion for tentatively referring these specimens to cf. *P. pusillus*. The paraconule and metaconule on the upper molars of cf. *P. pusillus* are distinct and are subequal in size. On M^2 and M^3, where the crowns are more transversely widened than on M^1, the conules usually bear small postparaconule and premetaconule wings. The conules are high and almost as large (girth) as those of *Alphadon marshi* and *A. wilsoni*. Only one complete M^4 is referable to cf. *P. pusillus*. On this molar there is a small cuspule situated between the metacone and metaconule, occupying the usual position of the premetaconule wing.

Several characters were noted by Clemens (ms.) for distinguishing upper molars of *Peradectes* from those of *Alphadon*. One character mentioned above is the slightly smaller size of the conules in *Peradectes*. Further, in *Peradectes* the paracone is slightly smaller than the metacone and the stylar cusps are relatively smaller than those of *Alphadon*. Both these characters are true for the specimens of cf. *P. pusillus* from loc. V74111. In addition, stylar cusps B, C, and D are aligned on M^{1-2}, while most M^1's and M^2's of *A. marshi* and *A. wilsoni* tend to have a slightly indented ectoflexus.

Stylar cusp B is the largest and highest stylar cusp on M^{1-3}. On M^4 from loc. V74111, stylar cusp A is higher but not larger than the ridgelike cusp B. On M^1, cusps C and D are subequal and are only slightly smaller than cusp B. On M^2, cusps C and D are subequal but are distinctly lower than cusp B. Cusp C is usually absent on M^3 and cusp D is small and ridgelike. M^{1-3} usually bear a small crista directed lingually from cusp D toward but not reaching the metacone. All the upper molars have a well-developed paracrista which becomes relatively longer with the expansion of the parastylar region going from M^1 through M^4. Stylar cusp A is small (except on M^4) but distinct on all the molars. The metacrista is well developed on all but M^4. There is usually no distinct stylar cusp E at the posterolabial termination of the metacrista.

The protocone is moderately bulbous on M^1 but becomes progressively less so on each successive molar.

Lower molars. Clemens (ms.) discussed the lower molar morphology of *Peradectes* and contrasted it to that seen in *Alphadon*. He considered the two most consistent differences between the genera to be derivations in *Peradectes* relative to *Alphadon*. These characters are: (1) the contact of the cristid obliqua with the trigonid shifts from below the protolophid notch in *Alphadon* to below the protoconid in *Peradectes*; and (2) the entoconid and hypoconulid are closely approximated in *Alphadon* and more distinctly separated in *Peradectes*. In most of the isolated lower molars of cf. *P. pusillus* from loc. V74111, the cristid obliqua contacts the trigonid below the protoconid.

The relative positions of the entoconid and hypoconulid are more difficult to assess.

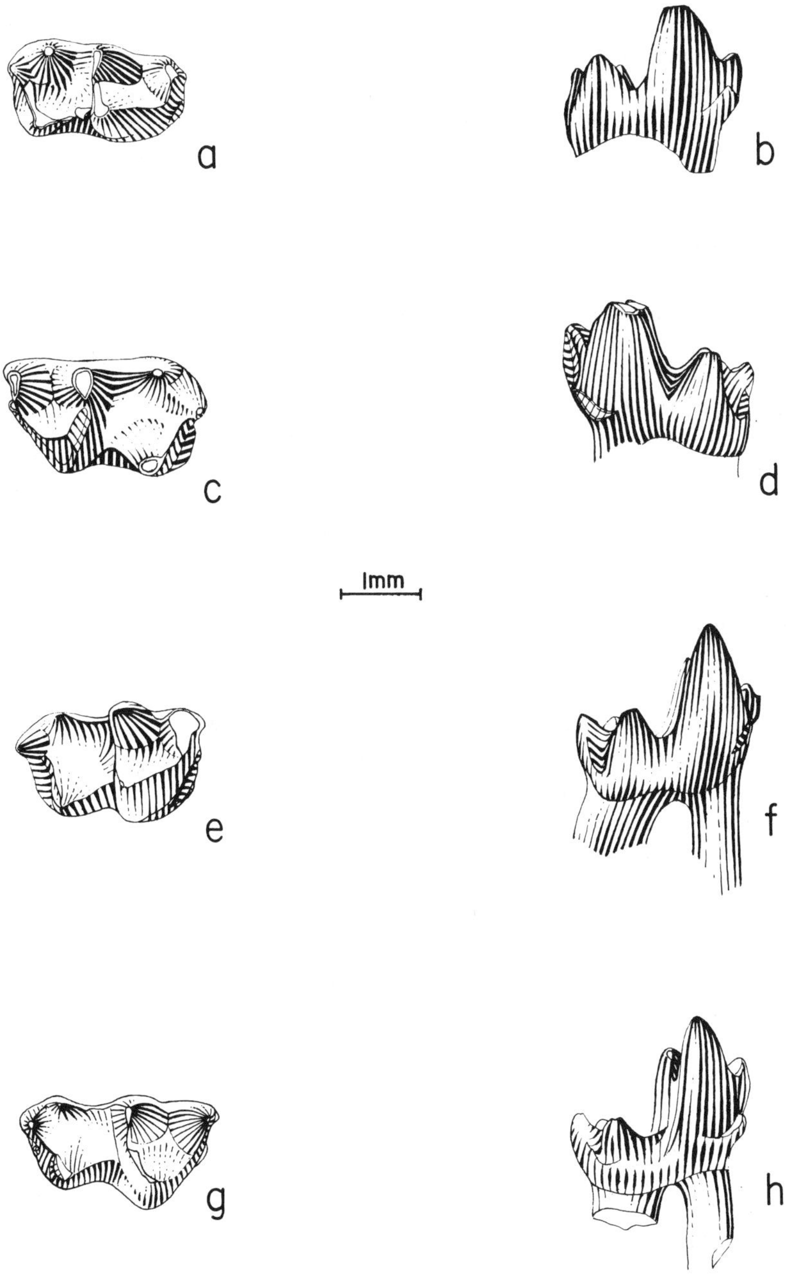

FIG. 43. *Peradectes* cf. *P. pusillus* (Matthew and Granger). Isolated lower molars, loc. V74111. Right M_1, UCMP 117776: (a) & (b). Left M_2, UCMP 117791: (c) & (d). Right M_3, UCMP 117796: (e) & (f). Right M_4, UCMP 117801: (g) & (h). Occlusal view: (a), (c), (e), & (g); labial view: (b), (d), (f), & (h). (A. L.)

Clemens (ms.) noted: "In *Peradectes* the entoconid is more distinctly separated from the hypoconulid [relative to its position in *Alphodon*] and situated farther mesially near the midpoint of the lingual margin of the talonid basin." In comparing the specimens of cf. *P. pusillus* from loc. V74111 with those of *A. marshi* and *A. wilsoni* from loc. V73087 I did not find a consistent difference in the position of the entoconid between these genera. This is one possible way in which cf. *P. pusillus* differs from other species of *Peradectes*. However, in comparing the occlusal views of lower molars of *P. pusillus* (Clemens, ms., figs. 1B, 4B, 6B) with those of various Lancian species of *Alphadon* (Clemens, 1966, figs. 58b, 61b, 62b, 63b, 64b; Lillegraven, 1969, figs. 14–1a, 18–2b, 18–2b, 3b, 4a) I could find no consistent difference in the placement of the entoconid. There may be some tendency in both genera for the entoconid to shift posteriorly relative to the hypoconid in going from M_1 through M_4. It should be emphasized that this second comparison was based on illustrations of *P. pusillus*, not on the actual specimens. Nevertheless, based on a comparison of cf. *P. pusillus* (loc. V74111) and *A. marshi* and *A. wilsoni* (loc. V73087) I do find that the hypoconulid is relatively taller and the hypoconulid and entoconid are somewhat more distinct in cf. *P. pusillus*. The greater distinctiveness of these cusps was recognized by Matthew and Granger (1921, p. 2) for *Thylacodon pusillus*. Clemens (ms.) has synonymized *Thylacodon* with *Peradectes* and has recognized this character as being present in all species of the genus.

Deciduous premolars. A fragmentary DP^3 and an isolated DP_3 from loc. V74111 are referred to cf. *Peradectes pusillus*. To my knowledge DP^3's have not been reported for this genus. The fragmentary DP^3, UCMP 117806, lacks the metacone and metastylar region. The paracone is the tallest preserved cusp. There may have been a well-developed, sharp-crested centrocrista joining the paracone and metacone. This cannot be determined with any certainty because of damage to this portion of the tooth. The parastylar lobe is well developed but positioned lower on the anterior side of the paracone. A small, distinct stylar cusp A is present. There appears to be no stylar shelf or cusp B labial to the paracone. This is rather curious, since on the upper molars cusp B is the largest stylar cusp. In Lancian marsupials for which DP^3 is at least tentatively identified, cusp B is usually present on both the DP^3 and upper molars (*Alphadon rhaister* Clemens, 1966, and *Glasbius twitchelli* n. sp., below) or is usually absent on both (*Pediomys hatcheri* and ?*Pediomys* sp. Clemens, 1964, and *P. krejcii* Lillegraven, 1969). The only other exception is *Didelphodon vorax* (Clemens, 1964), where cusp B is present on the upper molars but is absent on DP^3. It is possible that UCMP 117806 is a DP^3 of *Pediomys*; however, this is unlikely since no other teeth of this taxon have been recovered from loc. V74111. If this DP^3 does belong to cf. *Peradectes pusillus*, which seems likely, the lack of cusp B is regarded as a derived character relative to *Alphadon*. The protocone is well developed but is positioned low on the tooth and is not as transversely widened as on the upper molars. A small but distinct paraconule and metaconule are present. The crown was probably supported by three roots.

The lower DP_3, like the upper, is distinctly smaller than the molars. This evidence, plus a morphology that differs slightly from the lower molars, suggests that UCMP 117807 is a DP_3 of cf. *Peradectes pusillus*. The DP_3, like the DP^3, is almost fully molariform and differs from the lower molars in only a few characters. As in most of the lower molars, the cristid obliqua contacts the trigonid below the protoconid. The greatest differences between the DP_3 and the lower molars are in the proportions of the crown. Compared to the lower molars, the trigonid of the DP_3 is relatively longer than the talonid. This is due to a more anteriorly positioned paraconid and slight anterior canting of the trigonid. The DP_3

TABLE 24. Measurements of isolated upper molars and deciduous premolars of *Peradectes* cf. *P. pusillus* from UCMP loc. V74111.

		L	*W–A*	*W–P*	*L/W–A*	*L/W–P*	*W–A/W–P*
	N	8	8	9	8	8	8
M^1	OR	2.20 –2.45	2.18 –2.43	2.38 –2.78	0.96 –1.05	0.84 –0.97	0.83 –0.93
	$\bar{x}$	2.35	2.30	2.58	1.02	0.91	0.89
	N	8	9	9	8	8	9
M^2	OR	2.23 –2.61	2.61 –2.85	2.76 –3.06	0.85 –0.94	0.80 –0.88	0.89 –0.95
	$\bar{x}$	2.42	2.71	2.92	0.89	0.83	0.93
	N	3	5	3	3	3	3
M^3	OR	2.08 –2.43	2.86 –3.25	2.88 –3.35	0.74 –0.82	0.72 –0.82	0.97 –1.00
	$\bar{x}$	2.31	3.01	3.07	0.78	0.77	0.99
	N	2	1	3	1	1	1
M^4	OR	1.64 –1.97	— 2.81	2.16 –2.32	— 0.70	— 0.89	— 1.27
	$\bar{x}$	1.81	—	2.23	—	—	—
DP^3	N	—	1	—	—	—	—
	OR	—	1.39	—	—	—	—

is also narrower relative to length than the lower molars. This is seen in the high length/trigonid width and length/talonid width ratios (Table 25). Only M_4's have a relatively narrower talonid.

The DP_3 shows no trace of labial cingula below the paraconid and protoconid and hypoconid that are sometimes present on the molars. On the molars the hypoconid is always larger but not always taller than the entoconid and hypoconulid. On the DP_3 the hypoconid is the dominant cusp in girth and height. Parts of two subequal roots are preserved on the DP_3.

Clemens (ms., fig. 5) mentioned and figured a tooth of *P. pusillus*, AMNH 58385, that he felt is "either a DP_3 or, more likely, an M_1." His description of the trigonid proportions is similar to that given above for the DP_3 from loc. V74111. A comparison of his figure of this specimen (his fig. 5) with that of UCMP 117807 (Fig. 44) indicates that the specimens are very similar. He did not give measurements of the specimen or note its size relative to the molars. If it is distinctly smaller than the lower molars I suspect it is a DP_3 and not an M_1.

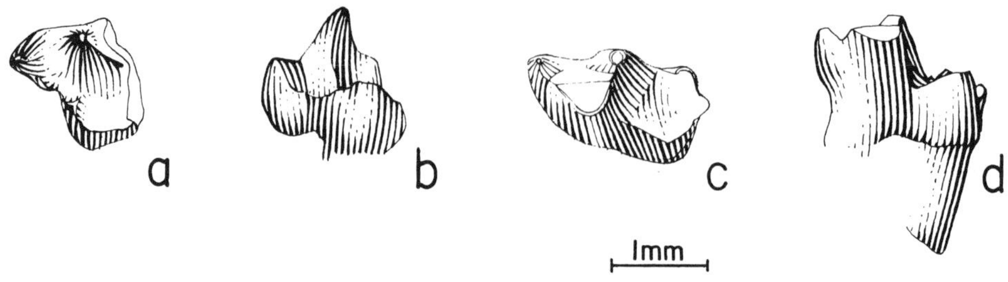

FIG. 44. *Peradectes* cf. *P. pusillus* (Matthew and Granger). Isolated deciduous premolars, loc. V74111. Left DP^3 fragment, UCMP 117806: (a) occlusal view; (b) lingual view. Left DP_3, UCMP 117807: (c) occlusal view; (d) labial view. (A. L.)

TABLE 25. Measurements of isolated lower molars and deciduous premolars of *Peradectes* cf. *P. pusillus* from UCMP loc. V74111.

		L	*W–TR*	*W–TA*	*L/W–TR*	*L/W–TA*	*W–TR/W–TA*
M_1	N	9	9	9	9	9	9
	OR	2.30 –2.49	1.13 –1.36	1.23 –1.47	1.74 –2.16	1.65 –1.92	0.88 –0.99
	$\bar{x}$	2.39	1.28	1.36	1.88	1.77	0.94
M_2	N	8	8	8	8	8	8
	OR	2.31 –2.55	1.22 –1.51	1.32 –1.56	1.59 –2.06	1.54 –1.86	0.90 –0.99
	$\bar{x}$	2.47	1.41	1.47	1.77	1.68	0.96
M_3	N	6	6	6	6	6	6
	OR	2.42 –2.61	1.43 –1.56	1.29 –1.52	1.57 –1.74	1.61 –1.89	1.02 –1.11
	$\bar{x}$	2.50	1.49	1.42	1.68	1.76	1.06
M_4	N	7	7	7	7	7	7
	OR	2.21 –2.50	1.30 –1.48	0.97 –1.24	1.63 –1.79	1.96 –2.48	1.19 –1.44
	$\bar{x}$	2.38	1.40	1.08	1.69	2.21	1.31
DP_3	N	1	1	1	1	1	1
	OR	1.96	0.89	0.98	2.20	2.00	0.91

Discussion. *Peradectes* cf. *P. pusillus* is common at loc. V74111. The only therian species represented by more teeth (and probably more individuals) is *Procerberus formicarum.* This differs from the slightly older Harbicht Hill locality of Sloan and Van Valen (1965). They do not report any marsupials at Harbicht Hill, although one unidentified small marsupial was recovered from Harbicht Hill by a UCMP–LACM crew. The discrepancy may be due to the smaller size of the sample from Harbicht Hill, or it may reflect a real difference in faunal composition due to dissimilarities in age, paleoecology, preservation, etc. There is presently no evidence to favor any one of these possibilities strongly, but age differences are probably at least partly the cause.

Subfamily GLASBIINAE Clemens, 1966

Glasbius Clemens, 1966

Glasbius Clemens, 1966, p. 24

Genotypic species. Glasbius intricatus Clemens, 1966.

Referred species. Glasbius twitchelli, n. sp.

Distribution. Lance Formation, Wyoming; Hell Creek Formation, Montana (both Lancian).

Revised diagnosis. Principal cusps on upper molars low relative to protofossa; metacone higher than paracone and metaconule larger than paraconule; stylar shelf broad; B large on M^{1-4}; D higher than B on M^{1-2}, smaller than B on M^3, absent on M^4; A and E small; C small or absent; M_2 longer than other molars; talonid (including labial cingulum) wider than trigonid, except on M_4; width of $M_3 > M_2 > M_1$; cristid obliqua contacts trigonid on back of protoconid; trigonid short anteroposteriorly, paraconid and metaconid closely approximated on M_{3-4}; difference in height of trigonid over talonid not great and decreases from M_1 to M_4; lingual side of crown higher than labial, except on M_{1-2}, where protoconid is subequal to metaconid; basal cingulum on anterior, labial, and posterior sides of M_{1-4} with variously developed cusps on some molars; M^4_4 much smaller than preceding molars; P_{1-3} double-rooted, increasing in size posteriorly; anteroposterior axis of P_1 rotated at 30–45° angle to long axis of dentary.

Comments. When Clemens (1966) proposed this genus he erected a monotypic didelphid subfamily, the Glasbiinae, to include it. The naming of a new species, *Glasbius twitchelli*, has necessitated the preceding revisions of the generic diagnosis. Besides additional morphological data provided by the new specimens, two minor reinterpretations are given. Clemens (1966, diagnosis on p. 25) notes that the labial side of the crown on the lower molars is higher than the lingual. My interpretation is that the lingual side is higher except for the protoconid of M_{1-2}, which is subequal to the metaconid (see Clemens, 1966, fig. 17b, and Figs. 46, 47, and 48 below). Clemens (1966, p. 26) noted that stylar cusp D is larger than B on M^1 but slightly smaller on M^2. Reexamination of his material and of additional specimens belonging to *G. twitchelli* suggests that stylar cusp D is larger than B on both M^1 and M^2 (see Clemens, 1966, figs. 14a and 16a, and Fig. 45 below).

Clemens (1966, pp. 29–30) tentatively referred two lower jaws, AMNH 58759 and PU 17286, to *Glasbius intricatus*. AMNH 58759, a dentary containing P_{2-3}, M_{1-3}, is here regarded as not belonging to *Glasbius*. PU 17286, a dentary containing P_3, M_{1-3}, is referred to *G. twitchelli*. These specimens are discussed in more detail below.

Glasbius twitchelli, n. sp.

(Figs. 45–49 & Tables 26–28)

Etymology. Named for the late Walter Twitchell and his family, whose hospitality greatly aided our fieldwork.

Type. UCMP 115853, an isolated right M^3.

Type locality. UCMP loc. V73087, Hell Creek Formation, Montana.

Referred specimens. Two worn maxillary fragments with M^{2-3}, UCMP 117553 and 117554; 1 DP^3; 4 M^1's; 2 M^2's; 3 M^3's; 1 M^4; dentary fragments with P_1 fragment, P_2, UCMP 117569, and with P_{2-3}, M_{1-2}, M_3 fragment, UCMP 117570; 1 DP_3; 4M_1's; 5 M_2's; 2 M_3's; and 1 M_4 from loc. V73087. Dentary fragments with P_3, M_{1-3}, PU 17286, loc. V73097; with M_2 fragment, M_3, UCMP 117540, loc. V72215; and with P_3 fragment, M_{1-2}, UCMP 117539, loc. V74116. Dentary fragment with M_{2-4}, UCMP 91728; 1 M^1, 3 M^2's, 1 M^4, and 3 M_1's from locs. V65127 and V70201 (Bug Creek Anthills localities). Worn maxillary fragment with M^{2-3}, UCMP 108055, loc. V72212. 1 M_1, loc. V73077. 1 M_3, loc. V73078.

Localities. See Referred specimens.

Distribution. Hell Creek Formation, Montana (Lancian). (Note: with the removal of PU 17286 from *G. intricatus* the distribution of this species is limited to the Lance Formation, Wyoming.)

Diagnosis. In comparison to *Glasbius intricatus*: stylar cusp C absent on all molars except for slight swelling on anterior margin of cusp D of M^1; labial (basal) cingulum of unworn M_2 bears cusp below and anterior to hypoconid; M_3 has similar cusp plus additional basal cusp below protoconid. Molars are slightly larger than their counterparts in *G. intricatus*.

Description. The bases for recognition of a new species of *Glasbius* from the Hell Creek Formation, Montana, are tenuous. Minor but consistent differences are present between the material Clemens (1966) referred to *G. intricatus* and the new material from Montana. The specimens from Montana could be near the morphological extremes of *G. intricatus* but, if included, would require some changes in the diagnosis of this species. Since the morphological differences are relatively consistent it was decided that the Montana and Wyoming samples should be separated at the species level. Comparison of Tables 26 and 28 with Clemens' table 6 (1966, p. 32) shows that the Montana specimens are generally larger. This may also be a species difference, but the samples are too small to warrant a statistical comparison.

Upper molars. Three maxillae, each preserving M^{2-3}, were recovered from the Hell Creek Formation. Unfortunately, all are heavily worn and all but one show some breakage. The least damaged maxilla, UCMP 117552, is the only measurable specimen and is included in Table 26 with the isolated teeth. The three maxillae can confidently be referred to *Glasbius*, based on molar shape.

The description of M^{1-3} of *Glasbius intricatus* given by Clemens (1966, pp. 25–26) fits the morphology of *G. twitchelli* equally well, with some minor exceptions. Clemens noted that stylar cusp C is small but present on M^{1-3} of *G. intricatus*. Almost all unworn upper molars of *G. twitchelli* lack any trace of this cusp. The only exception is M^1, which bears a small expansion on the anterior margin of cusp D in the position of cusp C (Fig. 45). Clemens also noted that on M^3 one large or two small cusps are present in the region of cusp D. All M^3's of *G. twitchelli* that preserve this region (5 specimens, including the type) have only one cusp D.

No M^4's are known for *Glasbius intricatus*. Two M^4's can tentatively be referred to *G. twitchelli*. As is shown in Table 26, the two molars vary considerably in size. This may be the result of a misallocation of one of the teeth to *G. twitchelli*. However, it appears that *Glasbius* was in the process of reducing M^4_4. It does not seem unusual that a tooth undergoing reduction should vary considerably in size and possibly in morphology. These

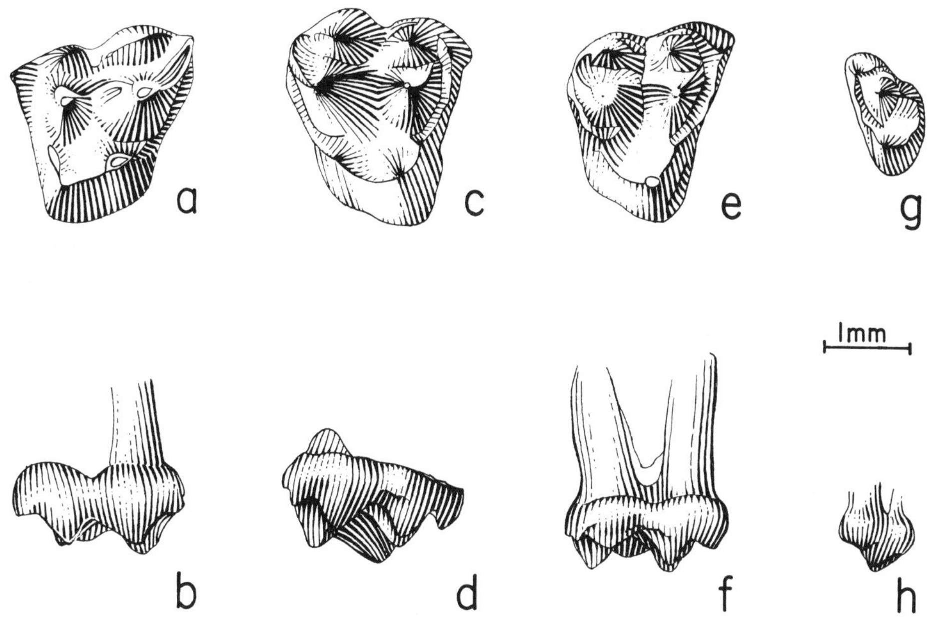

FIG. 45. *Glasbius twitchelli*, n. sp. Isolated upper molars, loc. V73087. Left M^1, UCMP 115852: (a) & (b). Right M^2, UCMP 117544: (c) & (d). Right M^3, UCMP 115853, type : (e) & (f). Left M^4, UCMP 117554: (g) & (h). Occlusal view: (a), (c), (e), & (g); labial view (b), (d), (f), & (h). (A. L.)

M^4's, UCMP 117554 (loc. V73087) and UCMP 102029 (loc. V70201), are referred to *G. twitchelli* because of their rounded, slightly bulbous shape and low principal cusps. The protocone is the larger cusp but is subequal in height to the paracone. The metacone is reduced to a short ridge projecting posteriorly from the paracone. The metaconule is absent on UCMP 117554 and a small swelling on the posterior of the protocone may be its remnant on UCMP 102029. A small crista on the lingual side of the paracone, slightly more prominent in UCMP 102029, is all that remains of the paraconule. The parastylar lobe of the stylar shelf is moderately large, while the metastylar lobe is absent. A low, anteroposteriorly oriented ridge is present on the parastylar lobe. Two small swellings on the ridge (the second being slightly larger) are probably much reduced cusps A and B. UCMP 102029 bears a cingulum from stylar cusp A to below the anterior side of the protocone. In UCMP 117554 the cingulum is not complete below the protocone. Both M^4's have two roots, a smaller one below the parastylar lobe of the stylar shelf and a second, larger one supporting the protocone and posterior edge of the crown.

Lower premolars. Two specimens, AMNH 58759 and PU 17286, that Clemens (1966) tentatively referred to *Glasbius intricatus* preserve portions of the lower premolar dentition. As will be discussed below, it is doubtful that AMNH 58759 is referrable to *Glasbius*, and PU 17286 is now considered to belong to *G. twitchelli.*

Four specimens of *Glasbius twitchelli* preserve portions of the premolar dentition or alveoli for these teeth. The most complete of these is UCMP 117570 (loc. V73087), which preserves the posterior margin of the canine alveolus, two alveoli for P_1, P_{2-3}, M_{1-2}, and

TABLE 26. Measurements of upper molars and deciduous premolars of *Glasbius twitchelli*, n. sp., from various localities.

		L	*W–A*	*W–P*
M^1	N	2	2	2
	OR	2.01 –2.08	1.96	2.16 –2.19
	$\bar{x}$	2.05	1.96	2.18
M^2	N	5	5	5
	OR	2.05 –2.11	1.99 –2.41	2.13 –2.45
	$\bar{x}$	2.07	2.24	2.31
M^3	N	5	5	5
	OR	1.68 –1.89	1.98 –2.26	1.80 –2.15
	$\bar{x}$	1.77	2.08	1.97
M^4	N	2	2	2
	OR	0.91 –1.24	1.32 –1.50	1.07 –1.26
	$\bar{x}$	1.08	1.41	1.17
DP^3	N	1	1	1
	OR	1.47	1.05	1.43
UCMP 115853 M^3, type		1.87	2.26	2.08

the trigonid of M_3 (Fig. 46). The P_{2-3} have suffered a good deal of postmortem abrasion which has removed much of the enamel. The morphology of the two premolars is similar; the major difference is the larger size of P_3. Both premolars are two-rooted and each possesses a single anterior principal cusp positioned over the anterior root. The convex anterior edge of the cusp slightly overhangs the root. Both premolars possess a small posterior heel that is wider than the anterior principal cusp. The surfaces of both heels are too worn to be certain of the morphology; however, the P_2 may have had a central ridge running down the back of the anterior principal cusp onto the posterior heel. The P_3 possesses two ridges that run along either side of the back of the principal cusp. These ridges define a central depression along the back of the cusp. It is possible that this represents the original morphology, but since most of the enamel is missing it seems equally reasonable to assign this to either dental attrition or postmortem abrasion.

The size of the alveoli for P_1 and the distance between them suggest that this tooth was slightly smaller than P_2. The posterior alveolus is slightly larger than the anterior. Although two single-rooted premolars could have been present, this seems unlikely. Between the anterior and posterior roots of lower premolars and molars in *Glasbius*, as in many other small mammals, the bone is formed into a slight ridge that is higher than the bone separating the roots of adjacent teeth. Such a ridge is present between the two alveoli anterior to P_2 in UCMP 117570. Additional evidence that the two alveoli are for a double-

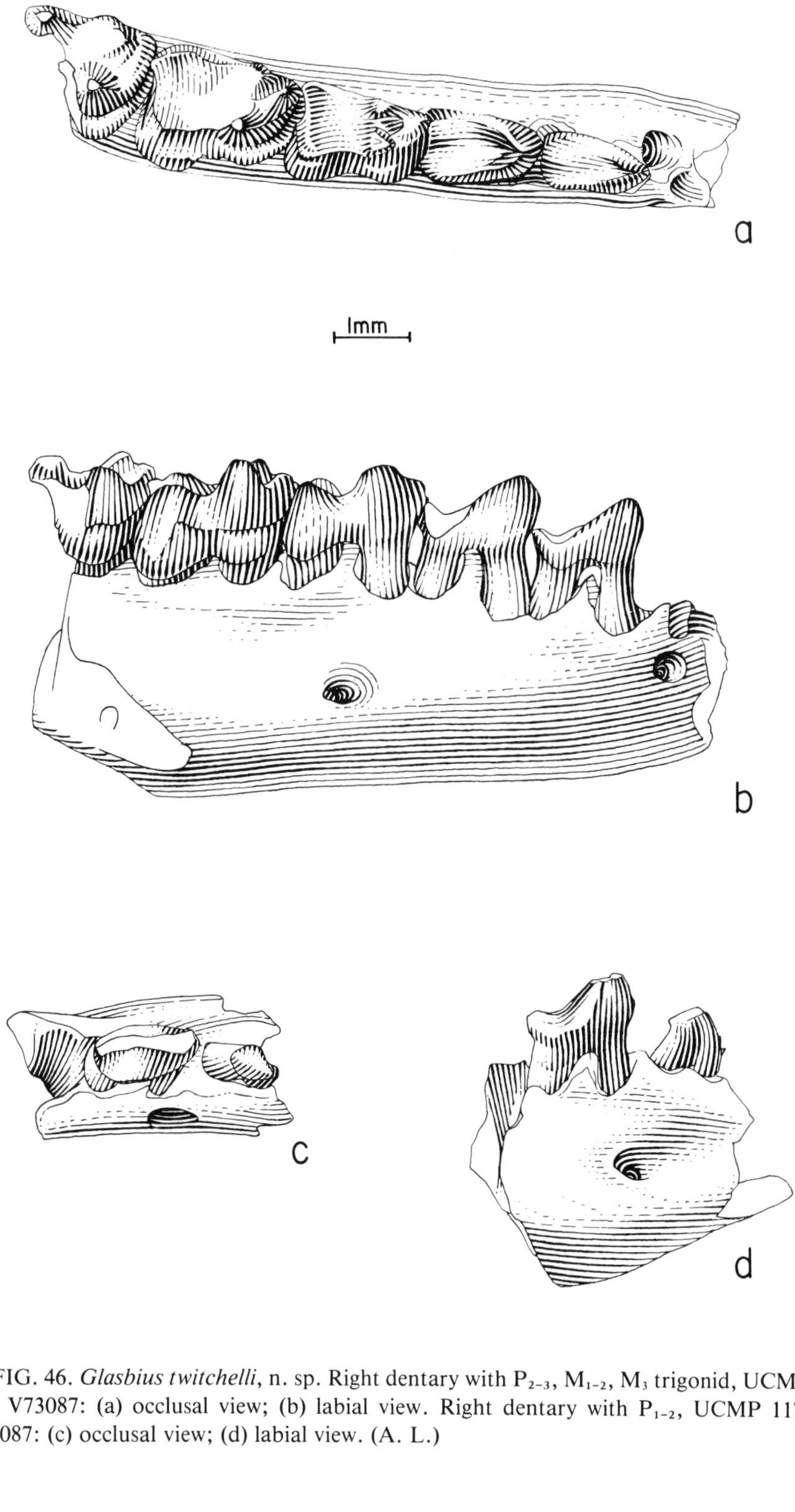

FIG. 46. *Glasbius twitchelli*, n. sp. Right dentary with P_{2-3}, M_{1-2}, M_3 trigonid, UCMP 117570, loc. V73087: (a) occlusal view; (b) labial view. Right dentary with P_{1-2}, UCMP 117569, loc. V73087: (c) occlusal view; (d) labial view. (A. L.)

rooted P_1 comes from UCMP 117569. This specimen preserves part of the alveolus for the canine, a damaged two-rooted P_1, and a slightly damaged two-rooted P_2. Referral of this specimen to *Glasbius* is tenuous, but one feature suggests that this allocation is correct. In UCMP 117569 the long axis of P_1 is set at an angle of about 30° to the long axis of the dentary, with the anterior end of the tooth directed anterolabially. The condition is similar in UCMP 117570, although the angle of the P_1 alveoli is closer to 45° in this specimen.

Two other dentaries, UCMP 91728 and UCMP 117539, also preserve P_1 alveoli that are oriented obliquely to the long axis of the low jaw. Although both of these dentaries have alveoli for all three premolars, only UCMP 117539 preserves portions of a crown, the talonid of P_3.

The premolar morphology of PU 17286 was described and figured by Clemens (1966, p. 30 and fig. 20). By the time I examined it, in 1976, the specimen had been slightly damaged: the crown of P_3 and the portion of the canine alveolus figured by Clemens (1966, fig. 20) were missing. Fortunately, the premolar region as described by Clemens is

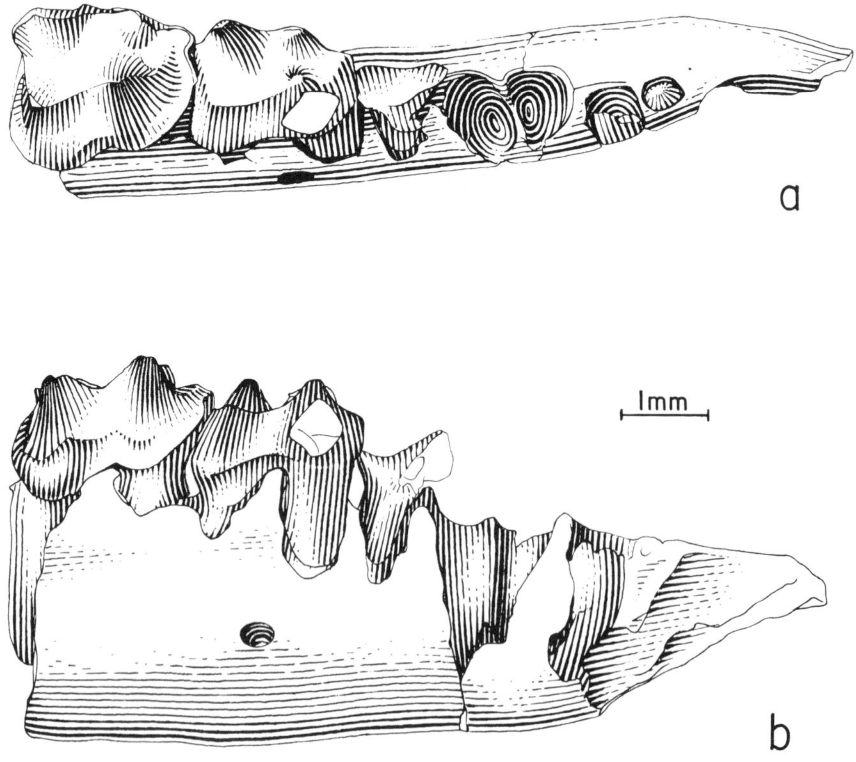

FIG. 47. *Glasbius twitchelli*, n. sp. Right dentary with P_3 fragment, M_{1-2}, UCMP 117539, loc. V74116: (a) occlusal view; (b) labial view. (A. L.)

still intact. Clemens suggested several alternatives for the number of premolars originally present in PU 17286. He noted that it may have possessed a premolar dentition similar to that of AMNH 58759 (1966, p. 30). AMNH 58759 possesses a two-rooted P_{2-3} and two alveoli that were probably for a two-rooted P_1. Comparison of PU 17286 with the specimens of *Glasbius twitchelli* described above suggest that this specimen probably had three two-rooted premolars, as in AMNH 58759, but the P_1 was probably set obliquely in the dentary. As can be seen in Clemens; fig. 20 (1966), the labial side of PU 17826 is not preserved anterior to the posterior root of P_1. A labially positioned anterior alveolus for P_1 was probably present on this missing portion.

Although the premolar dentition is known only for *Glasbius twitchelli*, its premolar character state is tentatively regarded as representative of the genus. Partly based on this evidence, AMNH 58759 is not included in the hypodigm of *Glasbius*. Unlike the specimens described above, the P_1 of this specimen is not set obliquely in its dentary. Further, in UCMP 117570, the only specimen that preserves P_{2-3} and can be referred to *Glasbius* confidently, the P_3 is longer than P_2. As was noted by Clemens (1966, p. 30), the P_2 is longer than the P_3 in AMNH 58759. Since the size relationship, $P_3 > P_2$, is known in only one specimen of *Glasbius*, it is impossible to determine whether this is the most common condition for the taxon.

Lower molars. Three of the four dentaries, plus PU 17286, mentioned in the preceding section preserve molars. Another dentary, UCMP 117540, also preserves molars (M_2 talonid and M_3). The molars in these dentaries (see Referred specimens and Table 27 for records of associated molars and premolars) and the seventeen isolated lower molars recovered from the Hell Creek Formation differ from those described by Clemens (1966) only in minor morphological details. Clemens noted that the development of cusps on the labial basal cingulum is variable in *Glasbius intricatus* (compare the M_3's in his figs. 17 and 18). The unworn lower molars of *G. twitchelli* are very consistent in the number of cusps carried on the basal cingulum. M_1 possesses a narrow labial cingulum that is discontinuous below the hypoconid and sometimes below the protoconid, but no basal cusps are present. On M_2 the labial cingulum is usually continuous from below the anterior side of the protoconid to below the posterior side of the hypoconid. All the M_2's bear one variously developed basal cusp below and slightly anterior to the hypoconid. The labial cingulum on M_3 is continuous, as in M_2. In addition to a cusp below and anterior to the hypoconid, the M_3's possess a prominence or a cusp on the labial cingulum below the protoconid. The two M_4's from the Hell Creek Formation do not differ from those described by Clemens (1966).

As in the upper molars, the lower molars of *Glasbius twitchelli* tend to be larger than those of *G. intricatus*. The sizes of the molars in PU 17286 are comparable to the other lower molars recovered from the Hell Creek Formation. PU 17286 preserves M_{1-3} in addition to fragments of the premolar dentation. Most of the enamel is lacking on M_{1-2}, but it can be determined that the M_2 and the well-preserved M_3 possess basal cingular cusps as in the M_{2-3} of *G. twitchelli*. Based on this evidence, PU 17286, first referred to *G. intricatus* by Clemens (1966, p. 30) is here transferred to *G. twitchelli*.

With the recovery of additional dentaries of *Glasbius* it is now reasonable to exclude AMNH 58759 from the hypodigm of this genus. Clemens (1966, p. 29) tentatively referred this specimen to *G. intricatus* but recognized that it differs from this species "in its distinctly smaller size, and molars with relatively larger paraconids, greater separation of paraconids, and metaconids and narrower labial cingula." Based on these differences and

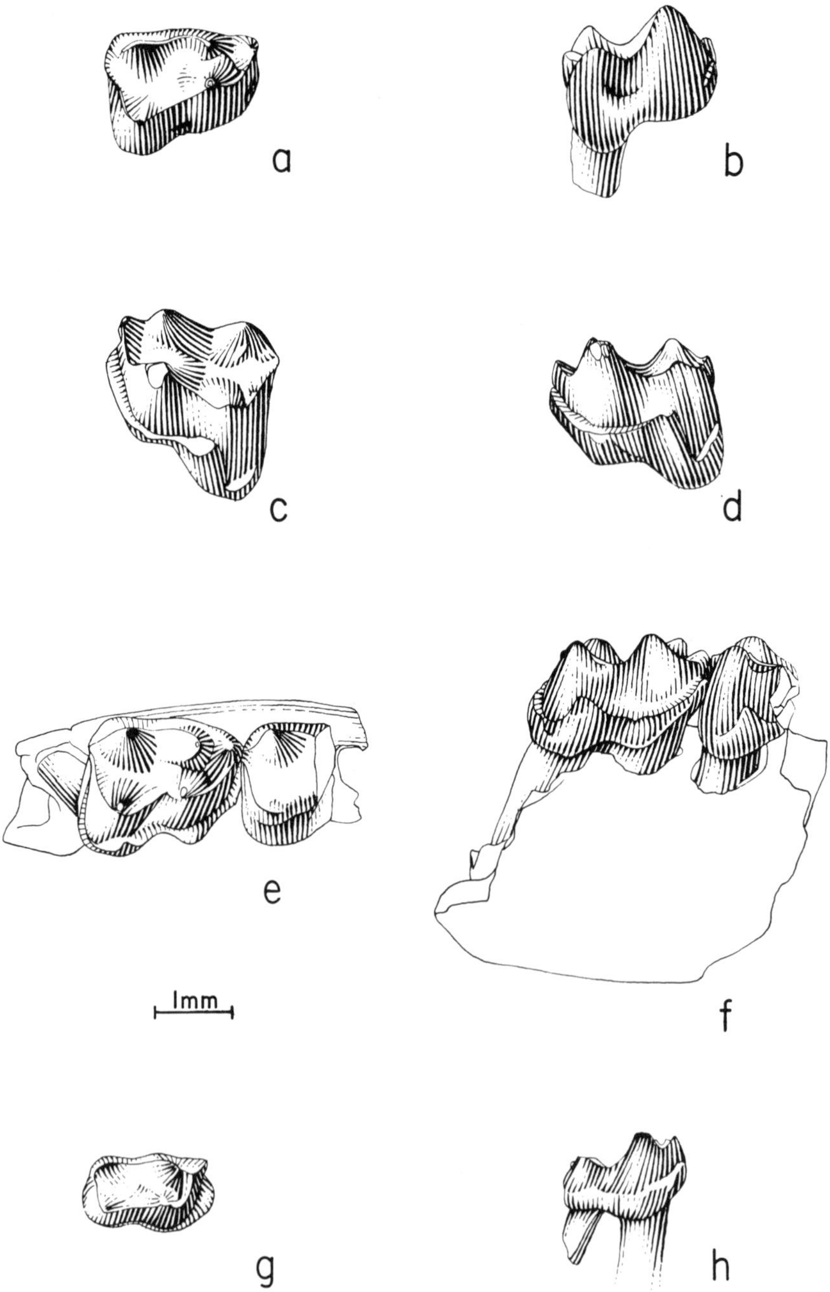

FIG. 48. *Glasbius twitchelli*, n. sp. Lower molars. UCMP 117540 is from loc. V72215; the remainder are from loc. V73087. Right M_1, UCMP 117558: (a) & (b). Left M_2, UCMP 117562: (c) & (d). Right dentary with M_2 fragment and M_3, UCMP 117540: (e) & (f). Right M_4, UCMP 117566: (g) & (h). Occlusal view: (a), (c), (e), & (g); labial view: (b), (d), (f), & (h). (A. L.)

those noted previously for the premolars, AMNH 58759 is removed from *Glasbius*. Its true affinities are unclear, but it could be a new genus and species of Late Cretaceous metatherian or, less likely, an aberrant specimen belonging to one of the smaller species of *Pediomys*.

The fossils recovered from the Hell Creek Formation add a few details to the description of the dentary given by Clemens (1966). He noted the presence of a mental foramen below the trigonid of M_1. This condition is found in dentaries of *Glasbius twitchelli*. This material also demonstrates the presence of a mental foramen below the posterior root of P_1. The symphyseal region is an elongate ovoid shape with its long axis oriented at a low angle to the horizontal. The posterior margin of this region extends below the posterior root of P_2.

Deciduous premolars. Two teeth recovered from loc. V73087 may be a DP^3 and a DP_3 of *Glasbius twitchelli*. The DP^3, UCMP 117567 (Fig. 49a, b), is very questionably referred to *G. twitchelli*. It seems equally likely that the tooth belongs to a small species of *Alphadon*. DP^3's that might belong to small species of this genus have not been reported previously. The tooth is anteroposteriorly elongate. The cusp A region is formed into an elongate lobe. A small cusp A is present and is approximately in line with the paracone and metacone, as is the case in the DP^3 of *A. rhaister* (Clemens, 1966, p. 13). A moderately large cusp B and a slightly larger and taller cusp D are present. No cusp C or E is present, although the stylar shelf in the region of E is well developed. The paracone is slightly larger than the metacone, but the metacone is the taller of the two cusps. The protocone is positioned quite low on the tooth. A small swelling can be seen in the paraconule region, and a small metaconule is present. Minute premetaconule and postparaconule wings are also present. The DP^3 is three-rooted.

The DP_3, UCMP 117568 (Fig. 49c, d), can more confidently be referred to *Glasbius twitchelli*. Unlike the molars of *Glasbius*, the trigonid of DP_3 is anteroposteriorly elongate. The protoconid is the largest trigonid cusp. This cusp has three well-developed crests originating from its apex. An anterior crest (paralophid) connects the protoconid with the paraconid, which is positioned low and lingual to the midline. The second crest (protolophid) joins the protoconid and the metaconid. The latter cusp is slightly taller than the paraconid and is positioned posterolingual to the protoconid. The third crest joins the protoconid and hypoconid. In the talonid the entoconid and smaller hypoconulid are twinned. The hypoconid is a large cusp but is lower than the lingual side of the talonid. Below the hypoconid, the labial side of the talonid forms a small bulbous protuberance reminiscent of the condition seen in molars of *Glasbius*. This feature is the strongest evidence for referring the DP_3 to *Glasbius*. The only suggestion of a labial cingulum is found below the anterior portion of the protoconid, where there is a slight expansion. No roots are preserved, but presumably two were present.

Discussion. Clemens (1966) discussed the possible affinities of *Glasbius*. He felt that its affinities were with the didelphids. The material recovered from the Hell Creek Formation does not offer any data to alter that view; however, it should be noted that *Glasbius* does show at least one similarity with *Pediomys* in the lower dentition. As Clemens (1966) has shown, the cristid obliqua contacts the back of the trigonid between the protoconid and metaconid in *Alphadon* and below the protoconid in *Pediomys*. *Glasbius* resembles *Pediomys* in this character. Further, I regard this character state as being derived over that seen in *Alphadon*. The problem arises in determining if *Glasbius* and *Pediomys* share this character state or if it was evolved separately. I could find no strong evidence to support

FIG. 49. *Glasbius twitchelli,* n. sp. Isolated deciduous premolars, loc. V73087. Right DP^3, UCMP 117567: (a) occlusal view; (b) labial view. Left DP_3, UCMP 117568: (c) occlusal view; (d) labial view. (A.L.)

either case, but my subjective judgment is that it was a parallel development related to a tendency toward bunodonty. Other marsupials considered to be didelphids, such as *Carolameghinia*, have developed bunodont dentitions (expansion of the talonid). If this development is regarded as a parallelism in *Glasbius* and *Carolameghinia*, as was done by Clemens (1966), then it is plausible that it evolved in parallel in *Pediomys* and *Glasbius*.

Glasbius shows an interesting distribution in the Late Cretaceous of North America. Clemens (1966) indicated that it was a rare element of the Lance local fauna of the type Lance Formation, Wyoming. Lillegraven (1969) reported that it was absent from the Trochu local fauna, Scollard Formation, Alberta. It is also interesting to note that Clemens found that in the Lance local fauna *Pediomys* is the most abundant marsupial and in the Trochu local fauna Lillegraven found *Alphadon* to be the most abundant marsupial. Complete faunal comparisons like those done by Estes (1964) and Estes and Berberian (1970) will be done elsewhere for the Hell Creek Formation material. Based on numbers of teeth and jaws, *Glasbius* is a moderately common element in the Hell Creek faunal-facies localities. As in the Trochu, *Alphadon* is the most common marsupial. *Glasbius* is next in abundance, and *Pediomys* is the least common of the three. Sloan and Van Valen (1965) reported *Glasbius* from their two stratigraphically lowest Bug Creek faunal-facies localities in McCone County (recorded in their table 1 as "undescribed genus and species"). According to their table 1, it is not as abundant as any of the species of *Pediomys* but at Bug Creek Anthills it is more abundant than *Alphadon*. Comparison of *Glasbius* from the Bug Creek faunal-facies of McCone County with the Hell Creek faunal-facies of Garfield County shows that *G. twitchelli* is present in both faunal-facies.

Family PEDIOMYIDAE (Simpson, 1927a) Clemens, 1966

Pediomys Marsh, 1889a

Comments. Pediomyids are more common than didelphids in the Lance local fauna (Clemens, 1966, p. 85 and table 16). The opposite is apparently true for the Trochu local fauna (Lillegraven, 1969, p. 50). Based on the figures given by Sloan and Van Valen (1965, table 1), marsupials comprise only slightly more than one percent of the mammals in the Bug Creek Anthills local fauna. Also, according to their figures, pediomyids (which they include in the Didelphidae) are by far more common than didelphids or stagodontids, representing about 85 percent of the marsupials found at the locality. Their Hell Creek faunal-facies localities were not rich, but even at these localities there is a suggestion that pediomyids are more common than didelphids.

TABLE 27. Measurements of associated lower premolars and molars of *Glasbius twitchelli*, n. sp.

Specimen	*UCMP*	P_1		P_2		P_3		M_1			M_2			M_3			M_4		
Number	*Locality*	*L*	*W*	*L*	*W*	*L*	*W*	*L*	*W–TR*	*W–TA*	*L*	*W–TR*	*W–TA*	*L*	*W–TR*	*W–TA*	*L*	*W–TR*	*W–TA*
UCMP 117569	V73087	0.98	0.52	1.47	0.73														
UCMP 117570	V73087			1.49	0.75	1.68	0.92	1.95	1.01	1.21	2.13	1.24	1.54	—	1.52	—			
UCMP 117539	V74116							2.04	1.10	1.34	2.21	1.38	1.71						
UCMP 117540	V72215										—	—	1.51	2.08	1.48	1.64			
PU 17286	V73097					1.40	0.70	2.00	1.00	1.20	2.20	1.28	1.55	2.09	1.42	1.75			
UCMP 91728	V70201										2.26	1.44	1.73	2.20	1.53	1.70	1.56	1.10	0.95

TABLE 28. Measurements of lower premolars and molars of *Glasbius twitchelli*, n. sp., from various localities, including associated teeth from Table 26.

		L	*W*	*W–TR*	*W–TA*
P_1	N	1	1	—	—
	OR	0.98	0.52		
P_2	N	2	2	—	—
	OR	1.47 –1.49	0.73 –0.75	— —	— —
	$\bar{x}$	1.48	0.74	—	—
P_3	N	2	2	—	—
	OR	1.40 –1.68	0.70 –0.92	— —	— —
	$\bar{x}$	1.54	0.81	—	—
M_1	N	11	—	11	11
	OR	1.95 –2.09	— —	1.00 –1.27	1.20 –1.50
	$\bar{x}$	2.02	—	1.13	1.38
	s	0.049	—	0.087	0.107
	CV	2.43	—	7.70	7.75
M_2	N	9	—	9	10
	OR	2.12 –2.26	— —	1.24 –1.44	1.51 –1.73
	$\bar{x}$	2.18	—	1.35	1.64
	s	—	—	—	0.081
	CV	—	—	—	4.94
M_3	N	6	—	7	6
	OR	1.98 –2.21	—	1.27 –1.53	1.35 –1.75
	$\bar{x}$	2.09	—	1.43	1.62
M_4	N	2	—	2	2
	OR	1.54 –1.56	—	0.85 –1.10	0.79 –0.95
	$\bar{x}$	1.55	—	0.98	0.87

		L	*W*	*W–TR*	*W–TA*
DP_3	N	1	—	1	1
	OR	1.46	—	0.77	0.82

Curiously, in Garfield County, the opposite appears to be true for the Hell Creek faunal-facies. At the richest and best-sampled of these localities, V73087, *Alphadon* and *Glasbius* (both didelphids) are more common than pediomyids. There is some suggestion, based on the sediment size and the fossil sample, that there is bias against larger mammals at V73087, so the rarity of the larger *Pediomys florencae* and *P. hatcheri* is not unexpected. Clemens (1966, table 16) found that two species of *Pediomys*, *P. elegans* and *P. cooki*, which are in the same size range as *A. marshi*, *A. lulli*, and *A. wilsoni*, are the two most common marsupials in the Lance local fauna at loc. V5620 and are the second and third most common at loc. V5711. *A. marshi* and *A. wilsoni* are not uncommon at loc. V73087, yet the similarly sized *P. elegans* and *P. cooki* are rare. It is thus doubtful that the absence of these small pediomyids is the result of depositional bias. The representational differences may represent original differences in the biocoenosis.

The few pediomyids recovered from the Hell Creek Formation, Garfield County, add to the geographic range of several of the species. However, these specimens add little morphological information to the descriptions and figures given by Clemens (1966) and Lillegraven (1969).

Pediomys elegans Marsh, 1889a
(Table 29)

Pediomys elegans Marsh, 1889a, p. 89

Type. YPM 11866, broken and heavily weathered upper molar (Marsh, 1889a, pl. IV, figs. 23–25).

Neotype. CM 11658, fragmentary right maxilla with M^{2-4}, designated as neotype by Simpson (1929b, p. 111).

Type locality. Mammal locality no. 1 of Lull (1915) = UCMP loc. V5003. Neotype somewhere in type Lance Formation, probably loc. V5003.

Referred specimens. RM^1, UCMP 117630; $RM_{3?}$, UCMP 17631; left dentary with P_1, M_1, and M_2 trigonid fragment, UCMP 117629; all from loc. V73087. RM_4, UCMP 108970, from loc. V73078.

Localities. See Referred specimens.

Distribution. Lance Formation, Wyoming; Scollard Formation, Alberta; Hell Creek Formation, Montana (all Lancian).

Revised diagnosis. Clemens, 1966, p. 35.

Description. Of the specimens referred above to *Pediomys elegans*, only the M^1 (UCMP 117630) and the dentary fragment (UCMP 118970) can be assigned with any confidence.

The M^1 is slightly worn, which obscures some of the details. However, the specimen agrees well in morphology and size with the description of the M^1 of *Pediomys elegans* given by Clemens (1966, p. 37). This specimen, as well as the others, was also compared directly with the material of *P. elegans* described by Clemens (1966).

The dentary fragment, UCMP 118970, is referred to *Pediomys elegans* based on size and a direct comparison with material from the Lance local fauna. The specimen preserves the posterior part of the canine alveolus, a worn P_1, two alveoli for P_2, the anterior root and posterior alveolus for P_3, a complete M_1, and a fragment of the M_2 trigonid. The only measurable tooth, the M_1, is slightly smaller than those measured by Clemens (1966, table 21). The only smaller species of *Pediomys* known from the Lancian is *P. krejcii*. A dentary fragment of this species was also recovered from loc. V73087. It is distinctly smaller than UCMP 118970, and there are morphological differences. Although Clemens (1966) observed that the dentary deepens with age in *P. elegans*, the dentary of this species is relatively more massive than in the slender-jawed *P. krejcii*. In addition, Clemens (1966, pp. 97–98) noted that the anterior root of P_1 in *P. elegans* is situated slightly labial to the canine. This is true for UCMP 118970. The P_1 is preserved in this specimen, unlike those available to Clemens. However, the P_1 is very worn and no details of the crown can be discovered. The anterior root is slightly smaller than the posterior root; this contrasts with the anterior dentition of *P. krejcii*, where the premolars are less crowded and the P_1 is not slanted labially as in *P. elegans*. These are probably consistent morphological differences, as they are seen in dentaries of *P. krejcii* from the Lance local fauna (Clemens, 1966), the Trochu local fauna (Lillegraven, 1969), and now the Flat Creek local fauna (loc. V73087), while *P. elegans* dentaries are known from the latter two local faunas.

In addition to an $M_{3?}$, loc. V73087, and an M_4, loc. V73078, which are tentatively referred to *Pediomys elegans* (see Referred specimens), there are three other isolated teeth which may belong either to *P. elegans* or *P. cooki*. The measurements of these specimens are included in Table 29 with those of *P. elegans*.

TABLE 29. Measurements of isolated molars of *Pediomys elegans* from various localities.

UCMP Number	*UCMP Locality*	*Tooth Site*	*L*	*W–A (uppers) or W–TR(lowers)*	*W–P (uppers) or W–TA (lowers)*
117629	V73087	M_1	1.67	1.20	1.30
117631	V73087	$M_{3?}$	1.99	1.35	—
117630	V73087	M^1	1.72	1.52	2.07
108970	V73078	M_4	1.86	1.13	1.10
*117612	V74116	M_x	—	—	1.67
*117633	V73087	M_1	1.76	1.06	1.18
*117634	V73087	M_1	1.91	1.16	1.24

*Lower molars belonging to *P. elegans* or cf. *P. cooki*.

Pediomys cf. *P. cooki* Clemens, 1966

Pediomys cooki Clemens, 1966, p. 38

Type. UCMP 47738, fragmentary right maxilla with P^3, M^{1-3} (Clemens, 1966, fig. 24).

Type locality. UCMP loc. V5711, Lance Formation, Wyoming.

Referred specimens. RM^3 fragment, UCMP 117632.

Locality. UCMP loc. V73087.

Distribution. Lance Formation, Wyoming; Hell Creek Formation, Montana (both Lancian); possibly St. Mary River Formation, Alberta, and Fruitland and Kirtland formations, New Mexico (both Edmontonian).

Diagnosis. Clemens, 1966, p. 39.

Description. Only an M^3 (L = 2.11 mm), lacking the protocone, is tentatively referred (indicated by "cf.") to this species, based on the criteria listed by Clemens (1966). As was noted in the description of *Pediomys elegans*, there are three isolated lower molars which may belong to either *P. elegans* or *P. cooki.*

Discussion. Clemens (1966, p. 40) tentatively identified an M^2 fragment, Hell Creek Formation, described by Simpson (1927a, fig. 5) as *Pediomys cooki.*

Sloan and Van Valen's description of the Bug Creek Anthills locality in 1965 preceded Clemens' description of *Pediomys cooki* in 1966. However, Sloan (1976) reports the occurrence of this species at Bug Creek Anthills, Bug Creek West, and Brownie Butte (= UCMP loc. V73097). Thus *P. cooki*, like *P. elegans*, is known from the two stratigraphically lower Bug Creek faunal-facies localities in McCone County (Bug Creek Anthills and Bug Creek West) and three Hell Creek faunal-facies localities in Garfield County (Crooked Creek of Simpson, 1927a; Brownie Butte, loc. V73097; Flat Creek 5, loc. V73087).

Pediomys krejcii Clemens, 1966
(Table 30)

Pediomys krejcii Clemens, 1966, p. 42

Type. UCMP 51390, fragmentary left maxilla with P^3, M^{1-3} (Clemens, 1966, fig. 26).

Type locality. UCMP loc. V5711, Lance Formation, Wyoming.

Referred specimens. 3 fragmentary upper molars; 1 DP^3 fragment; 1 M_4; right dentary fragment with P_{2-3}, M_3, UCMP 117623; all from UCMP loc. V73087. 3 isolated upper molars from other localities.

Localities. UCMP locs. V73087, V74116, and V76167.

Distribution. Lance Formation, Wyoming; Scollard Formation, Alberta; Hell Creek Formation, Montana (all Lancian); possibly St. Mary River Formation, Alberta (Edmontonian).

Diagnosis. Clemens, 1966, p. 43.

Description. The Hell Creek Formation specimens referred to *Pediomys krejcii* differ from those described by Clemens (1966) and Lillegraven (1969) in only a few details.

Clemens (1966, p. 43) noted that none of the upper molars from the Lance local fauna possessed pre- and postcingula. Subsequently, Lillegraven (1969, p. 45) found that some upper molars of *P. krejcii* in the Trochu local fauna do possess pre- and postcingula. None of the six upper molars in the Hell Creek Formation sample has these cingula.

The dentary fragment, UCMP 117623, is almost identical to those described by Lillegraven (1969) and Clemens (1966). This specimen preserves P_{2-3} and M_3.

As was discussed above in the description of *Pediomys elegans*, this species can be distinguished from *P. krejcii* on the basis of the premolar dentition. In the smaller, more slender-jawed *P. krejcii* the premolars are relatively less crowded and in an approximately straight line behind the canine. In *P. elegans* the premolars are relatively more crowded and the anterior alveolus of P_1 is situated labial to a line drawn through the canine and more posterior premolar alveoli.

Lillegraven (1969, p. 47) described a DP^3 of *Pediomys krejcii* from the Trochu local fauna. A DP^3, UCMP 117627, from the Flat Creek local fauna (loc. V73087) can also be referred to this species. Only the posterior width of the tooth could be measured; the parastylar region is missing. This measurement (Table 30) agrees well with that given by Lillegraven (1969, table 6). UCMP 117627 differs in two ways from the DP^3 described by Lillegraven. First, he noted that the paracone is the tallest and most robust cusp. On UCMP 117627 the paracone is distinctly taller than the metacone but the two cusps are equal in girth. Second, and possibly more important, the DP^3 described by Lillegraven has a well-developed stylar cusp D. UCMP 117627 does not have cusp D, but a ridge almost as well-developed as the metacrista occupies the position of this cusp. Since only two DP^3's are known, these differences are best considered, at least for the time being, as intraspecific variation.

TABLE 30. Measurements of premolars and molars of *Pediomys krejcil* from various localities.

UCMP Number	*UCMP Locality*	*Tooth Site*	*L*	*W–A (uppers) or W–TR(lowers)*	*W–P (uppers) or W–TA (lowers)*
117624	V73087	$M_{4?}$	1.68	0.93	0.84
117625	V73087	$M^{1?}$	—	—	1.39
117626	V73087	$M^{2?}$	1.43	1.62	2.03
117627	V73087	DP^3	—	—	1.07
117611	V74116	M^2 or M^3	—	2.00	—
117614	V74116	M^2	1.46	1.61	2.10
117616	V76167	$M^{3?}$	—	—	2.34

Right dentary fragment, UCMP 117623, loc. V73087

P_2		P_3		M_3		
L	*W*	*L*	*W*	*L*	*W–TR*	*W–TA*
1.01	0.48	1.13	0.65	1.44	1.04	0.96

Pediomys hatcheri (Osborn, 1898) Simpson, 1927b

(Table 31)

Protolambda hatcheri Osborn, 1898, p. 172
(See Clemens, 1966, p. 45, for synonymies.)

Type. AMNH 2202, fragmentary right maxilla with M^3, designated as a lectotype by Clemens, 1966, p.45 (Osborn, 1898, fig. 1A, center); AMNH 2203, isolated M^2, designated as a syntype by Clemens, 1966, p. 45.

Type locality. "Lance Creek area, Lance Formation, Wyoming" (Clemens, 1966, p. 45).

Referred specimens. 1 M^x fragment, UCMP 117635, loc. V73087. M^2 or M^3 fragment, UCMP 110630, loc. V72208. 1 M_4, UCMP 108979, loc. V73076.

Localities. See Referred specimens.

Distribution. Lance Formation, Wyoming; Scollard Formation, Alberta; Hell Creek Formation, Montana; North Horn Formation, Utah (all Lancian).

Revised diagnosis. Clemens, 1966, p. 45.

Comments. *Pediomys hatcheri* is a rare element in the Hell Formation of Garfield County. It is also rare in the Trochu local fauna (Lillegraven, 1969) but is relatively common in the Lance local fauna (Clemens, 1966). Sloan and Van Valen (1965, table 1) indicate that *P. hatcheri* is the third most common marsupial in the Bug Creek Anthills local fauna, while the larger *P. florencae* is not present. Estes and Berberian (1970) indicate that both species are present in the Bug Creek Anthills local fauna. Thus it is reasonable to suspect that Sloan and Van Valen's relative abundance figures include both *P. hatcheri* and *P. florencae*.

The Hell Creek Formation sample of *Pediomys hatcheri* adds no morphological data to that given by Clemens (1966).

TABLE 31. Measurements of isolated molars of *Pediomys hatcheri* from various localities.

UCMP Number	*UCMP Locality*	*Tooth Site*	*L*	*W–A (uppers) or W–TR(lowers)*	*W–P (uppers) or W–TA (lowers)*
110630	V72208	M^2 or M^3	—	—	3.93
108979	V73076	M_4	3.73	2.06	1.85

Pediomys florencae Clemens, 1966

(Table 32)

Pediomys florencae Clemens, 1966, p. 50

Type. UCMP 51440, fragmentary left maxilla with M^{2-3} (Clemens, 1966, fig. 31).

Type locality. UCMP loc. V5820, Lance Formation, Wyoming.

Referred specimens. 10 teeth and tooth fragments from various UCMP localities.

Localities. UCMP locs. V72213, V73087, V74106, V74116, V75178, and V76161.

Distribution. Lance Formation, Wyoming, and Hell Creek Formation, Montana (both Lancian).

Diagnosis. Clemens, 1966, p. 50.

Description. Half of the isolated teeth (5 specimens) referable to *Pediomys florencae* were recovered from loc. V75178. The other five localities yielded one specimen each. Except for one tooth, this referred material does not add anything to the description of Clemens (1966).

The exception is a fragment of a right DP^3 lacking the protocone. The specimen, UCMP 117636, strongly resembles the DP^3 that Clemens identified as ?*Pediomys* sp. (1966, fig. 30). UCMP 117636 is referred to *P. florencae* because of its large size and its reduced stylar shelf, especially labial to the paracone.

According to Estes and Berberian (1970), *Pediomys florencae* also occurs in the Bug Creek local fauna.

TABLE 32. Measurements of isolated premolars and molars of *Pediomys florencae* from various localities.

UCMP Number	*UCMP Locality*	*Tooth Site*	*L*	*W*	*W–A (uppers) or W–TR(lowers)*	*W–P (uppers) or W–TA (lowers)*
117636	V73087	DP^3	3.20	—	—	—
117610	V74106	M_2 or M_3	4.35	—	2.62	3.06
117617	V75178	M_2 or M_3	5.14	—	3.02	3.29
117620	V75178	M_4	4.49	—	2.60	2.33
117621	V75178	$P^{3?}$	3.84	2.28	—	—

?*Pediomys* cf. *P. florencae* Clemens, 1966

(Fig. 50)

A large, edentulous dentary fragment, UCMP 110541, was recovered from loc. V73084 (Flat Creek local fauna) in 1973. Morphologic and biostratigraphic criteria suggest that the specimen is a marsupial that can questionably be referred to *Pediomys florencae* (indicated by "?" and "cf.").

In Lancian eutherians the angular process of the dentary can be lingually inflected, but to my knowledge none is labially reflected. In Lancian marsupials the angular process is usually both lingually inflected and labially reflected; this is the case in UCMP 110541.

The size of UCMP 110541 (max. length = 60 mm) approaches that of *Didelphodon vorax*, the largest known Lancian marsupial. However, *D. vorax* can easily be eliminated as a possibility. Even a small, possibly sub-adult dentary of *D. vorax* is decidedly more robust than UCMP 110541, making reference to this taxon highly unlikely.

The dentary of the next largest Lancian marsupial, *Pediomys florencae*, is about the same size or slightly smaller than UCMP 110541, based on the reconstruction of *P. florencae* by Clemens (1966, fig. 74). The probable size relationships of the teeth (based on the alveoli) and the shape of the symphysis also suggest that this specimen may be tentatively referred to *P. florencae*.

A large section of bone is missing on the labial side in the middle of the dentary, and a smaller lingual part is missing at the anterior end. It is possible to determine the number and approximate size and shape of the alveoli. In the middle of the dentary, where the labial side is missing, eight alveoli are present. The alveoli decrease in size going anteriorly. All but the posteromost one were probably round in outline. The last one is distinctly anteroposteriorly elongate. All eight alveoli are oriented approximately vertically. A small section anterior to the last eight alveoli preserves both labial and lingual sides of the dentary. This section has two subequal ovoid alveoli. The two alveoli are similar in size to the most posterior one in the dentary and are larger than the seven intervening alveoli. Anterior to this region the lingual side of the mandible is lacking. In the posterior part of this region two small, possibly round, alveoli are present. Anterior to these there are a minimum of three concavities that could be alveoli for one or more anteromost premolars. At the anterior edge of the dentary a large, almost horizontally oriented concavity probably contained part of the canine root. Lingual to this a very small concavity, probably for an incisor, is present. Not counting the incisor and canine alveoli and the three small concavities which may be for the anteromost premolar(s), the dentary preserves twelve alveoli. Based on a comparison with other Lancian therians, these alveoli probably are for

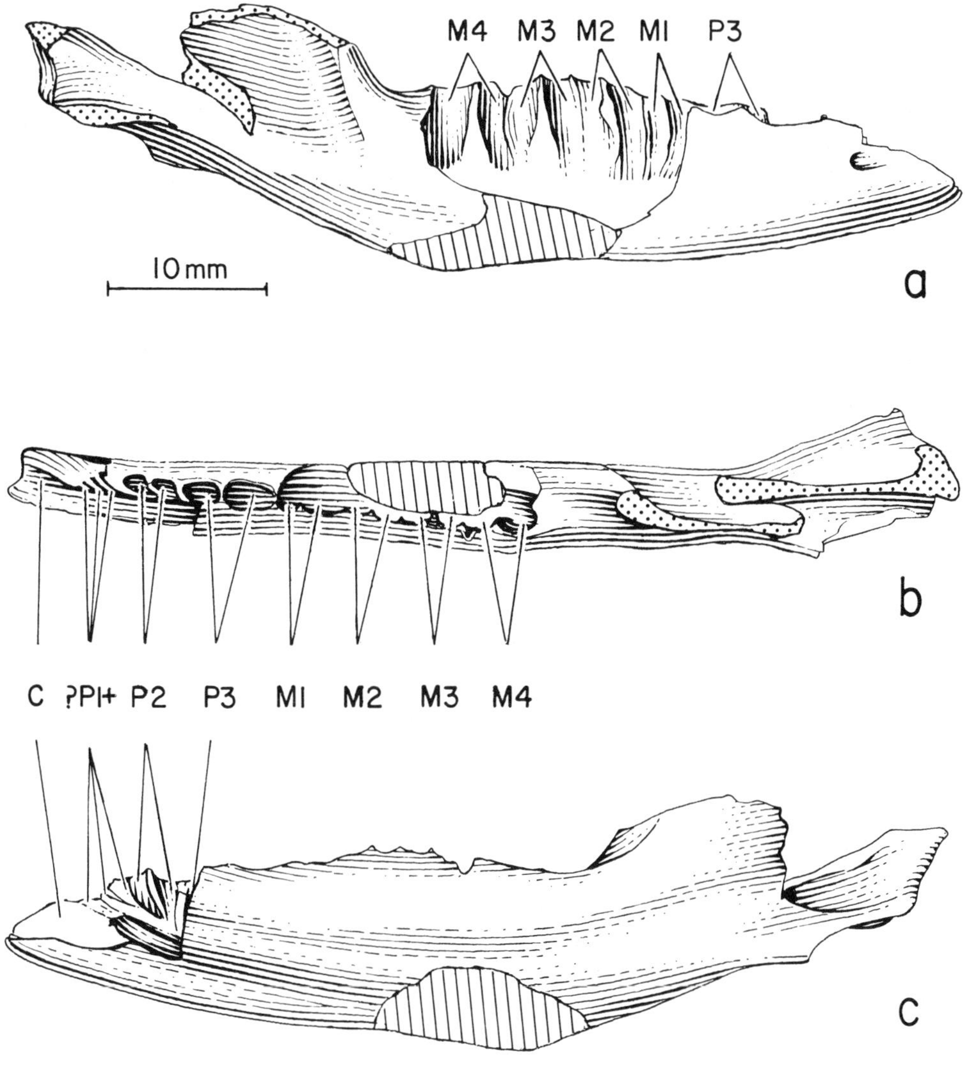

FIG. 50. ?*Pediomys* cf. *P. florencae* Clemens. Edentulous dentary, UCMP 110541, loc. V73084: (a) labial view; (b) occlusal view; (c) lingual view. Lines indicate plaster; stippling indicates broken surfaces. (A. L.)

six two-rooted teeth. The total length of these twelve alveoli is 26 mm. This compares to approximately 24 mm for the length of the twelve posteromost alveoli (for P_{2-3}, M_{1-4}) of *Pediomys florencae* based on Clemens' (1966, fig. 74) reconstruction of the dentary.

In Clemens' reconstruction of *Pediomys florencae* the molars become progressively smaller from M_4 through M_1. (M_3 and M_4 may have been similar in length, but M_4 was probably narrower, according to Clemens, 1966, p.100). The P_3 is an elongate tooth, similar in length to the M_3 (and M_4?). The shorter P_2 is similar in length to the M_1. This size pattern of teeth is essentially the same as that described for the alveoli of UCMP 110541 in the preceding paragraph. The size relationships of the alveoli in UCMP 110541 and *P. florencae* are not known in any other Lancian marsupial or placental. The closest approach to this pattern is in *P. hatcheri*; however, in this species the P_3 is not as elongate (Clemens, 1966, fig. 70) and the disparity in size of the molar series is not as great (Clemens, 1966; compare his tables 23 and 24 and figs. 71, 72, and 74). In addition, Clemens (1966, p. 105) noted that in *P. florencae* the symphysis is elongated, while in *P. hatcheri* it is deep and heavy. The symphyseal region of UCMP 110541 is similar to the condition in *P. florencae*.

UCMP 110541 also bears a striking resemblance to *Eodelphis browni*. UCMP 110541 was compared to the type specimen of *E. browni* (AMNH 14169). UCMP 110541 and the type of *E. browni* are almost the same size. The P_{2-3}, M_{1-4} alveoli measure slightly more than 26 mm in length in both specimens. The same size relationships of the teeth or alveoli are present in the type of *E. browni* as was observed above for *Pediomys florencae* and UCMP 110541. The P_2 is shorter than the elongated, but not bulbous, P_3. The M_1 is distinctly shorter than P_3. The molars show a marked size gradient, increasing from M_1 through M_3 and M_4.

The dentary of the type of *Eodelphis browni* is slender, as in UCMP 110541 and *Pediomys florencae*. However, the symphyseal region is slightly deeper and more robust in *E. browni* than in *P. florencae* and UCMP 110541. The more slender dentary of *E. browni* is one of the characters that Clemens (1966) used to reestablish *E. browni* as separate from the heavier-jawed *E. cutleri* (=*Boreodon matutinis*, according to Sahni, 1972, but *B. matutinus* is a *nomen dubium* according to Clemens, 1979). Fox (written comm.) agrees with Clemens (1966, 1979) that there are two groups of *Eodelphis* based on size, but the two groups overlap. However, Fox also believes that the type specimen of *E. browni* (AMNH 14169) is most likely a dentary of *E. cutleri*; in the former the teeth have been reduced in size because of postmortem corrosion of the enamel. Due to these varying opinions of *Eodelphis*, comparisons of dentary size of this genus to UCMP 110541 and *P. florencae* cannot be done with certainty.

Similarly, opinions vary concerning the number of roots possessed by the P_1 in the various specimens referable to *Eodelphis*. However, comments on this topic are not germane to the present discussion because the region between the alveoli for the canine and P_2 is not well preserved in UCMP 110541. The alveoli for what are presumed to be P_2 and P_3 are present in this specimen and do allow comparisons. In the type of *E. browni* the two roots for each of the P_2 and P_3 diverge ventrally (see Simpson, 1929a, fig. 48, and Clemens, 1966, pp. 58–60). In contrast, the roots of P_{2-3} in *Pediomys florencae* and UCMP 110541 are parallel and vertically implanted.

In the type of *Eodelphis browni*, the anterior margin of the ascending ramus forms approximately a right angle with the corpus of the dentary. In UCMP 110541, the ascending ramus is not well preserved, but it appears to form an angle of less then 90° with the corpus. According to Clemens' reconstruction of *Pediomys florencae* (1966, fig. 74), the anterior margin of the ascending ramus begins at an angle of less than 90° but quickly assumes a near-vertical orientation.

The positions of the mental foramina vary somewhat in these three forms. In the type of *Eodelphis browni* a large foramen is present below the anterior root of P_3 and a second, smaller one is present below the posterior root of M_1. In Clemens' reconstruction of *Pediomys florencae*, two small mental foramina are present. One is below and between the posterior root of P_1 and the anterior root of P_2. The second is below the posterior root of M_1, as in *E. browni*. UCMP 110541 preserves only one large foramen below P_1. The anterior mental foramen is situated low on the dentary in *E. browni*. In the other two forms it is placed distinctly higher.

Referral of UCMP 110541 to either *Eodelphis browni* or *Pediomys florencae* is plausible. However, *Eodelphis* is not known from the Lancian. *E. browni* occurs in the Judithian, and the genus is questionably known from the Edmontonian (Sloan and Russell, 1974). UCMP 110541 is tentatively referred to ?*Pediomys* cf. *P. florencae* based in part on this negative biostratigraphic evidence.

Family STAGODONTIDAE Marsh, 1889b

Didelphodon Marsh, 1889a

Didelphodon vorax Marsh, 1889a

(Table 33)

Didelphodon vorax Marsh, 1889a, p. 88
(See Clemens, 1966, p. 60, for synonymies.)

Type. YPM 11827, left M^2 (Marsh, 1889a, pl. IV, figs. 1–3).

Type locality. Mammal locality no. 1 of Lull (1915), UCMP loc. V5003, Lance Formation, Wyoming.

Referred specimens. 10 isolated teeth (LACM and UCMP specimens) from various LACM and UCMP localities; UCMP 117670, right dentary fragment with P_3; five edentulous jaw fragments from various LACM and UCMP localities.

Localities. UCMP locs. V72207, V72208, V72212, V73087, V73090, V73103, V75173, V75178, and V76161; LACM locs. 3083 and 3116.

Distribution. Lance Formation, Wyoming; Scollard Formation, Alberta; and Hell Creek Formation, Montana and South Dakota (all Lancian).

Revised diagnosis. Clemens, 1966, p. 60, with modifications by Clemens, 1968, pp. 4–10.

Description. The postcanine dentition of *Didelphodon* was reviewed by Clemens in 1966. In this paper he also described the ear region of *D. vorax*. Later (1968) he described a well-preserved dentary of *D. vorax* from the Hell Creek Formation, Garfield County. In this study he modified his earlier ideas on the orientation of the upper premolars (see Clemens, 1968, for discussion) and reconstructed the lower dentition of *D. vorax*, including the canine. In 1969, Lillegraven described isolated lower canines that he referred to *D. vorax*. In 1973, Clemens described a newly discovered maxilla of *D. vorax* from the Lance Formation, Wyoming, that preserved M^{1-3} and alveoli for P^3 and M^4.

None of the new material from the Hell Creek Formation adds to the descriptions given by Clemens and Lillegraven. However, two specimens in the collection tend to support these authors' ideas concerning the morphology and orientation of the upper premolars.

Upper premolars referred to *Didelphodon vorax* by Clemens (1966) possess a large main cusp and an expanded lobe. Clemens (1966) tentatively oriented the premolars with the lobe on the labial side. Subsequently (1968), he argued that the lobe was probably on the lingual side. Lillegraven (1969) described the first known upper P^3 of *D. vorax* that preserved the roots. One root is transversely widened, supporting the accessory lobe and part of the main cusp. The second root is much smaller, supporting only a part of the main cusp. The maxilla of *D. vorax* described by Clemens in 1973 preserves part of an alveolus for P^3. Based on the maxilla, Clemens noted that the only possible configuration was with a large root placed posteriorly and the small root placed at the anterolabial corner. In this orientation the accessory lobe in lingual. Two P^3's (see Table 33) from the Hell Creek Formation each preserve two roots similar to the roots of P^3 described by Lillegraven (1969, p. 48 and fig. 25–3d). However, one of the P^3's, UCMP 117669, is quite small and could be P^2. These specimens support Lillegraven's interpretation of the P^3 roots and Clemens' orientation of the upper premolars in *D. vorax*.

TABLE 33. Measurements of isolated premolars and molars of *Didelphodon vorax* from various localities.

Specimen Number	*Locality*	*Tooth Site*	*L*	*W (max.)*
UCMP 117699	UCMP V72208	$P^{3?}$	5.27	5.72
UCMP 109027	UCMP V73087	DP^3	4.04	3.88
UCMP 117670	UCMP V75173	P_3	5.95	6.32
UCMP 117671	UCMP V75178	P^3	6.23	6.45
LACM 55732	LACM 3083	M_4	6.74	4.98
LACM 55740	LACM 3116	M_2	5.99	4.67
LACM 55748	LACM 3116	P_3	7.75	5.71

Discussion. In 1966 (p. 58) Clemens provisionally restricted *Didelphodon padanicus* "to include only the individuals (or individual) represented by the type specimens [AMNH 3013]." The specimens probably came from the Hell Creek Formation, South Dakota (Wilson, 1965). All material of this genus from the Lance Formation, Wyoming, was referred to *D. vorax* by Clemens (1966).

During the course of the present study comparisons were made between the type of *Didelphodon padanicus*, the dentary of *D. vorax* (LACM 15433) from the Hell Creek Formation, Garfield County, described by Clemens (1968), and the type of *Stagodon validus* (USNM 2136) described by Marsh (1892). The latter is from the type Lance Formation and includes alveoli for three incisors and the base of the canine and fragments or alveoli for P_{1-3}. The specimen has been figured by Marsh (1892, pl. VIII, fig. 7) and Clemens (1966, fig. 36b, outline drawing). The crown of P_3 is almost obliterated, but the outline of the tooth is preserved, showing that it was very large and subquadrate in shape. It is similar in size and shape to the P_3 of the *D. padanicus* type specimen.

The P_3 in the Hell Creek Formation specimen of *Didelphodon vorax* (Clemens, 1968, fig. 1) is also large but is relatively much smaller than in the two other specimens. It is possible that USNM 2136, Marsh's type of *Stagodon validus*, is a specimen of *D. padanicus* from the type Lance Formation. However, little is known of the premolar variation in this genus, so this is regarded as an unsupported supposition. It should be added that if future finds indicate that *D. padanicus* is present in the type Lance Formation, this species name (Cope, September 1892) would be a junior synonym of *Didelphodon* ("*Stagodon*") *validus* (Marsh, March 1892).

Infraclass EUTHERIA

The eutherians in the Hell Creek and lowermost Tullock formations of Garfield County are as taxonomically diverse as the multituberculates and more so than the metatherians. However, with a few exceptions (e.g., *Procerberus*), the eutherians do not rival the multituberculates in numerical abundance.

Like the multituberculates and metatherians, the eutherians lack any species in common between the pre-mantuan and the Lancian Hell Creek faunal-facies localities in Garfield County. Sloan and Van Valen (1965) found that their stratigraphically lowest Bug Creek faunal-facies locality, Bug Creek Anthills, yielded a number of Hell Creek faunal-facies mammals. However, these authors noted that Hell Creek faunal-facies or typical Cretaceous mammals are progressively lost going upward in their Bug Creek faunal-facies sequence (Van Valen and Sloan, 1977a; Sloan, 1976).

The Hell's Hollow local fauna includes the richest pre-mantuan locality, V74111 (as well as V74110). In this local fauna the most taxonomically diverse group is that of the condylarths. The only other eutherian in this local fauna is *Procerberus,* which is the single most abundant eutherian in the Hell's Hollow local fauna.

None of the above eutherians is found in the Hell Creek faunal-facies in either Garfield or McCone County. Within this faunal-facies, palaeoryctids are the most taxonomically diverse eutherian group. The leptictoid *Gypsonictops* is also present. All the Hell Creek faunal-facies eutherians are rare. This rarity is in part due to collecting biases (i.e., the lack of rich Hell Creek faunal-facies localities), but also appears to represent a real scarcity of eutherians. This is similar to the situation in the Lance local fauna (Clemens, 1973; Estes and Berberian, 1970), where eutherians are relatively rare, but contrasts with the Trochu local fauna (Lillegraven, 1969) where eutherians rival metatherians in abundance and taxonomic diversity.

The classification of eutherians used in this study to some extent reflects recent ideas concerning phylogenetic relationships within the infraclass. Probably the most stimulating and controversial paper dealing with eutherian relationships to appear in recent years is that of McKenna (1975). This is the first major attempt since Simpson (1945) to review the higher relationships among mammals. Although I applaud this attempt at showing cladistic relationships among mammals, I do not subscribe to the proposed cladistic classification (see the chapter on Techniques of Study).

Two major groupings of eutherians, proteutherians and Condylarthra, are recognized in this paper.

The term proteutherian is used in the sense of Butler (1972). He suggested this group be of ordinal rank and include: "ancestral placentals"; "forms which have diverged comparatively little from the basic placental stock"; and short-lived, specialized Cretaceous and earliest Tertiary lines (p. 263). He fully realized that this would be a "waste-basket" taxon to replace the Insectivora, which would disappear from usage or be restricted to the Lipotyphla. I agree with Butler that Proteutheria is a "waste-basket" taxon and therefore use the term only informally in this study. In the same paper Butler also suggested that the proteutherians might be placed in Eutheria, *incertae sedis*. This is the usage followed in the present study for formal classification.

The remaining eutherians from the Hell Creek Formation are included in the order Condylarthra. Unlike the proteutherians, the condylarths are considered to be a monophyletic group (possibly with a few exceptions) relative to other latest Cretaceous and earliest Tertiary eutherians. At the same time it is realized that the Condylarthra is a grade that gave rise to myriad later groups (e.g., subungulates, ungulates, cetaceans, etc.). The order may eventually be eliminated or highly restricted (as was done by McKenna, 1975, but without his proliferation of categories).

Infraclass EUTHERIA, *incertae sedis*

Proteutherians (*sensu* Butler, 1972)

Comments. The proteutherians recovered from the Hell Creek Formation, Garfield County, are few and fragmentary (except for *Procerberus*) and add little to the morphological data given by Sloan and Van Valen (1965), Lillegraven (1969), and Clemens (1973). Only additions are given here; the reader is directed to these authors for detailed morphological descriptions. The major addition in this study is the recognition of taxa that have not previously been reported from the Hell Creek Formation of Montana.

Superfamily LEPTICTOIDEA (Gill, 1872) Butler, 1956

Leptictidae Gill, 1872, p. 19
Leptictoidea Butler, 1956, p. 476

Family, *incertae sedis*

Gypsonictops Simpson, 1927a

Gypsonictops Simpson, 1927a, p. 6
Euangelistes Simpson, 1929b, p. 107

Gypsonictops cf. *G. illuminatus* Lillegraven, 1969

(Table 34)

Gypsonictops illuminatus Lillegraven, 1969, p. 51

Type. UA 2447, right maxillary fragment with P^{3-4}, M^{1-3} (Lillegraven, 1969, fig. 27–6).

Type locality. Loc. KUA-1, Scollard Formation, Alberta.

Referred specimens. 4 upper molars; 1 P_4; 3 lower molars from UCMP loc. V73087. 10 teeth and tooth fragments from other localities.

Localities. UCMP locs. V72212, V73076, V73077, V73078, V73087, V73097, and ?V74116.

Distribution. Scollard Formation, Alberta, and Hell Creek Formation, Montana (both Lancian).

Diagnosis. Lillegraven, 1969, p. 51.

Description. Most of the specimens referable to *Gypsonictops* are broken or worn, making species assignment difficult. The specimens are intermediate in size between *G. hypoconus* (Clemens, 1973, table 1) and *G. illuminatus* (Lillegraven, 1969, tables 9–11). Lillegraven (1969, p. 51) noted several characters differentiating *G. illuminatus* from the smaller *G. hypoconus*. It is only possible to compare two of these characters in the Hell Creek Formation sample.

First, Lillegraven noted that the upper molars (presumably referring to the protocone) are more anteriorly recumbent in *G. illuminatus*. The Hell Creek Formation specimens were compared to upper molars from the Lance local fauna and a cast of the type of *G. illuminatus* (UA 2447). There is some suggestion that the Hell Creek upper molars are slightly more recumbent than those from the Lance local fauna.

Second, Lillegraven noted that the trigonids of lower molars of *Gypsonictops illuminatus* are "proportionately slightly lower" than in *G. hypoconus*. All but one lower molar (UCMP 108945) in the Hell Creek sample are worn, making this comparison difficult. However, the trigonids in this sample may be somewhat lower than in *G. hypoconus*.

These comparisons are tenuous and referral of the Hell Creek Formation specimens to *Gypsonictops illuminatus* is at best tentative (indicated by "cf.").

An M^2, AMNH 21984, from the Hell Creek Formation (near Forsyth, Montana), figured by Simpson (1927a, fig. 6) and Clemens (1973, fig. 10) may belong to *Gypsonictops illuminatus*.

TABLE 34. Measurements of isolated teeth of *Gypsonictops* cf. *G. illuminatus* from various localities. Tentative identification of the isolated lower molars is done after Clemens (1973, fig. 7).

UCMP Number	*UCMP Locality*	*Tooth Site*	*L*	*W*	*W–A (uppers) or W–TR(lowers)*	*W–P (uppers) or W–TA (lowers)*
117640	V73087	$M^{2?}$	a2.20		a3.68	3.48
117646	V73098	$M^{1?}$	a1.80		a2.96	a2.84
117644	V74116	P_3	2.12	1.20		
108057	V72212	P_4	2.25		1.45	1.26
117637	V73087	P_4	2.32		1.44	1.16
109028	V73087	$M_{1?}$	2.29		1.70	1.73
117638	V73087	$M_{1?}$	2.60		a1.64	a1.80
108945	V73077	$M_{2?}$	2.50		1.90	1.73
108946	V73077	$M_{2?}$	2.46		1.83	1.65
108975	V73078	$M_{2?}$	2.24		1.74	1.55

a Approximate measurement.

Comments. Sloan and Van Valen (1965; Sloan, 1976) reported the occurrence of two species of *Gypsonictops, G. hypoconus* and *G. petersoni,* from the Hell Creek Formation, McCone County. Sloan (1976) reported that the two species are present in both the Hell Creek and Bug Creek faunal-facies (approximately the *Triceratops* and *Protungulatum-Stygimys* communities of Sloan). Clemens (1973) synonymized *Euangelistes petersoni* with *G. hypoconus*. He noted that if a second species of *Gypsonictops* is present in the Hell Creek Formation a new species name should be proposed to replace *Gypsonictops* ("*Euangelistes*") *petersoni*.

Clearly, a comparison of the specimens of *Gypsonictops* from the Hell Creek Formation of Garfield and McCone counties should be made, in order to establish which species of this genus are present.

Novacek (1977) recently reviewed Paleocene and Eocene leptictids from North America. He has argued, based on a cladistic analysis, that *Gypsonictops* should be removed from the Leptictidae but that it is the "most closely related sister group" (1977, p. 12). However, he did not suggest a familial assignment. A possible solution would be to elevate Van Valen's (1967) Gypsonictopinae to familial status. This should probably be postponed until a study, similar to Novacek's, has been done of *Gypsonictops* and also of such forms as *Kennalestes* (Kielan-Jaworowska, 1969, 1975a; Fox, 1977; Kielan-Jaworowska and others, 1979). Accordingly, a formal change in rank is not proposed here.

Infraclass EUTHERIA, *incertae sedis*
Family PALAEORYCTIDAE (Winge, 1917) Simpson, 1931

Comments. The taxon Palaeoryctidae is used as a grade in the present study. There is some evidence (e.g., McKenna, 1975) that the Cretaceous and Tertiary forms included in the family are not a monophyletic group (*sensu* Hennig, 1975). The subfamily Didelphodontinae Matthew, 1918, also appears to be an artificial grouping and this category is not used.

The taxonomy of the Lancian representatives of this group at the species level is well understood largely due to the efforts of Lillegraven (1969) and Clemens (1973). However, a thorough review of the family is needed above the species level.

Except for a large sample of mostly isolated teeth of *Procerberus*, palaeoryctids are rare elements in the Hell Creek Formation. In some instances only one tooth of a species is known, but fortunately the morphology is distinct and referral to a particular species is relatively easy.

Cimolestes Marsh, 1889a
Cimolestes incisus Marsh, 1889a

Cimolestes incisus Marsh, 1889a, p. 89
Nyssodon punctidens Simpson, 1927b, p. 124

Type. YPM 11775, left M_1 (Marsh, 1889a, pl. IV, figs. 12–15; Simpson, 1929a, fig. 50, no. 1; Slaughter, 1965, fig. 6).

Type locality. Mammal locality no. 1 of Lull (1915), UCMP loc. V5003, Lance Formation, Wyoming.

Referred specimens. LM^1 fragment, UCMP 117654; RM^2 fragment, UCMP 117655; LDP^3 fragment, UCMP 117656, all from UCMP loc. V73087.

Localities. See Referred specimens.

Distribution. Lance Formation, Wyoming, and Hell Creek Formation, Montana (both Lancian).

Revised diagnosis. Clemens, 1973, p. 45.

Description. Four fragmentary teeth recovered from the Hell Creek Formation, Garfield County, are within the approximate size range of *Cimolestes incisus* and *C. cerberoides*. Assignment of the four teeth to one or another of these species is somewhat arbitrary. Referral of the upper molar fragments was aided by Lillegraven's observation (1969, p. 63) that the P^4 and M^{1-3} are slightly less anteroposteriorly compressed in *C. cerberoides* and the protocone is less transversely extended. Based on these criteria, one upper molar is referred to cf. *C. cerberoides* (see below). Another, more anteroposteriorly compressed, upper molar fragment, UCMP 117655, is referred to *C. incisus*. A third fragment, UCMP 117654, which does not preserve the protocone, is referred to *C. incisus* based on comparisons with material from the Lance local fauna and a cast of the type specimen of *C. cerberoides*, UA 2973. A moderately deep ectoflexus and small paracrista suggest that the tooth is an M^2 fragment. Also, the length of the fragment (L = 3.52 mm) is within the observed range of M^2 lengths of *C. incisus* (Clemens, 1973, table 2) but is greater than that seen in *C. cerberoides* M^2's (Lillegraven, 1969, table 13).

A DP^3, UCMP 117656, that lacks the parastyle and part of the protocone is indistinguishable from the DP^3 of *Cimolestes incisus* figured by Clemens (1973, fig. 18i & j).

Cimolestes cf. *C. cerberoides* Lillegraven, 1969

Cimolestes cerberoides Lillegraven, 1969, p. 61

Type. UA 2973, fragment of left side of skull with P^{3-4}, M^1, M^3 (Lillegraven, 1969, fig. 31-3).

Type locality. Loc. KUA-1, Scollard Formation, Alberta.

Referred specimens. LM^2 lacking parastylar lobe and part of the paracone, UCMP 117661.

Locality. UCMP loc. V72207.

Distribution. Scollard Formation, Alberta, and Hell Creek Formation, Montana (both Lancian).

Description. One upper molar fragment, UCMP 117661, from the Hell Creek Formation is tentatively (indicated by "cf.") referred to *Cimolestes* cf. *C. cerberoides*. The parastylar lobe is missing, but the indented ectoflexus and the well-developed metastylar lobe suggest that the tooth is an M^2. The paracone is also broken off near its base, but when present it was definitely larger and taller than the metacone. In both *C. cerberoides* and *C. incisus* the disparity in size between the paracone and metacone increases going from M^1 through M^3.

Two weak characters favor a tentative assignment of this tooth to *Cimolestes* cf. *C. cerberoides*. First, as in *C. cerberoides*, the protocone region is relatively less anteroposteriorly compressed than in *C. incisus*. The second character is size. Only the posterior width of UCMP 117661 can be measured. This dimension (W-P = 4.32 mm) is very close to the mean posterior width of *C. cerberoides* (Lillegraven, 1969, table 13). The value is also within the range of the posterior width of *C. incisus* (Clemens, 1973, table 2) but is at the extreme lower end.

If this tentative referral of UCMP 117661 to *Cimolestes cerberoides* is correct, it constitutes a geographic range extension of the species.

Cimolestes stirtoni Clemens, 1973

(Table 35)

Cimolestes stirtoni Clemens, 1973, p. 60

Type. UCMP 48591, left upper molar, probably M^2 (Clemens, 1973, fig. 21a-c).

Type locality. UCMP loc. V5711, Lance Formation, Wyoming.

Referred specimens. LM^1, UCMP 117660; RM_2, UCMP 117659; LDP_4, UCMP 117658, all from UCMP loc. V73087.

Locality. See Referred specimens.

Distribution. Lance Formation, Wyoming, and Hell Creek Formation, Montana (both Lancian).

Diagnosis. Clemens, 1973, p. 60.

Description. Three isolated teeth from loc. V73087, Flat Creek local fauna, are morphologically very similar to the teeth of *Cimolestes stirtoni* from the Lance local fauna (Clemens, 1973). The only consistent difference is in size: the teeth from loc. V73087 are slightly smaller in all dimensions than their counterparts in *C. stirtoni* from the Lance local fauna but are larger than those of *C. incisus*. Both samples of *C. stirtoni* are far too small for statistical comparison. Speaking subjectively, it is doubtful whether these size differences, although consistent, are significant.

UCMP 117660 is an M^1 of *Cimolestes stirtoni*. The tooth is complete except for the paracone. The metastylar lobe is better developed than the parastylar lobe, as would be expected on an M^1. The two M^1's Clemens (1973, p. 63 and fig. 21h) referred to *C. stirtoni* are not as complete as UCMP 117660. The stylar shelf opposite the paracone (the parastylar lobe) is not as wide as in the M^1's described by Clemens. The stylocone is small but distinct and, except for a smaller parastyle, is the only stylar cusp. A weak paracrista is present at the base of the broken paracone but does not extend completely to the stylocone or parastyle. The metacrista is formed into a crestlike cusp. A distinct notch is present between the metacrista and the metacone. UCMP 117660 has well-developed but narrow pre- and postcingula. Clemens (1973, p. 63) did not observe cingula on the one M^1 preserving the protocone. However, he did note the presence of rugosities in the position of the precingulum on the M^1, and the type specimen, an M^2, possesses both pre- and postcingula.

UCMP 117659 is an unworn M_2 of *Cimolestes stirtoni*. As in the M_2's from the Lance local fauna, the trigonid is much wider than the talonid and the small paraconid is closely approximated to the metaconid.

A large DP_4, UCMP 117658, showing considerable dental attrition is almost identical to the DP_4 figured by Clemens (1973, fig. 23a–c). The trigonid and talonid cusp apices as well as the interconnecting cristids and lophids show considerable wear. In addition, major elongate wear facets are present on the labial side of the tooth below the paralophid and the cristid obliqua. Smaller facets are present on the labial side below the paraconid and on the posterior side below the hypoconid and hypoconulid.

The preceding three teeth represent the first reported occurrence of *Cimolestes stirtoni* from the Hell Creek Formation, Montana. The previous geographic range included only the type Lance Formation, Wyoming.

TABLE 35. Measurements of isolated teeth of *Cimolestes stirtoni* from UCMP loc. V73087.

UCMP Number	*Tooth Site*	*L*	*W–A (uppers) or W–TA(lowers)*	*W–P (uppers) or W–TA (lowers)*
117660	M^1	3.92	5.44	5.84
117659	M_2	4.24	3.28	2.48
117658	DP_4	3.76	1.84	1.92

Cimolestes propalaeoryctes Lillegraven, 1969

Cimolestes propalaeoryctes Lillegraven, 1969, p. 69

Type. UA 3756, right dentary with 2 incisors, canine, P_{1-4}, M_{1-3}.

Type locality. Loc. KUA-1, Scollard Formation, Alberta.

Referred specimens. LM_1 or LM_2, lacking the tip of the metaconid, UCMP 117653, UCMP loc. V73087.

Locality. See Referred specimens.

Distribution. Scollard Formation, Alberta, and Hell Creek Formation, Montana (both Lancian).

Diagnosis. Lillegraven, 1969, p. 69.

Description. A single lower molar, UCMP 117653, from the Flat Creek local fauna (loc. V73087) can confidently be referred to *Cimolestes propalaeoryctes*. The specimen is not morphologically distinguishable from the molars in a cast of the type specimen, UA 3756. As was noted by Lillegraven (1969), the trigonid is high, the protoconid and metaconid are fused throughout much of their height, the paraconid tends to be reduced, and the talonid is narrow and anteroposteriorly short. It was not possible to determine whether UCMP 117653 is an M_1 or an M_2, but based on the trigonid width/talonid width ratio (=1.38) identification as an M_2 seems more likely. UCMP 117653 is slightly smaller (L=2.20; W–TR=1.60; W–TA=1.16) than the M_1's and M_2's Lillegraven referred to this species (1969, table 16).

The occurrence of this species in the Hell Creek Formation, Montana, constitutes a geographic range extension. *Cimolestes propalaeoryctes* was previously known only from the Trochu local fauna.

Comments. Lillegraven (1969) demonstrated the similarities between *Cimolestes propalaeoryctes* and *Palaeoryctes puercensis* from the Torrejonian. He considered *C. propalaeoryctes* to be a possible ancestor of *P. puercensis* based on these similarities.

Lillegraven (1969, p. 73) discussed what he considered to be nomenclatural problems involving these two genera. He felt that "All but the most ardent 'splitters' " would include his new species (i.e., *Cimolestes propalaeoryctes*) in *Cimolestes*, based on molar morphology. Yet, he felt that the premolars were essentially the same as in *Palaeoryctes*. He argued that it would be possible to establish a new genus for *C. propalaeoryctes* based on the differences in evolutionary rate between the premolar and molar series. However, he felt this ". . . would be without scientific or practical value." He concluded that the name *Cimolestes propalaeoryctes* has a distinct advantage because "Such a phylogenetically based name emphasizes a common origin with other species of *Cimolestes*, yet points out its probable descendent [*Palaeoryctes*]."

Having a slightly different philosophical bias toward phylogenetic reconstruction and classification, I feel that a third possible nomenclature, not mentioned by Lillegraven, should be considered. This third approach would be to refer *Cimolestes propalaeoryctes* to *Palaeoryctes*. If it is true (and it does seem likely) that *C. propalaeoryctes* and *Palaeoryctes* possess shared derived characters relative to *Cimolestes*, then the classification should reflect this relationship. There may be other options besides referring *C. propaleoryctes* to *Palaeoryctes*; however, it is not appropriate to maintain the former species in *Cimolestes* if the relationship is based on shared primitive characters. No formal changes are suggested here because such changes should await a full cladistic analysis of the family.

Cimolestes magnus Clemens and Russell, 1965

Comments. Teeth of this species have not been recovered from the Hell Creek Formation of Garfield County. However, Estes and Berberian (1970, p. 8) report this taxon from the Bug Creek Anthills local fauna, McCone County. Also, Clemens (1973, fig. 24c) mentioned and figured a maxillary fragment of this species from the Hell Creek Formation of Carter County, in the southeastern corner of Montana. All other species of Lancian palaeoryctids are at least tentatively recognized from the Hell Creek Formation, Garfield County, although they are a rare element (except *Procerberus*). The lack of *Cimolestes magnus* is probably a result of collecting and/or preservational biases.

Batodon Marsh, 1892

Batodon tenuis Marsh, 1892

(Table 36)

Batodon tenuis Marsh, 1892, p. 258

Type. USNM 2139, fragmentary left dentary, with part of the canine, alveolus of P_1, and P_{2-4} (Marsh, 1892, pl. XI, fig. 5; Simpson, 1929a, fig. 55; Clemens, 1973, fig. 25d & e).

Type locality. Mammal locality no. 1 of Lull (1915); UCMP loc. V5003, Lance Formation, Wyoming.

Referred specimens. RM^2, lacking the parastylar lobe, UCMP 117649; RP_4, UCMP 117650; LM_2, UCMP 117651; M_x talonid, UCMP 117652, all from UCMP loc. V73087.

Locality. See Referred specimens.

Distribution. Lance Formation, Wyoming; Scollard Formation, Alberta; Hell Creek Formation, Montana (all Lancian).

Revised diagnosis. Clemens, 1973, pp. 67–68.

Description. The single upper molar of *Batodon tenuis*, UCMP 117649, from loc. V73087 was identified as an M^2 based on a comparison with the M^2 figured and described by Lillegraven (1969, fig. 38 and p. 82). Since the parastylar lobe is missing on UCMP 117649, the morphology of this region is unknown. Lillegraven (p. 82) noted: "Precingula and postcingula are strong and variable in construction." On one molar he found the cingula to be connected lingually below the protocone; on the other, the cingula are not connected. In UCMP 117649 the cingula are well developed but are not connected lingually below the protocone. The tooth shows no visible wear.

UCMP 117650 is a P_4 of *Batodon tenuis*. It resembles the two P_4's described by Lillegraven (1969) in possessing a distinct metaconid. UCMP 117650 differs from the P_4 figured by Lillegraven (fig. 39–1) in the development of the anterior accessory cusp (paraconid). In UCMP 117650 an anterior accessory cingulum is present but a distinct cusp is lacking.

An M_2, UCMP 117651, lacking the tips of the trigonid cusps, and a talonid, UCMP 117652, are referable to *Batodon tenuis*.

This is the first occurrence of *Batodon* from the Hell Creek Formation, Montana. Sloan and Van Valen (1965) did not indicate the presence of this taxon at the Bug Creek Anthills locality. However, Clemens (pers. comm.) has found *Batodon* in the UCMP collections from Bug Creek Anthills.

TABLE 36. Measurements of isolated teeth of *Batodon tenuis* from UCMP loc. V73087.

UCMP Number	*Tooth Site*	*L*	*W*	*W-A (uppers) or W-TR(lowers)*	*W-P (uppers) or W-TA (lowers)*
117649	M^2	—		—	1.56
117650	P_4	1.20	0.68		
117651	M_2	1.28		0.80	0.68
117652	M_x	—		—	0.72

Procerberus Sloan and Van Valen, 1965

Procerberus formicarum Sloan and Van Valen, 1965

(Fig. 51 & Tables 37–39)

Procerberus formicarum Sloan and Van Valen, 1965, p. 225

Type. UMVP 1460, dentary fragment with P_{3-4}, M_1 (type specimen not figured by Sloan and Van Valen, 1965; isolated teeth figured by Sloan and Van Valen, 1965, fig. 6; see also Lillegraven, 1969, fig. 32–1,2,3, and Van Valen, 1969, fig. 3).

Type locality. Bug Creek Anthills, Hell Creek Formation, Montana.

Referred specimens. 17 P^3's; 7 P^4's; 13 M^1's; 15 M^2's; 14 M^3's; left dentary fragment with P_2 roots, P_{3-4}, M_1 trigonid, UCMP 117890; 12 P_4's; 8 M_1's; 15 M_2's; 17 M_3's; 5 tooth fragments, all from UCMP loc. V74111. 5 isolated teeth from UCMP loc. V74110.

Localities. See Referred specimens.

Distribution. Hell Creek Formation (Lancian) and lowermost Tullock Formation (premantuan), Montana.

Diagnosis. Sloan and Van Valen, 1965, pp. 225–226, plus Lillegraven's (1969, pp. 63–69) referral of the species to the Palaeoryctidae.

Description. Sloan and Van Valen (1965) and Lillegraven (1969) have adequately described the dental morphology of *Procerberus formicarum*. Only additions to these authors' descriptions are given here.

Procerberus formicarum is very common in the Hell's Hollow local fauna, second only to *Mesodma garfieldensis* in abundance.

The upper molars and premolars resemble those shown by Lillegraven (1969, fig. 32–1,2,3) and Van Valen (1969, fig. 3). In fig. 3 of Van Valen (1969) the P^2 and P^3 are isolated teeth. He shows a space between P^2 and P^3 and between P^3 and P^4. The figure is clearly a composite of three photographs and it is not clear whether the spaces are intended to represent diastemata. However, there is one specimen, UCMP 117823, from loc. V74111 that suggests there is a slight diastema between P^2 and P^3 but not between P^3 and P^4. The specimen is a right maxillary fragment preserving P^3, part of the anterior alveolus of P^4, and part of the posterior alveolus of P^2. The anterior alveolus of P^4 is situated very close to the posterior root of P^3, while there is a distinct space or diastema between the posterior alveolus and P^2 and the anterior root of P^3.

A left dentary fragment, UCMP 117890, preserves two roots for P_2, P_{3-4}, and M_1 trigonid, and alveoli for the other molars. The P_2 was present in the specimen when found but was later lost. It was similar to the P_3 preserved in the jaw, except for its slightly smaller size. As was noted by Sloan and Van Valen (1965, p. 225) the P_3 (and P_2) is simple, with a low anterior basal cusp (~ paraconid), a tall protoconid, and a one-cusped heel. The P_2 and P_3 are labiolingually compressed. The dentary is laterally narrow and is not exceedingly deep relative to the height of the molars. The depth of the jaw below the alveoli of the posterior molars is 3.7 mm and gradually diminishes to 3.0 mm below the roots of P_2. A small depression on the labial side of the dentary below P_4 bears a mental foramen. An elongated triangular-shaped surface is present low on the anterolingual surface of the dentary. The apex of the triangular surface terminates below the anterior root of M_1. The surface is smooth but is probably the symphyseal region; if so, the symphysis is very dorsoventrally compressed but anteroposteriorly elongated, extending beneath the M_1.

A right dentary fragment, UF 22896, of *Procerberus formicarum* preserving part of P_4, M_{1-3}, and portions of the previously undescribed ascending ramus was recovered from the

TABLE 37. Measurements of isolated upper premolars and molars of *Procerberus formicarum* from UCMP loc. V74111.

		L	*W*	*W–A*	*L/P*
P^3	N	12	15		
	OR	2.2–2.6	1.6–2.0		
	$\bar{x}$	2.4	1.7		
	s	0.15	0.11		
	CV	6.3	6.5		
P^4	N	7		7	7
	OR	2.2–2.7		2.1–2.6	2.4–2.9
	s	2.5		2.3	2.7
M^1	N	10		10	12
	OR	2.3–3.0		3.0–3.7	2.9–3.8
	$\bar{x}$	2.6		3.3	3.3
	s	0.22		0.23	0.21
	CV	8.5		7.0	6.4
M^2	N	8		7	12
	OR	2.2–2.9		3.4–3.8	3.4–3.9
	$\bar{x}$	2.5		3.6	3.6
	s	—		—	0.20
	CV	—		—	5.6
M^3	N	9		8	7
	OR	1.8–2.6		3.1–3.8	2.6–3.0
	$\bar{x}$	2.1		3.4	2.7

Bug Creek Anthills locality by G. MacDonald. Mr. MacDonald and the researchers currently working on the Bug Creek Anthills local fauna, Drs. R. Sloan and L. Van Valen, have kindly allowed me to describe the specimen. The lower dentition of *P. formicarum* has been described elsewhere (Sloan and Van Valen, 1965). Only the dentary will be described and discussed here (Fig. 51).

TABLE 38. Measurements of left dentary with P_{3-4}, M_1 trigonid, UCMP 117890, of *Procerberus formicarum* from UCMP loc. V74111.

P_3		*P_4*			*M_1 trigonid*
L	*W*	*L*	*W–TR*	*W–TA*	*W–TR*
2.3	0.9	2.6	1.2	1.1	1.5

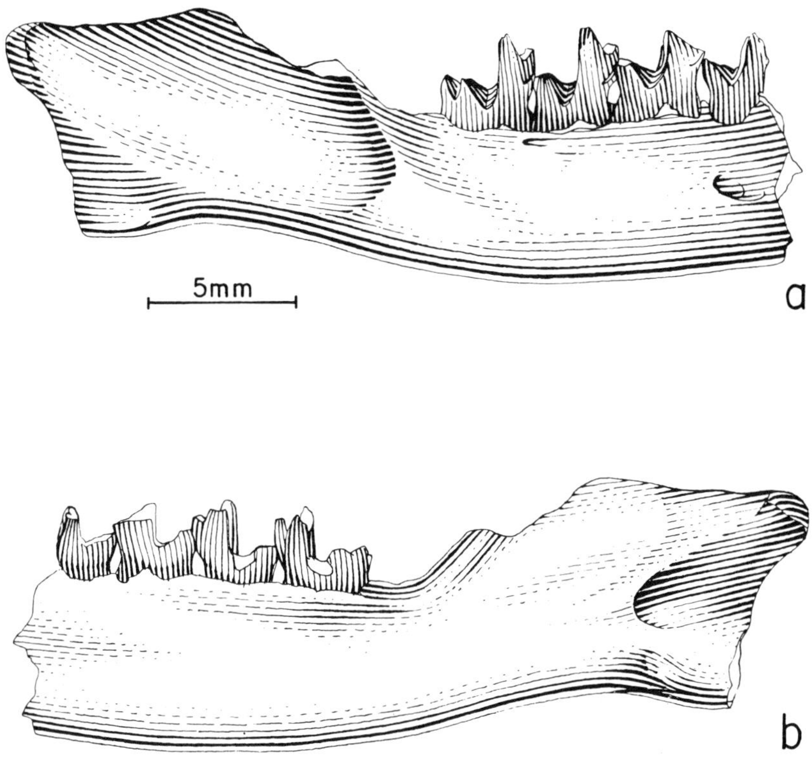

FIG. 51. *Procerberus formicarum* Sloan and Van Valen. UF 22896. Right dentary with P_4 fragment, M_{1-3}, Bug Creek Anthills local fauna: (a) labial view; (b) lingual view. (A. L.)

The corpus of the dentary in UF 22896 is of approximately uniform depth (approx. 4.3 mm) from the portion preserved below P_4 to posteroventral to M_3. The depth of the corpus slightly decreases until it merges with the ascending ramus. The corpus is transversely narrow. A small mental foramen is present labioventral to P_4. The narrowness of the dentary and the position of the mental foramen are similar to that seen in UCMP 117890, loc. V74111. The depth of the dentary is greater in UF 22896. The teeth show more dental attrition in UF 22896, suggesting that the mandible may have deepened with age as in some marsupials (e.g., Clemens, 1973) and eutherians (e.g., Lillegraven, 1969). However, the teeth are slightly larger in UF 22896. Further, UF 22896 and UCMP 117890 are from two localities almost certainly differing in age. A comparison of dentaries from the richer Bug Creek Anthills local fauna will be necessary to test this thesis.

TABLE 39. Measurements of isolated lower premolars and molars, and UCMP 117890, dentary fragment with P_{3-4}, M_1 trigonid, of *Procerberus formicarum* from UCMP loc. V74111.

		L	*W–TR*	*W–TA*
P_4	N	12	12	12
	OR	2.4 –2.9	1.0 –1.4	1.0 –1.3
	$\bar{x}$	2.6	1.2	1.1
	s	0.12	0.13	0.10
	CV	4.6	10.8	9.1
M_1	N	6	8	6
	OR	2.6 –2.9	1.5 –1.8	1.5 –1.6
	$\bar{x}$	2.7	1.7	1.5
M_2	N	12	14	13
	OR	2.3 –3.0	1.5 –2.3	1.3 –1.8
	$\bar{x}$	2.6	1.8	1.5
	s	0.19	0.19	0.12
	CV	7.3	10.5	8.0
M_3	N	15	15	17
	OR	2.4 –3.0	1.3 –2.0	1.0 –1.5
	$\bar{x}$	2.8	1.7	1.3
	s	0.19	0.18	0.13
	CV	6.8	4.7	10.0

In UF 22896 the mandibular condyle is intact. The surface of the condyle is slightly convex and is ovoid in outline. The long axis of the ovoid runs labiolingually. The condylar surface faces posterodorsally and its long axis is tilted slightly toward the lingual side. The condyle is at a level just above the apices of the trigonid cusps. In the only other Lancian palaeoryctid in which the condyle is known, *Cimolestes magnus*, the condyle is just below the level of the apices of the trigonid cusps (Lillegraven, 1969, p. 80). This contrasts with the condition seen in *Protungulatum donnae* (Sloan and Van Valen, 1965, fig. 6) and *Oxyprimus erikseni* (this study, Fig. 57), where the condyle is situated well above the apices of the trigonid cusps. The condyle in UF 22896 also appears to be positioned more posteriorly relative to the tooth row than in the two species of *condylarths.*

Most of the ascending ramus is lacking on UF 22896. However, some structures are preserved. The thickened anterior margin of the ascending ramus is oriented at an angle of about 63° to the alveolar border. This is 10°–14° less than is seen in *Protungulatum donnae* and *Oxyprimus erikseni* (see the discussion of *O. erikseni*). The anterior margin of the ascending ramus is not well preserved in most other Lancian palaeoryctids, but the angle in *Cimolestes magnus* (Lillegraven, 1969, fig. 37–3a, 3c) appears to be less than that seen in species of *Protungulatum.*

The angular process in UF 22896 is not complete. However, the portion preserved shows a slight but distinct inflection like that which Lillegraven (1969) found in other Lancian palaeoryctids. The angular process in dentaries of *Protungulatum donnae* does not appear to be inflected.

A small, posteriorly facing mandibular foramen is present on the lingual side of the dentary just anteroventral to the mandibular condyle.

Discussion. The characters in the posterior portion of the dentary of *Procerberus formicarum* are similar to those seen in other palaeoryctids. However, without further comparisons it is not possible to determine if these are shared derived or shared primitive characters (or, as is less likely, parallel developments). At least one character, the inflected angular process, can be assumed to be a primitive character, based on a cursory comparison with other Lancian therians. An inflected angle is found in palaeoryctids (i.e., *Cimolestes*), leptictoids (i.e., *Gypsonictops*), and Lancian marsupials. The commonality of this character suggests that it is primitive for therians, as has been noted by various other authors (e.g., Lillegraven, 1969).

Order CONDYLARTHRA

In Garfield County, condylarths have been recovered from four localities in the Hell Creek Formation and five localities in the lowermost Tullock Formation. Two of the Hell Creek Formation localities are possibly latest Cretaceous in age and thus would be Lancian, Bug Creek faunal-facies occurrences. These two localilties have produced an edentulous dentary and a dentary with one damaged M_2 (loc. V73098) and an edentulous dentary (loc. V73099). The other two localities are just below the Z coal within the uppermost Hell Creek Formation and are earliest Paleocene in age; these two localities are probably pre-mantuan. One (loc. V77129) has produced two molars and bone fragments, and the other (loc. V77128) has produced a number of condylarths currently being studied by W. A. Clemens.

In the lowermost Tullock Formation two localities of pre-mantuan, earliest Paleocene age (locs. V74110 and V74111, the Hell's Hollow local fauna) have produced most of the condylarths, representing four or five taxa. Another locality (loc. V72211) has produced one condylarth dentary and is pre-mantuan in age. Finally, two localities (loc. V72210 = LACM 3103, and loc. V77087) that have produced four condylarth taxa are mantuan in age.

In addition to the description of the condylarths from these localities, two upper dentitions from the lower Nacimiento Formation (Puercan), New Mexico, are described. These two upper dentitions are tentatively assigned to two species of *Oxyacodon* and this genus is transferred to the Anisonchinae (as was also done by Van Valen, 1978).

The majority of the condylarths described in the following section are referable, or at least questionably referable, to new taxa recently described by Van Valen (1978). He generously allowed me to make comparisons with the material he has been studying and provided an early draft of his descriptions. However, most of our conclusions regarding the phyletic relationships among the earliest condylarths were reached separately. This is particularly true in regard to early periptychid evolution. Also, as is shown by his figs. 2 and 7, he has in part accepted my unpublished views regarding the relationship of *Oxyprimus* to hyopsodontids (Van Valen, 1978). Although the future research may show both of us to be wrong regarding some of these relationships, the fact that basically similar views were reached by individuals with somewhat different perspectives helps strengthen the case for these phyletic relationships.

Family ARCTOCYONIDAE (Giebel, 1855) Murray, 1866

Subfamily OXYCLAENINAE (Scott, 1892) Matthew, 1937

Protungulatum Sloan and Van Valen, 1965

Comments. The UCMP Bug Creek Anthills collection (UCMP locs. V65127 and V70201) contains a number of mandibular and maxillary fragments reerable to *Protungulatum donnae.* The measurements for these specimens are given in Appendices 4 and 5. These specimens were used for comparisons in the following descriptions.

Protungulatum cf. *P. donnae* Sloan and Van Valen, 1965

(Fig. 52 & Table 40)

Referred specimens. 2 M_1's or M_2's; 2 M^1's; 2 M^2's; all from UCMP loc. V74111.

Locality. UCMP loc. V74111.

Distribution. Lowermost Tullock Formation, Montana (pre-mantuan)

Description. In the sample from loc. V74111 there are six isolated teeth which cannot be allocated to the three or four other species from loc. V74111 described on the following pages. These six teeth are morphologically most similar to *Protungulatum donnae* from Bug Creek Anthills.

M_x. Based on their morphology and size, UCMP 116498 and UCMP 116497 can be tentatively identified as an M_1 and M_2, respectively. They are similar in size to M_1's and M_2's of *Protungulatum donnae* from Bug Creek Anthills. The only possible morphological difference is a slight increase in talonid length of the molars from loc. V74111, indicated by the lower L–TR/L–TA ratio.

M^1. There are two M^1's from loc. V74111 that are slightly smaller than M^1's of

Protungulatum donnae. The only morphological difference is a very slight labial shift in the protocone in the loc. V74111 M^1's. (Note A/W-A in Table 40 and Appendix 4.)

M^2. Only one complete M^2 (UCMP 116501) and an M^2 fragment are referable to cf. *Protungulatum donnae.* UCMP 116501 is noticeably smaller (Table 40) than the six M^2's of *P. donnae* listed in Appendix 4. Like the M^1's, the only morphological difference between the one complete M^2 from loc. V74111 and M^2's of *P. donnae* is a slight labial shift in the protocone. (Note A/W-A in Table 40 and Appendix 4.)

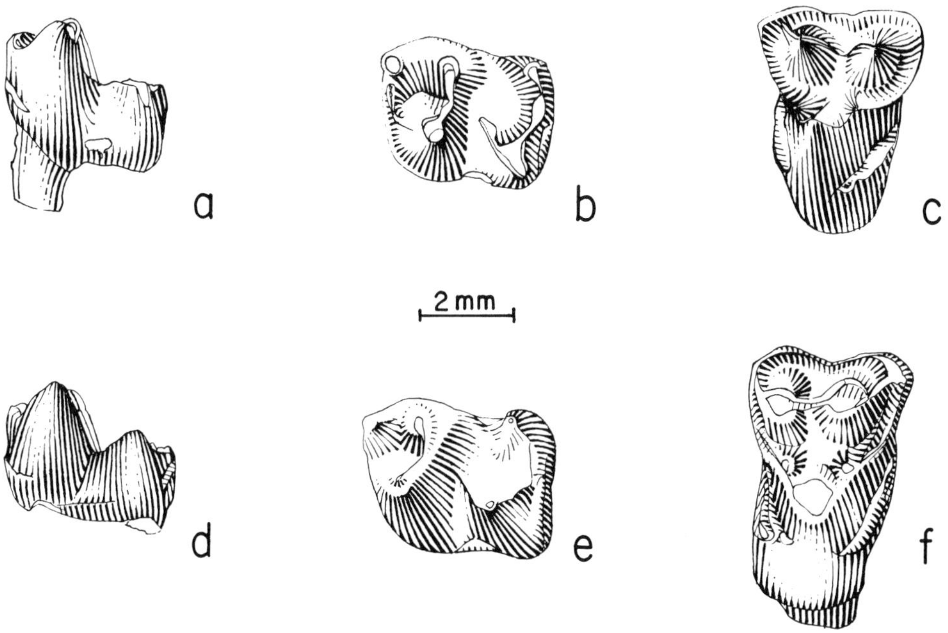

FIG. 52. *Protungulatum* cf. *P. donnae* Sloan and Van Valen. Isolated molars, loc. V74111. UCMP 116498, left M_1: (a) labial view; (b) occlusal view. UCMP 116499, left M^1: (c) occlusal view. UCMP 116497, left M_2: (d) labial view; (e) occlusal view. UCMP 116501, right M^2: (f) occlusal view. (A. L.)

Protungulatum mckeeveri, n. sp.

(Fig. 53 & Table 41)

Etymology. Named for Mr. Frank McKeever and family for the hospitality and help they gave our field crews.

Type. UCMP 121782, left dentary with P_4, M_{1-3}, only known specimen.

Type locality. UCMP loc. V72210 (=LACM loc. 3103).

Distribution. Lowermost Tullock Formation (mantuan).

TABLE 40. Measurements of isolated molars of *Protungulatum* cf. *P. donnae* from UCMP loc. V74111.

UCMP Number	*Position*	*L*	*W–A*	*W–P*	*A*	*B*	*C*	*D*	$\frac{L}{W-P}$	$\frac{L}{W-A}$	$\frac{A}{W-A}$	$\frac{B}{W-A}$	$\frac{C}{W-A}$	$\frac{D}{W-A}$
116499	M^1	3.79	4.63	4.94	1.92	0.42	1.43	0.86	0.77	0.82	0.41	0.09	0.31	0.19
116500	M^1	3.56	4.79	4.98	1.93	0.55	1.41	0.90	0.71	0.74	0.40	0.11	0.29	0.19
116501	M^2	3.64	5.33	5.55	2.16	0.63	1.47	1.07	0.66	0.68	0.41	0.19	0.28	0.20

UCMP Number	*Position*	*L*	*L–TR*	*L–TA*	*W–TR*	*W–TA*	*P–M*	$\frac{L-TR}{L-TA}$	$\frac{W-TR}{W-TA}$
116498	M_1	3.69	1.85	1.84	2.91	2.79	<0.10	1.01	1.04
116497	M_2	3.96	1.91	2.05	3.20	3.02	<0.10	0.93	1.06

Diagnosis. Similar in size to *Protungulatum donnae* and possibly *P. sloani,* but smaller than *P. gorgun*; in contrast to *P. donnae* the P_4, M_{1-2} have become lengthened relative to width, the talonid is slightly expanded (length and width), the length of the M_3 talonid is decreased relative to trigonid length, and the paraconid is slightly more reduced; the talonid is less expanded than in *P. sloani.*

Comments. The type of *Protungulatum mckeeveri*, UCMP 121782, was recovered in 1978 after the majority of the present study had been completed. Although in most aspects UCMP 121782 resembles *P. donnae*, the characters listed in the diagnosis indicate that the observed ranges of variation of *P. donnae* (which are known from large sample sizes) would have to be significantly modified to include this specimen. *P. mckeeveri* cannot be separated from *P. sloani* with as much certainty. Comparisons have not been made between specimens representing these two species; however, based on the photograph of the type of *P. sloani* (Van Valen, 1978, pl. 1, fig. 3), this species has noticeable more expanded talonid basins on M_{1-2} than are found in *P. mckeeveri*. It is not possible to tell from this photograh how the size of the paraconid on M_{1-2} compares to those in *P. donnae* or *P. mckeeveri*. I am reluctant to establish a new species based on a single specimen, but because the type is a dentary with associated teeth the observed differences between this specimen and previously described species of *Protungulatum* seem to warrant such a course.

Description. Since *Protungulatum mckeeveri* is quite similar to *P. donnae*, for the most part only the differences between these species will be described. The dentary, UCMP 121782, preserves the posterior alveolus for P_3 and complete, relatively unworn P_4, M_{1-3}. The corpus below these teeth is intact. Portions of the ascending ramus are also present but are poorly preserved. The anterior border of the ascending ramus is at about 73° to the alveolar border. This compares to an angle of about 77° in *P. donnae*. The length of M_{1-3} in UCMP 121782 is about 13 mm. This is almost identical to the M_{1-3} length of a specimen of *P. donnae* (UCMP 100896) noted in Appendix 5 but is larger than the 12.4 mm figure (mean plus one standard deviation) given by Sloan and Van Valen (1965) for a sample of *P. donnae* and the 10.7 mm figure given by Van Valen (1978) for the type of *P. sloani*. The measurements of the teeth in UCMP 121782 are all larger than the averages given for the sample of associated premolars and molars of *P. donnae* in Appendix 5, except for the P_4 width and the M_3 talonid length. (A statistical comparison is not warranted because of the small sample size of *P. donnae*, except possibly for M_2, where nine specimens are included.) Of the ten length measurements made of UCMP 121782 (one for the P_4 and three each for the molars), seven are greater than the observed ranges of *P. donnae* given in Appendix 5. Of the seven width measurements made of UCMP 121782 (one for the P_4 and two each for the molars) all fall within the observed ranges of *P. donnae* except for the width of P_4 which is below the observed range of *P. donnae*. Thus, on a qualitative basis, the P_4, M_{1-2} in UCMP 121782 appear to be longer relative to maximum width than in *P. donnae*. The one exception is the M_3 of UCMP 121782, where the total length is decreased by the relative reduction in talonid length of M_3. This is seen by the high trigonid length/talonid length ratio of this M_3 in comparison to M_3's of *P. donnae*. Also, the talonids of M_{1-2} in UCMP 121782 seem to be enlarged relative to the trigonids, in comparison to the M_{1-2} of *P. donnae* (based on a qualitative comparison of the trigonid length/talonid length ratios and trigonid width/talonid width ratios of M_{1-2}). In almost all other characters UCMP 121782 is indistinguishable from *P. donnae*. The one additional difference is in the paraconid. On the M_{1-3} of UCMP 121782 the paraconid is slightly smaller and slightly more approximated to the metaconid, but it is still lingual in position.

TABLE 41. Measurements of lower teeth in UCMP 121782, type, *Protungulatum mckeeveri*, n . sp., UCMP loc. V72210 (=LACM loc. 3103).

Position	*L*	*W*	*L–TR*	*L–TA*	*W–TR*	*W–TA*	*P–M*	*L–TR*/*L–TA*	*W–TR*/*W–TA*
P_4	3.75	2.10							
M_1	4.00		2.10	1.90	2.65	2.70	a0.20	1.11	0.98
M_2	4.40		2.25	2.15	3.30	3.05	a6.16	1.05	1.08
M_3	4.90		2.25	2.65	3.10	2.40	a0.12	0.85	1.30

a Approximate measurement.

Discussion. Separately, the above characters are insufficient to warrant a species-level distinction between *Protungulatum donnae* and UCMP 121782. However, combined they do suggest that UCMP 121782 is a separate species which in comparison to *P. donnae* has lengthened P_4, M_{1-2} relative to width, slightly expanded (length and width) the talonid, decreased the length of the M_3 talonid relative to trigonid length, and reduced the paraconid on M_{1-3}. It is possible that further study may show that *P. mckeeveri* is a junior synonym of *P. sloani*; however, based on a comparison with the photograph of the type specimen of *P. sloani* (Van Valen, 1978, pl. 1, fig. 3), *P. mckeeveri* has distinctly less expanded molar talonids. Also, the type of *P. mckeeveri* has an M_{1-3} length measurement of about 13 mm, in contrast to the 10.7 mm M_{1-3} length given by Van Valen (1978) for *P. sloani.* This may be significant in terms of a species distinction, but the difference in the measurements could be due to differences in measuring techniques. One final possibility is that *P. mckeeveri* could belong to a species of *Oxyprimus.* However, in a comparison with material referable to *O. erikseni* (described below), *P. mckeeveri* is distinctly larger and its molar paraconids are not as closely appressed to the metaconids. As reported by Van Valen (1978), the type of the largest species he referred to *Oxyprimus*, *O. albertensis*, has an M_2 length of 3.8 mm. The M_2 length in the type of *P. mckeeveri* is 4.40 mm.

Subfamily LOXOLOPHINAE Van Valen, 1978

Comments. Van Valen (1978) has proposed a new arctocyonid subfamily, the Loxolophinae. I am not convinced of the validity of this new subfamily, but because it represents the most recent view I have followed this usage.

Ragnarok Van Valen, 1978

Ragnarok harbichti Van Valen, 1978

(Fig. 54 & Table 42)

Ragnarok harbichti Van Valen, 1978, p. 56
Type. AMNH 35983, right mandible with M_{1-3} (Van Valen, 1978, pl. 3, fig. 4).
Type locality. Harbicht Hill, Hell Creek Formation, Montana.

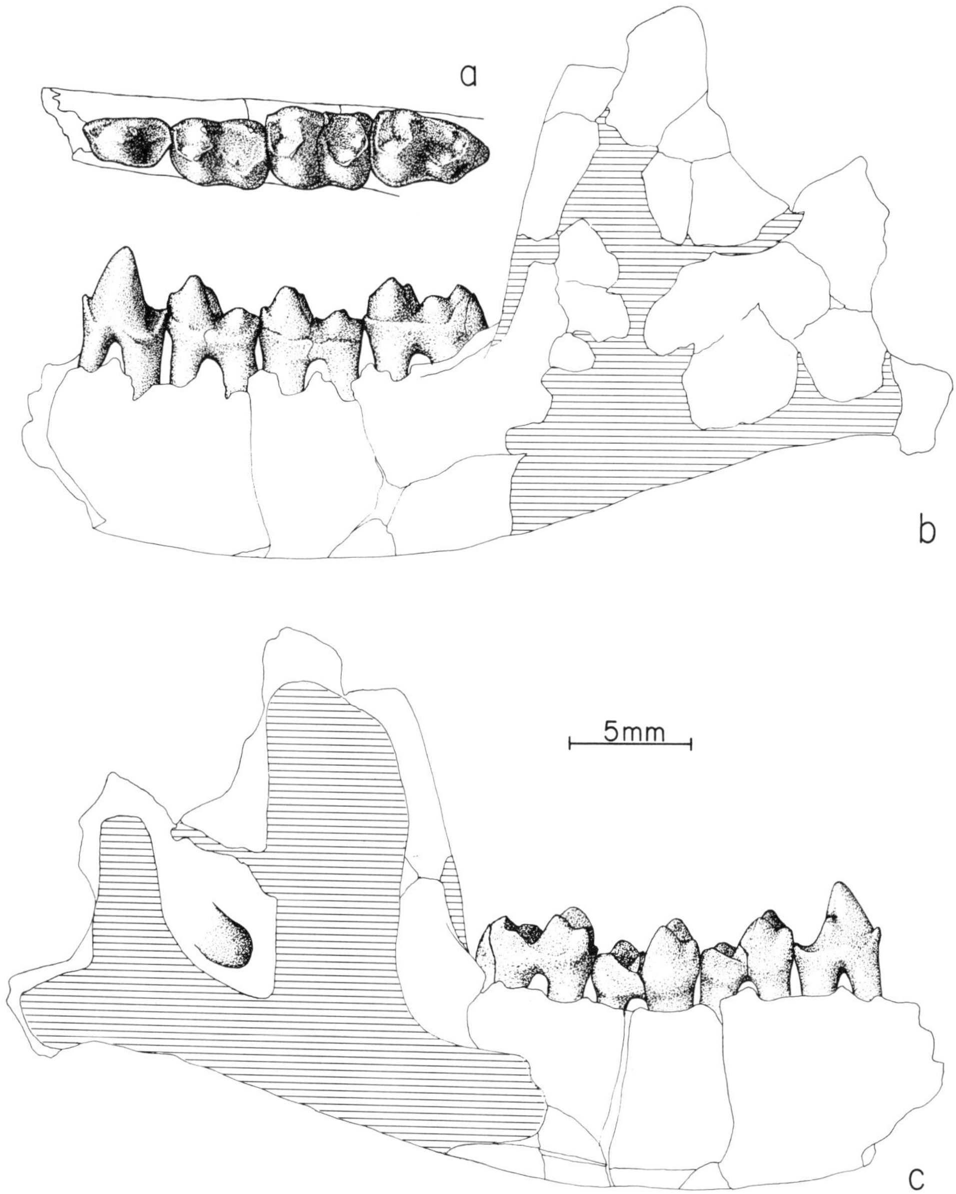

FIG. 53. *Protungulatum mckeeveri,* n. sp., type. Left dentary with P_4, M_{1-3}, UCMP 121782, loc. V72210 (= LACM loc. 3103): (a) occlusal view; (b) labial view; (c) lingual view. Lines indicate wax.

Referred specimens. 2 left molars, possibly M_{1-2}, questionably belonging to the same individual, UCMP 121791.

Locality. UCMP loc. V77129.

Distribution. Hell Creek Formation, Montana (Lancian and pre-mantuan), and possibly lowermost Tullock Formation (pre-mantuan; see discussion of *R. engdahli.*)

Diagnosis. Van Valen, 1978, p. 56.

Description. The material catalogued under UCMP 121791 includes a lower left molar and the posterior root of the preceding tooth held together by a dentary fragment, a larger isolated lower left molar, one definite edentulous dentary fragment, and two more questionable dentary fragments. None of the fragments can unequivocally be pieced together. The larger molar is almost certainly an M_2. The smaller molar may be the M_1 of the same individual or an M_2 of a smaller species. The former conclusion is supported by the fact that the dental wear and preservation of the teeth are similar, and the two teeth fit together moderately well. The latter conclusion is supported by the fact that the supposed M_1 seems slightly too small to go with the M_2, but comparisons of the size relationships of associated M_1's and M_2's in specimens referable to *Ragnarok harbichti* and *R. nordicum* suggest that the two molars could go together. The type of *R. harbichti* lacks the M_1 trigonid, but based on the photograph of the type specimen (Van Valen, 1978, pl. 3, fig. 4), the width of the M_1 talonid appears to be about 0.75 times the width of the M_2 trigonid and 0.85 times the width of the M_2 talonid. The comparable ratios for the supposed M_1 and the M_2 in UCMP 121791 are about 0.70 and 0.80. Comparisons were also made with the photograph of a dentary, PU 16720, referred to *R. nordicum* by Van Valen (1978, pl. 3, fig. 5) and another dentary, LACM 112902, that is referred to the same species. In LACM 112902 the M_1 is 0.86 times the length and 0.83 times the maximum width of the M_2. Based on the photograph of PU 16720, the comparable ratios are approximately 0.88 and 0.81. In UCMP 121791, if the two molars are M_{1-2}, the same relationships are 0.85 for length and 0.81 for maximum width. These various ratios suggest that the two molars catalogued under UCMP 121791 could belong to the same species and, less certainly, to the same individual.

In the diagnosis of *Ragnarok,* Van Valen (1978, p. 55) noted that the molars are moderately low-crowned and bulbous. Although this is not sufficient to differentiate *Ragnarok* from a number of other condylarths, it is adequate to distinguish the genus from other condylarths known from the Hell Creek and lowermost Tullock formations, Montana. In addition, *R. harbichti* is the largest condylarth known from the Hell Creek Formation and has (at least in M_2) the widest lower molars (particularly the trigonid) relative to length, except for a new species of *Ragnarok* described below.

The $M_{1?}$ and M_2 of UCMP 121791 are similar in morphology, although the $M_{1?}$ is distinctly longer relative to its width and has a relatively wider talonid. Both molars show apical wear and wear on the trigonid and talonid cristids, and the anterolabial corner of the $M_{1?}$ is damaged. In both molars the cusps are more bulbous (particularly the protoconid and metaconid on M_2) and the trigonid is relatively lower compared to the molars in all species of *Protungulatum*. The paraconid on both molars appears to be relatively lingual in position but could be slightly more labial than is seen in *Protungulatum*. The labial cingulum is very narrow and discontinuous.

Van Valen (1978) distinguished *Ragnarok harbichti* from *R. nordicum* by the smaller talonids in the former species. However, based on the information provided by Van Valen (1978), *R. harbichti* appears to be only slightly smaller or possibly the same size as *R. nordicum*. It is on the basis of the relatively smaller talonid that UCMP 121791 is referred to

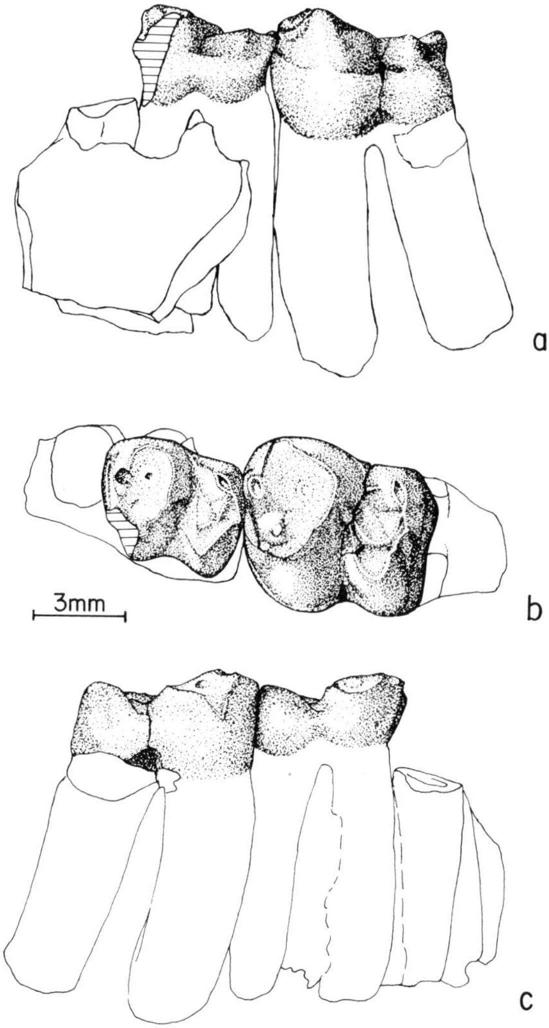

FIG. 54. *Ragnarok harbichti* Van Valen. UCMP 121791. Possibly associated left dentary fragment with posterior root for $P_{4?}$ and slightly damaged $M_{1?}$ and an isolated left M_2, loc. V77129: (a) labial view; (b) occlusal view; (c) lingual view. Lines indicate worn and broken surface.

R. harbichti. A dentary described just below has relatively wider molar talonids and thus has been referred to *R. nordicum.* Although it would be more desirable to have hypodigms for comparison, the limited material from Garfield County tends to support the species distinctions between *R. harbichti* and *R. nordicum*.

TABLE 42. Measurements of possibly associated lower teeth of *Ragnarok harbichti,* UCMP 121791, from UCMP loc. V77129.

Position	*L*	*L–TR*	*L–TA*	*W–TR*	*W–TA*	*P–M*	*L–TR* / *L–TA*	*W–TR* / *W–TA*
$M_{1?}$	5.10	2.80	2.30	3.95	3.85	0.15	1.22	1.03
M_2	5.90	3.25	2.65	5.60	4.90	0.10	1.23	1.14

Discussion. It should be noted that UCMP 121791 is the only mammalian specimen so far recovered from UCMP loc. V77129. As noted in the geology section, this locality occurs almost immediately below the lowest Z coal; thus, by definition the locality lies within the Hell Creek Formation. However, dinosaurs have not been recovered from the locality (Clemens, pers. comm.) and this is the basis for referring the locality to the earliest Paleocene (pre-mantuan). A laterally equivalent locality (loc. V77128) has produced more specimens and may be able to establish with more certainty the age of loc. V77129. The next specimen to be described is a dentary referable to *Ragnarok nordicum.* The specimen comes from a locality immediately overlying the lowest Z coal. Although I am not completely convinced, based on the material from Garfield County, that *R. harbichti* and *R. nordicum* are separate species, it should be emphasized that this material was referred to these two species of *Ragnarok* based on the morphological criteria recognized by Van Valen (1978) and not on the basis of stratigraphic position.

Ragnarok nordicum (Jepsen, 1930b) Van Valen, 1978

(Fig. 55 & Table 43)

Loxolophus nordicus Jepson, 1930b, p. 501

Type. PU 13285, right maxilla with M^{1-3} (Jepsen, 1930b, pl. IV, fig. 1).

Type locality. Mantua lentil, Polecat Bench Formation, Wyoming.

Referred specimen. Right dentary with P_3 alveoli, P_4, M_{1-2}, trigonid of M_3, LACM 112902.

Locality. LACM loc. 3103 (= UCMP loc. V72210).

Distribution. Lowermost Tullock Formation, Montana, and Mantua lentil, Polecat Bench Formation, Wyoming (both mantuan).

Diagnosis. Jepsen, 1930b, p. 501, plus Van Valen's comment (1978, p. 56) that the molar talonids of *Ragnarok harbichti* are smaller than those of *R. nordicum*.

Comments. Van Valen (1978) referred a dentary, PU 16720, to *Ragnarok nordicum,* and it is on this basis that the dentary, LACM 112902, is assigned to this species.

Description. Except for an M_3 of a ?periptychid from the same locality, LACM 112902 is the largest condylarth considered in this study. The dentary (LACM 112902) is moderately deep and robust relative to the size of the teeth. One mental foramen is preserved below the anterior root of P_4. The P_4 is noticeably shorter than the M_1 (approximately 87 percent of the M_1 length). As in the molars, the occlusal surface of the P_4 is quite worn by dental attrition and postdepositional damage. Also like the molars, but to a lesser degree, the P_4 shows some inflation. All three trigonid cusps are easily discernible in spite of the amount of wear. The preserved base of the metaconid suggests that this cusp was probably relatively larger than its counterpart in *Protungulatum donnae*. The paraconid is moderately large but was probably slightly smaller than the metaconid, and is positioned low and lingual to the midline. There is a very narrow, discontinuous labial cingulum at the antero- and posterolabial corners of the P_4. None of the occlusal surface of the talonid is preserved, but it was probably basined. The talonid is quite short but is close to two-thirds of the width of the trigonid at its widest point.

The occlusal surfaces of the molars preserve few details; however, the M_2 retains enough of the trigonid cusps to show that the paraconid is almost lingual in position, unreduced, and more closely appressed to the metaconid than in *Protungulatum donnae*. A small piece of paraconid preserved on the M_3 trigonid and the occlusal outline of the M_1 trigonid suggest a similar placement of the paraconid. Other than these cusps, no details are preserved on the occlusal surfaces. The molars are quite bulbous and when unworn, the trigonids were not as high relative to the talonids as in the *P. donnae*. On the M_2 a narrow labial cingulum runs from the anterolabial margin of the protoconid to below the cristid obliqua. The labial cingulum is almost nonexistent on M_1 and is discontinuous on M_3.

TABLE 43. Measurements of associated lower teeth of *Ragnarok nordicum*, LACM 112902, from LACM loc. 3103 (= UCMP loc. V72210).

Position	*L*	*W*	*L-TR*	*L-TA*	*W-TR*	*W-TA*	*P-M*	*L-TR*/*L-TA*	*W-TR*/*W-TA*
P_4	4.62	3.38							
M_1	5.29		a2.73	a2.56	4.08	4.15	<0.10?	1.07	0.98
M_2	6.13		3.27	2.86	5.47	5.09	0.18	1.14	1.07
M_3	—		a2.96	—	a4.42	—	—	—	—

a Approximate measurement.

Discussion. Although LACM 112902 is damaged, enough details are preserved to allow comparison with other condylarths. LACM 112902 was compared with PU 16720, a dentary from Mantua lentil that Van Valen has referred to *Ragnarok nordicum* (Van Valen, 1978, pl. 3, fig. 5). LACM 112902 was also compared with various arctocyonids such as *Oxyclaenus, Loxolophus,* and *Protogonodon* (a junior synonym of *Loxolophus*, according to Van Valen, 1978). Based on these comparisons, LACM 112902 most resembles *R. nordicum* and has been referred to this taxon. There are differences which are noted below, but based on limited sample sizes such differences could be attributable to intraspecific variation.

In comparing LACM 112902 to *Ragnarok nordicum* (i.e., PU 16720), LACM 112902 is slightly larger in most tooth dimensions (averaging approximately 6 percent larger) and has a slightly more robust dentary. In molar proportions, LACM 112902 appears to have

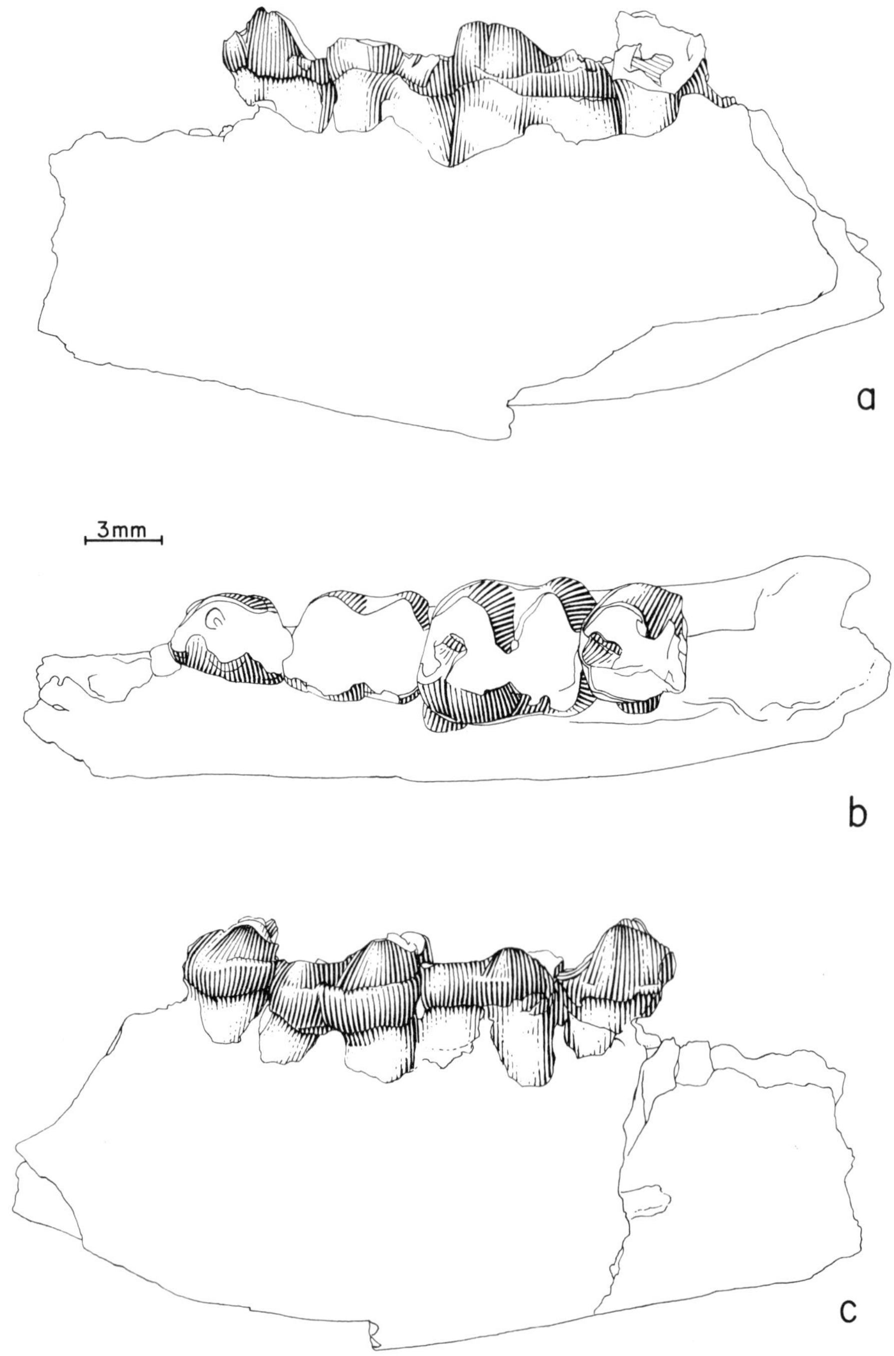

FIG. 55. *Ragnarok nordicum* (Jepsen). LACM 112902. Right dentary with P_4, M_{1-2}, M_3 trigonid, LACM loc. 3103 (=UCMP loc. V72210): (a) lingual view; (b) occlusal view; (c) labial view. (A. L.)

slightly narrower but longer talonids on M_{1-2} (talonid is lacking on M_3 of LACM 112902) relative to the trigonids. However, in comparing molar length to trigonid width, the M_{1-2} of LACM 112902 are distinctly wider. Thus, the apparent narrowness of the talonids in LACM 112902 is a function of the general increase in trigonid width relative to molar length in this specimen. The talonids are also wider relative to molar length in LACM 112902, but to a much lesser degree. This same relationship is present between the P_4 of LACM 112902 and PU 16720 (*R. nordicum*). The P_4 in the former specimen is distinctly wider relative to length (i.e., it is more massive). The slightly more massive (greater width relative to length) overall appearance of LACM 112902 may be enough to warrant separation at the specific level. However, with only two specimens to compare, I prefer to regard this as morphologic and geographic (LACM 112902 is from Montana and PU 16720 is from Wyoming) variation in one species.

LACM 112902 was also compared to the $M_{1?}$ and M_2 (UMCP 121791) previously referred to *Ragnarok harbichti*. LACM 112902 does have distinctly larger talonids. This was the criterion for Van Valen's (1978) separation of *R. nordicum* from *R. harbichti*, and on this basis LACM 112902 was referred to the former species and UCMP 121791 was referred to the latter. In most other details LACM 112902 and UCMP 121791 are very similar.

In a comparison with other early Paleocene arctocyonids, the molars in LACM 112902 are quite primitive in most aspects, except for their inflation, their slight increase in relative talonid size, and their closer approximation of the paraconid and metaconid. The inflation of the molars is shared with most other oxyclaenines (and loxolophines, in the sense of Van Valen, 1978). Based on Matthew's (1937) description and specimens in UCMP collections from the Nacimiento Formation, LACM 112902 most resembles *Loxolophus hyattianus* and *Protogonodon pentacus* in the degree of talonid expansion on M_{1-2}. The P_4 of LACM 112902 is distinctly different from these two species because of an enlarged metaconid and moderately large, lingually positioned paraconid. In this last feature there is some resemblance to *Loxolophus*. The morphology of the P_4 and the comparatively smaller (especially narrower) molar talonids in LACM 112902 are probably primitive retentions from *Protungulatum donnae*. However, *Ragnarok* does appear to be morphologically intermediate between *Protungulatum* and some other arctocyonids, particularly *Loxolophus*, and thus supports at least the earlier portion of the provisional phylogeny presented by Van Valen (1978, fig. 3).

Ragnarok engdahli, n. sp.

(Fig. 56 & Tables 44 & 45)

Etymology. Named for the Lester and Robert Engdahl families, whose help and friendship inestimably aided this project.

Type. UCMP 116541, isolated right molar.

Type locality. UCMP loc. V74110, lowermost Tullock Formation, Montana.

Referred specimens. 2 P_4's; 1 $M_{1?}$; 1 M_2; 1 M_3; 1 M^2 from UCMP loc V74110. Questionably, 1 M^3 from UCMP loc. V74111.

Localities. UCMP locs. V74110 and, questionably, V74111.

Distribution. Lowermost Tullock Formation, Montana (pre-mantuan).

Diagnosis. Smallest species of *Ragnarok*; P_4 metaconid is small (it is large in the only species, *R. nordicum*, in which the P_4 has been figured); M_2 talonids approximately similar in relative size to those of *R. nordicum* but relatively larger than those of *R. har-*

bichti; M^2 most similar to M^2 of *R. nordicum* (upper molars of *R. harbichti* have not been figured), but in *R. nordicum* protoconal region is more expanded and paracone and metacone are closer; as in *R. nordicum*, M^2 lacks lingually projecting hypocone of *R. wovokae.*

Comments. Among the sample of condylarths from the Hell's Hollow local fauna (locs. V74110 and V74111) there is a small group of isolated teeth that cannot be referred to any of the other species represented in the local fauna, nor can the specimens be referred to any other published species of condylarth without significantly altering the definitions of these taxa. Therefore, a new species is established for these specimens, but not without doubt as to its validity. The rationale and the doubts concerning this taxon are discussed below.

Description. P_4. The protoconid on P_4 is tall and bulbous, and is relatively larger than in P_4's of the other condylarths from the Hell's Hollow local fauna. The paraconid is distinct, but is low compared to the protoconid, and is lingual to the midline. A metaconid is present, but is very small and closely appressed to the protoconid. The talonid is reduced in length and width compared to *Protungulatum donnae.* There is a ridge running from a small posterior talonid cusp to the posterior side of the protoconid. The ridge makes a slight labial bend near its center. A second small ridge runs lingually from the posterior talonid cusp. These two structures, plus a narrow internal cingulum below the talonid, form a very small lingually sloping basin. A very narrow labial cingulum is present below the paraconid and the anterior margins of the protoconid and the talonid. This morphology bears a striking resemblance to that seen in the P_4 in *Protogonodon pentacus.* Except for its much larger size, the only differences are a smaller paraconid, a complete but very narrow labial cingulum, and the crenulation of the enamel in the P_4 of *P. pentacus.*

Lower molars. The molars are larger and somewhat more bulbous than in *Protungulatum donnae.* The paraconids appear to be slightly more labially placed (based on two M_2's and one M_3) in M_{2-3}. In a molar questionably identified as an M_1 the paraconid is lingually placed (i.e., in line with the metaconid and entoconid). The paraconid and metaconid are more closely appressed, particularly on M_2, than on the molars of any species of *Protungulatum.* The M_{1-2} are wider relative to length than are those of species of *Protungulatum.* Also, compared to *P. donnae* the talonids of M_{1-2} are relatively longer and the talonid of M_3 is relatively shorter. The talonids of M_{1-3} also appear to be relatively wider, primarily because of an enlarged hypoconid. The trigonids are slightly lower, as in other species of *Ragnarok, Loxolophus,* and *Oxyclaenus,* but not as low as in *Protogonodon.*

Upper molars. Only two upper molars from the Hell's Hollow local fauna are allocated to this taxon. An M^2, UCMP 116537, loc. V74110, is very similar to *Ragnarok nordicum* (Jepsen, 1930b) except for the former tooth's slightly smaller size. Jepsen (1930b) notes the presence of a slight enlargement in the mesostylar region of the ectocingulum which is not present on UCMP 116537, although on UCMP 116537 the portion of the ectocingulum labial to the metacone is slightly enlarged. Based on Jepsen's figure of *R. nordicum* (1930b, pl. IV, fig. 1), the stylar shelf appears to be slightly wider, the paracone and metacone are more widely spaced, and the protoconal region is less expanded in UCMP 116537.

The M^3, UCMP 116544, is the only tooth from loc. V74111 that has been allocated to this taxon. Most of the crown morphology has been obliterated by dental wear, making close comparison with other teeth impossible. However, narrow pre- and postcingula are

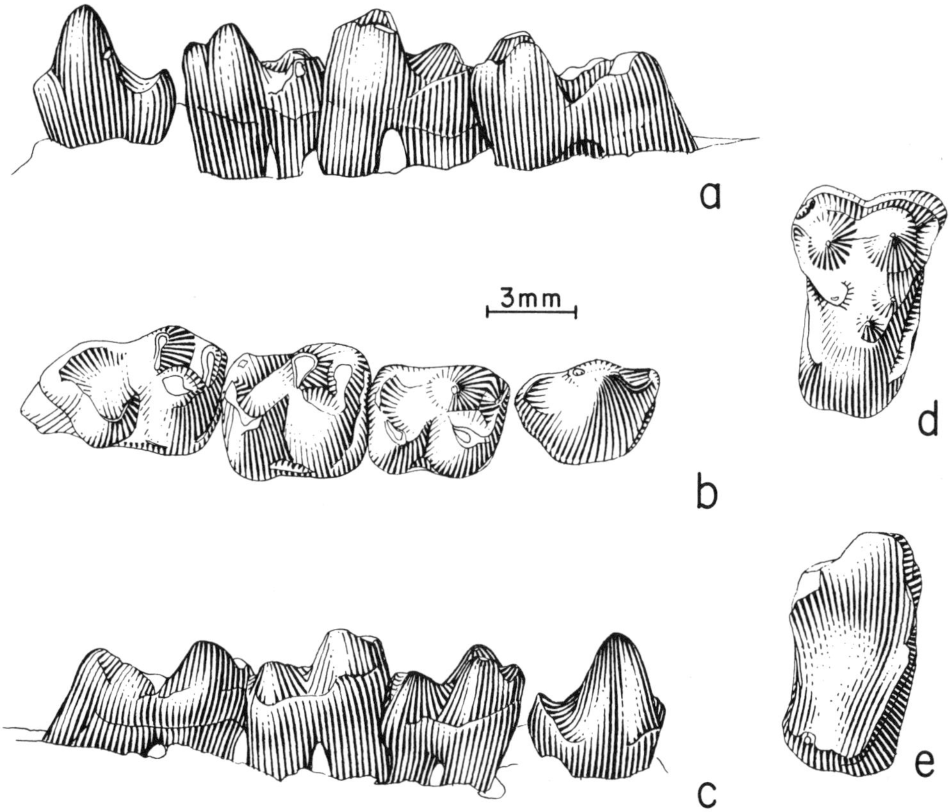

FIG. 56. *Ragnarok engdahli*, n. sp. Isolated lower teeth, loc. V74110. UCMP 116538, right P_4. UCMP 116540, right M_1. UCMP 116541, right M_2, type. UCMP 116543, right M_3: (a) lingual view; (b) occlusal view; (c) labial view. Upper molars, UCMP 116537, loc. V74110, right M^2: (d) occlusal view; UCMP 116544, loc. V74111, right M^3: (e) occlusal view. (A. L.)

preserved. The tooth is quite anteroposteriorly short and transversely wide. This tooth is smaller anteroposteriorly than UCMP 116537, but is wider transversely. The latter relationship is not typical in condylarths and suggests that this M^3 may belong to another, larger species.

Discussion. All the isolated teeth referred to *Ragnarok engdahli* have confidently been identified to their correct tooth site except for the M_1. However, the same cannot be said regarding their referral to a single taxon. This is particularly true for the worn M^3, UCMP 116537. The tooth is too large to be confidently referred to this species; it is equally probable that it belongs to a larger species such as *R. harbichti*, or less likely to *R. nordicum*, or to an as yet unrecognized taxon. The questionable M_1, UCMP 116540, could also be an M_2 based on the proportions of the crown. If this is the case it probably does not belong to *R. engdahli*, considering its different trigonid morphology. It could belong to the largest species of *Protungulatum, P. gorgun*, or possibly to *P. mckeeveri*. The general crown morphology of UCMP 116540 does not preclude either of these possibilities, but, as will be discussed below, the relatively greater width of UCMP 116540 suggests that it does not belong to any recognized species of *Protungulatum*.

The M^2 and particularly the M_2 suggest that this species is referable to *Ragnarok*. Based on a comparison with the figure of the type specimen of *R. nordicum* (Jepsen, 1930b, pl. IV, fig. 1), the M^2 of *R. engdahli* shares several features with those of *Protungulatum donnae* and *R. nordicum*. The single M^2, UCMP 116537, of *R. engdahli* is intermediate in size between *P. donnae* and *R. nordicum*. It shares with *P. donnae* a less closely spaced paracone and metacone and a less transversely widened crown, while its slightly more expanded protoconal region approaches the condition seen in *R. nordicum*.

The M_2 of *R. engdahli* more closely resembles the M_2 of *Ragnarok*, particularly *R. nordicum*, than it does any species of *Protungulatum*. As in both *R. nordicum* and *R. harbichti*, the paraconid and metaconid are relatively closely appressed. Although the M_2 of *R. engdahli* is smaller, it has, like these other two species of *Ragnarok*, a relatively wider crown. This relationship is shown in Table 45. The length to trigonid width (L/W–TR) and length to talonid width (L/W–TA) ratios of M_{1-2} were compared in the species of *Protungulatum* and *Ragnarok*. As can be seen in Table 45, the species of *Ragnarok*, including *R. engdahli*, almost consistently have wider M_{1-2}'s relative to length. The one notable exception is the quite narrow trigonid in the M_1 of PU 16720, *R. nordicum*. Sample sizes are too small to warrant meaningful statistical comparisons; however, assuming that the sample of *P. donnae* ($N=6$ for M_1; $N=9$ for M_2) approximates the range of variation observable for one species, none of the L/W–TR and L/W–TA ratios for the six M_1's overlapped those for the species of *Ragnarok* (except the L/W–TR of PU 16720, *R. nordicum)*. Five of the nine L/W–TR ratios for M_2's of *P. donnae* equaled or were less than the maximum L/W–TR ratio recorded for a species of *Ragnarok* (PU 16720, *R. nordicum)*. Only two of nine L/W–TA ratios for M_2's of *P. donnae* equaled or were less than the maximum L/W–TA ratio recorded for a species of *Ragnarok* (AMNH 35987, *R. harbichti*). The qualitative conclusion is that the relatively wider molar crown (M_{1-2}) distinguishes *Ragnarok* (including *R. engdahli*) from *Protungulatum*.

The $M_{1?}$ UCMP 116540, referred to *Ragnarok engdahli* has a relatively wider crown than M_1's referred to species of *Protungulatum*. If UCMP 116540 is an M_1 it probably belongs to *R. engdahli*. However, the tooth is slightly unusual for an M_1 in having the trigonid wider than the talonid. This relationship does occur in M_1's, particularly in *P. donnae*, but is more characteristic of M_2's.

Like the M^2, the two P_4's referred to *Ragnarok engdahli* are intermediate in morphology between *Protungulatum donnae* and *R. nordicum*. In *R. engdahli,* as in *P. donnae*, the P_4 metaconid is small, while in *R. nordicum* it is noticeably larger. The condition in the former two taxa is assumed to be the primitive condition. In contrast, both *R. engdahli* and *R. nordicum* show considerable inflation of the P_4, a derived condition.

The evidence, particularly from the M_2, clearly indicates that a species of *Ragnarok* is present in the Hell's Hollow local fauna. The morphology of the M_2 most closely resembles *R. nordicum* (i.e., it has a relatively larger talonid). The specimens from Hell's Hollow local fauna could represent extremely small specimens of *R. nordicum*; however, this seems unlikely when the morphology of the presumed P_4 and M^2 of *R. engdahli* are also considered. In addition, the presence of a large M^3 (UCMP 116544), questionably referred to *R. engdahli*, suggests that a larger taxon such as *R. nordicum* or *R. harbichti* is probably in the Hell's Hollow local fauna.

TABLE 44. Measurements of teeth of *Ragnarok engdahli*, n. sp. All teeth are from UCMP loc. V74110, except for UCMP 116544, which is from UCMP loc. V74111.

UCMP Number	*Position*	*L*	*W–A*	*W–P*	*A*	*B*	*C*	*D*	*L/W–P*	*L/W–A*	*A/W–A*	*B/W–A*	*C/W–A*	*D/W–A*
116537	M^2	4.75	6.60	6.84	2.78	0.64	1.87	1.31	0.69	0.72	0.42	0.10	0.28	0.20
116544	M^3	4.11	7.25	a6.08	—	—	—	—	0.68	0.57	—	—	—	—

a Approximate measurement.

UCMP Number	*Position*	*L*	*W*	*L–TR*	*L–TA*	*W–TR*	*W–TA*	*P–M*	*L–TR/L–TA*	*W–TR/W–TA*
116538	P_4	4.56	3.47							
116539	P_4	4.91	3.10							
116540	$M_{1?}$	4.43		2.25	2.08	3.68	3.45	<0.10	1.08	1.07
116541 (type)	M_2	4.79		2.42	2.37	4.15	3.81	0.22	1.02	1.09
116542	M_2	4.75		2.41	2.34	3.96	3.80	0.40	1.03	1.04
116543	M_3	5.99		2.77	3.22	3.88	3.16	0.15	0.86	1.23

TABLE 45. Comparisons of molar length to widths in M_{1-2} of all species of *Protungulatum* and *Ragnarok* for which lower dentitions are known.

	M_1		M_2	
	L/W–TR	*L/W–TA*	*L/W–TR*	*L/W–TA*
Protungulatum donnae[1]	1.40	1.41	1.24	1.36
Protungulatum gorgun[2]	—	—	1.45	1.49
Protungulatum mckeeveri	1.51	1.48	1.33	1.44
Protungulatum sloani[3]	1.44	1.35	1.29	1.29
Ragnarok harbichti				
(type, AMNH 35987)[3]	—	—	1.14	1.29
(UCMP 121791)	1.29	1.32	1.05	1.20
Ragnarok nordicum				
(PU 16720)[4]	1.47	1.29	1.22	1.22
(LACM 112902)	1.29	1.27	1.12	1.20
Ragnarok engdahli[5]	1.20	1.28	1.18	1.26

[1] Averages determined from the specimens recorded in Appendix 5.
[2] Determined from measurements given by Van Valen (1978, p.52). (Note: Van Valen gives a different value for the length of M_2 on p. 73, caption; using this figure the ratios given above would be 1.42 and 1.46.)
[3] Determined from measurements given by Van Valen (1978, p. 56).
[4] Estimated from Van Valen (1978, pl. 3, fig. 5).
[5] Ratios for M_2 are averages from UCMP 116541 and UCMP 116542. See text for discussion.

Family ?HYOPSODONTIDAE (Trouessart, 1879) Lydekker, 1889
Subfamily ?HYOPSODONTINAE (Trouessart, 1879) Matthew, 1937

Comments. One of the genera named by Van Valen (1978), *Oxyprimus*, shares with hyopsodontids certain characters which I interpret to be derived. In unpublished comments I said that *Oxyprimus* was probably related to hyopsodontids, but I was much less certain about the affinities. However, combining these earlier data (shown by Van Valen, 1978, figs. 2 and 7) with Van Valen's descriptions of new mantuan species of *Oxyprimus*, I feel the case is clearer for referring this genus to the Hyopsodontidae, rather than the Arctocyonidae, as was done by Van Valen (see discussion.)

The genus is only questionably referred to the family for several reasons. First, although Van Valen (1978, p. 60) has accepted my views concerning this genus, he apparently is not sure whether all the species he referred to *Oxyprimus* can possibly be placed in the Hyopsodontidae. Since I have only studied in detail the material from Garfield County, I cannot make a more definitive statement regarding the genus as a whole. Second, the Hyopsodontidae (and Mioclaenidae) have not been recently reviewed and the tentative comparisons of *O. erikseni* with hyopsodontids are based largely on the older literature. However, as will be discussed below, the morphology of *O. erikseni* suggests a closer affinity to hyopsodontids than to mioclaenids. This suggests an early divergence of hyopsodontids and mioclaenids and thus tends to support Van Valen's (1978) resurrection of the Mioclaenidae as a separate family.

Oxyprimus Van Valen, 1978
Oxyprimus erikseni Van Valen, 1978
(Figs. 57–59 & Table 46)

Oxyprimus erikseni Van Valen, 1978, p. 53

Type. UMVP 1561, right M_1 (Van Valen, 1978, pl. 1, fig 5).

Type locality. Harbicht Hill, Hell Creek Formation, Montana.

Referred specimens. 1 DP^4; 1 $M^{1?}$; 1 $M^{2?}$; 1 M^3; left mandible with P_{3-4}, M_{1-3}, UCMP 116503; 1 P_3; 2 P_4's; 1 M_1; 2 M_2's; 1 M_3 fragment; and isolated molar fragments; all from UCMP loc. V74111. 1 $M^{2?}$ from UCMP loc. V74110.

Localities. UCMP locs. V74110 and V74111.

Distribution. Hell Creek Formation (Lancian) and lowermost Tullock Formation (pre-mantuan), Montana.

Diagnosis. Van Valen, 1978, p. 53.

Comments. In his description of *Oxyprimus erikseni,* Van Valen (1978) only noted the type specimen, an M_1. The referral of the preceding specimens to *O. erikseni* was based on a comparison with specimens under study by Van Valen. However, sample sizes are limited and it is possible that the material from locs. V74111 and V74110 could belong to another species of *Oxyprimus* from Mantua lentil.

Description. *Mandible.* One mandibular fragment, UCMP 116503, preserves the posterior root of P_2, P_{3-4}, and M_{1-3} (Fig. 57). Anterior to P_3 little of the mandible is preserved; however, posterior to M_3 all but portions of the condyle and angular process are present. The jaw depth is comparable to that seen in *Protungulatum donnae* (see Sloan and Van Valen, 1965, fig. 7). UCMP 116503 is relatively narrower (labiolingually) than in *P. donnae.* In the former there is a slight but definite S-curve to the corpus when viewed dorsally, with the posterior portion bowing lingually and the anterior portion bowing labially. One mental foramen is preserved below and slightly posterior to the anterior root of P_3. Of the two jaw fragments of *P. donnae* in the UCMP Bug Creek Anthills collection that preserve the posterior mental foramen, one (UCMP 112002) has the foramen in a position similar to that of UCMP 116503. The other jaw fragment (UCMP 112110) preserves the posterior mental foramen below the posterior root of P_3, as in the holotype of *P. donnae* (SPSM 62-2028, Sloan and Van Valen, 1965, fig. 7). The holotype of *P. donnae* does not preserve the complete ascending ramus; however, the portion present shows a great similarity to the complete ascending ramus of UCMP 116503. In UCMP 116503 the ascending ramus is quite high, rising about 12 mm above the alveolar border, and is at about a 73° angle to the latter (compared to an angle of about 77° in *P. donnae*). The ascending ramus terminates dorsally at a straight margin set at right angles to the anterior margin. As in *P. donnae*, the condyle is distinctly higher than the molar apices. The dorsal aspect of the condyle is not preserved in UCMP 116503, but it presumably resembled the convex, ovoid surface seen in *P. donnae.* Most of the angular process is not preserved in UCMP 116503, but the portion present indicates that the angular process extended posteriorly beyond the level of the condyle, as in *P. donnae*, and that is was not inflected. The large, triangularly shaped masseteric fossa covers most of the labial surface of the ascending ramus. The bone is very thin in most of the fossa, which is bounded anteriorly by a thickened ridge. Posteroventrally, the fossa gradually shallows and is bordered by a low ridge which continues posteriorly toward the angular process. Posterodorsally, the fossa is bounded by a low ridge running from the condyle toward the anterodorsal aspect of the ascending ramus. There are six to eight minuscule foramina in the anteroventral corner of the fossa. On the lingual side of the mandible, anteroventral to the condyle, there is a large mandibular foramen set in a teardrop-shaped depression.

In contrast to the mandible of *Procerberus formicarum* (UF 22896) previously described (Fig. 51), the articular condyles of *Oxyprimus erikseni* and *Protungulatum donnae* are relatively higher than and closer (anteroposteriorly) to the tooth row. The condition seen in *P. donnae* and *O. erikseni* is regarded as the derived state.

P_{3-4}. The two-rooted P_3 of *Oxyprimus erikseni* is similar to the P_3 of *Protungulatum donnae*, although like the other teeth of *O. erikseni* it is distinctly smaller and less robust. There is a prominent central cusp, a much lower anterior cusp, and a posterior heel bearing a small cusp at its posterior edge. The two-rooted P_4 of *O. erikseni* is almost as long as the P_4 of *P. donnae*, but is distinctly narrower. Based on one specimen, UCMP 116503, the P_4 of *O. erikseni* is longer than the M_1, while in *P. donnae* the two teeth are subequal in length. As in *P. donnae*, the protoconid of P_4 is the highest and largest cusp. Unlike *P. donnae*, the metaconid of *O. erikseni* is taller and more prominent. The paraconid is also larger but is still well below the level of the other two trigonid cusps. The paraconid is lingual to the midline of the P_4, possibly more so than in *P. donnae*. The P_4 talonid of UCMP 116503 is heavily worn, but two other P_4's (UCMP 116505 and UCMP 116506) show that the talonid is wide and trenchant, with one posterior cusp and a "cristid obliqua" crest running along the midline of the talonid. The wear facet on the talonid of the P_4 in UCMP 116503 has obliterated all surface features and forms a steeply inclined surface sloping labially. The trigonid is less worn, with facets on the apices of the protoconid and metaconid and along the anterior margin of the paralophid.

There is usually a slight disatema between the canine-P_1, P_{1-2}, and P_{2-3} in *Protungulatum donnae.* Neither P_1 nor P_2 is preserved in UCMP 116503; however, the close placement of the posterior root of P_2 to P_3 suggests that this diastema may not have been present in *Oxyprimus erikseni.*

Lower molars. The lower molars of *Oxyprimus erikseni* do not differ greatly from those of *Protungulatum donnae* except for the smaller size and less bulbous outline in the former species. The less bulbous shape may well be due to allometry. Based on the very few molars of *O. erikseni* (a maximum of three complete molars at any molar site), this species may have relatively wider and longer talonids than *P. donnae*. The exception is the talonid of the one complete M_3 in UCMP 116503, which is relatively shorter than any of the M_3's in the jaws of *P. donnae* shown in Appendix 5. *P. donnae* generally has an incomplete labial cingulum anterior to the paraconid, in the hypoflexid region, and posterior to the hypoconid. The labial cingulum in the hypoflexid region may carry a small ectostylid. In *O. erikseni* this labial cingulum is usually reduced to a narrow ledge only anterior to the paraconid. In *P. donnae* the paraconids of M_{2-3} are lingual to the midline and are aligned with the metaconid and entoconid. In M_1 the paraconid is also lingual to the midline but is labial to a line through the metaconid and entoconid. Too few molars of *O. erikseni* are represented in the loc. V74111 sample to allow a close comparison; however, the available information suggests that *O. erikseni* is generally similar to *P. donnae* in paraconid size. The paraconids in *O. erikseni* are more closely appressed to the metaconids.

Upper molars. No upper molars or premolars have been found in association with lower molars at loc. V74111 or loc. V74110. Based on morphology and size, one DP^4 and four isolated molars from these two localities are assigned to *Oxyprimus erikseni.* One specimen, UCMP 116511, loc. V74111, may be an M^1 of *O. erikseni.*The most distinctive feature of this tooth is its narrowness in the transverse dimension giving it a distinct

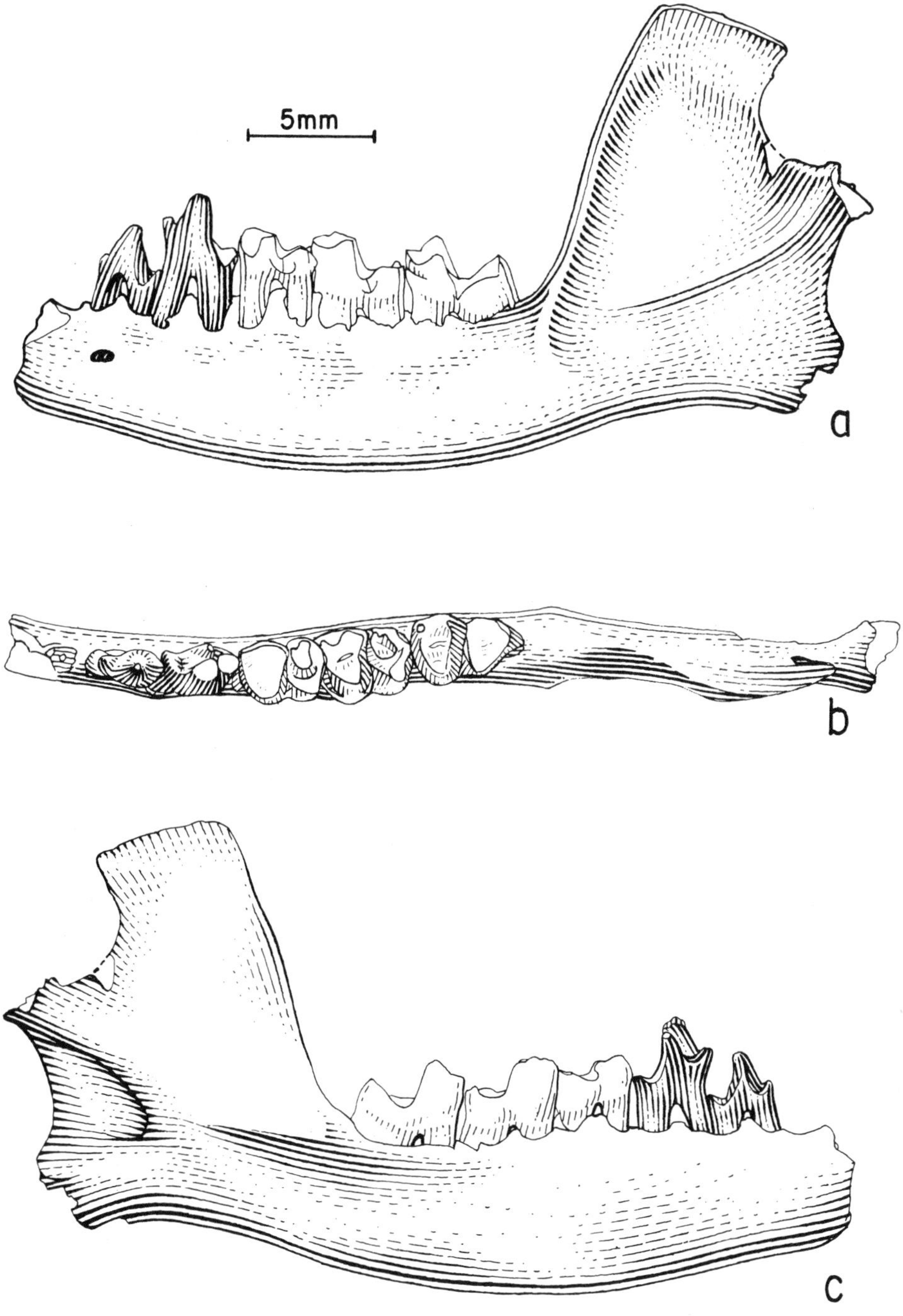

FIG. 57. *Oxyprimus erikseni* Van Valen. UCMP 116503. Left dentary with P_{3-4}, M_{1-3}, loc. V74111: (a) labial view; (b) occlusal view; (c) lingual view. (A. L.)

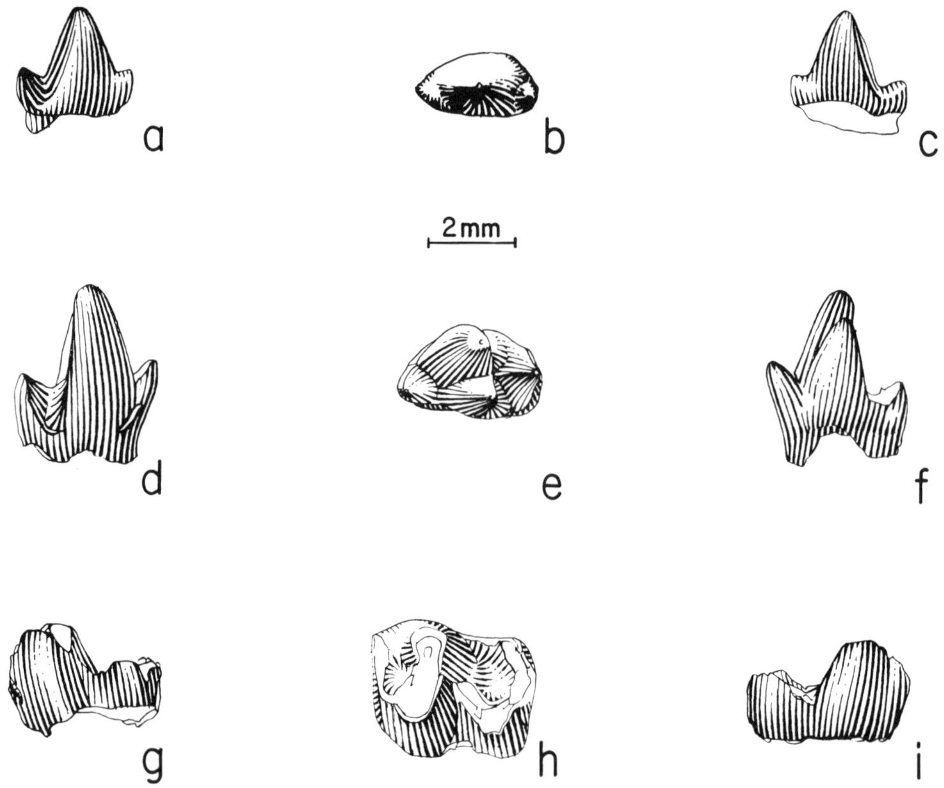

FIG. 58. *Oxyprimus erikseni* Van Valen. Isolated lower premolars and molars, loc. V74111. UCMP 116504, right P_3: (a), (b), & (c). UCMP 116505, right P_4: (d), (e), & (f). UCMP 116508, left M_2: (g), (h), & (i). Labial view: (a), (d), & (g); occlusal view: (b), (e), & (h); lingual view: (c), (f), & (i). (All except b, A.L.)

triangular outline. It also possesses a large cingular hypocone, unlike M^1's of *Protungulatum donnae.* The postparaconule and premetaconule wings have been accentuated by wear but may have been prominent in an unworn condition. This differs from the small or absent postparaconule and premetaconule wings on the M^1's of *P. donnae.* Other than these differences, UCMP 116511 resembles M^1's of *P. donnae.* This tooth shows an even stronger resemblance to four isolated teeth in the CM Bug Creek Anthills collection which are probably DP^4's of *P. donnae* (CM 37161 through CM 37164). Like UCMP 116511, these DP^4's are narrow transversely and triangular in outline, and they possess large cingular hypocones. It is possible that UCMP 116511 is a DP^4 of *O. erikseni*, *P.* cf. *P. donnae*, or *Mimatuta morgoth* (described just below). Unlike these four DP^4's, UCMP 116511 possesses roots. Further, its coloration does not differ from other teeth from loc. V74111, as is sometimes the case for deciduous teeth. Another tooth, UCMP 116512, from loc. V74111 even more strongly resembles the DP^4's mentioned above. More like the DP^4's but less like UCMP 116511, UCMP 116512 is quite narrow transversely, has a well-developed, anterolabially positioned parastylar region but a narrow stylar shelf labial to the paracone and a more widely spaced paracone and metacone, and it lacks roots. The narrowness of UCMP 116512 can clearly be seen in the L/W–A ratio where, unlike "true" molars, the ratio is over one (i.e., the tooth length is greater than its anterior width; see Table 46). The relatively greater length is due in part to the narrowness of the tooth, but the anterolabially directed parastylar region also adds to the overall length. The evidence suggests that UCMP 116512 is a DP^4, while UCMP 116511 is an M^1. Because of its small size, UCMP 116511 has been allocated to *O. erikseni*. However, UCMP 116512 is only questionably assigned to *O. erikseni*, because it is similar in size to the DP^4's of *P. donnae* mentioned above. This tooth could just as easily belong to *P.* cf. *P. donnae*, or even possibly to *Mimatuta morgoth.*

Two isolated molars, UCMP 116515 from loc. V74110 and UCMP 116513 from loc. V74111, are probably M^2's of *Oxyprimus erikseni.* They are almost indistinguishable from M^2's of *Protungulatum donnae* except for their smaller size. The parastylar region of UCMP 116515 is missing, but the portion of the stylar shelf preserved indicates that the ectoflexus was moderately deep, as in *P. donnae.* UCMP 116513 is heavily worn due to post-depositional chemical wear, but it appears that the ectoflexus is not as deep as in UCMP 116515. UCMP 116513 could also be an M^1 of *P.* cf. *P. donnae.*

An M^3, UCMP 116514, is assigned to *Oxyprimus erikseni* because of its small size but could also belong to *Protungulatum* cf. *P. donnae.* None of the upper molars shows extensive dental wear. The earliest stages of attrition are confined to the apices of the three primary cusps and then are followed by wear in the protofossa and along the centrocrista, metacrista, conular wings, and lingual cingula.

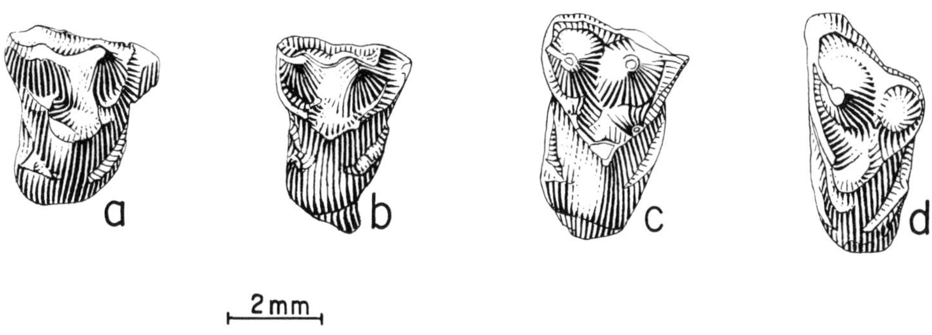

FIG. 59. *Oxyprimus erikseni* Van Valen. Isolated upper teeth. UCMP 116512, loc. V74111, right DP^4?: (a) occlusal view. UCMP 116511, loc. V74111, right M^1: (b) occlusal view. UCMP 116515, loc. V74110, right M^2: (c) occlusal view. UCMP 116514, loc. V74111, left M^3: (d) occlusal view. (A. L.)

TABLE 46. Measurements of teeth of *Oxyprimus erikseni*, from UCMP loc. V74111, except UCMP 116515, which is from UCMP loc. V74110.

UCMP Number	*Position*	*L*	*W–A*	*W–P*	*A*	*B*	*C*	*D*	*L/W–P*	*L/W–A*	*A/W–A*	*B/W–A*	*C/W–A*	*D/W–A*
116511	$M^{1?}$	2.95	3.30	3.60	1.04	a0.50	a1.04	0.72	0.82	0.89	0.32	a0.15	a0.32	0.22
116512	DP^4	3.45	3.39	3.82	1.27	0.40	1.23	0.49	0.90	1.02	0.37	0.12	0.36	0.14
116513	M^2	3.43	4.15	4.50	1.63	a0.41	a1.20	0.91	0.74	0.80	0.39	a0.10	a0.30	0.22
116515	M^2	a3.22	—	4.57	1.69	0.52	1.33	—	0.70	—	—	—	—	—
116514	M^3	3.08	4.82	4.13	1.42	0.66	1.31	1.49	0.75	0.64	0.29	0.12	0.27	0.31

UCMP Number	*Position*	*L*	*W*	*L–TR*	*L–TA*	*W–TR*	*W–TA*	*P–M*	*L–TR/L–TA*	*W–TR/W–TA*
*116503	P_3	2.52	1.25							
116504	P_3	2.75	1.28							
*116503	P_4	3.17	1.95							
116505	P_4	3.35	2.00							
116506	P_4	3.42	1.68							
	$\bar{x}=$	3.31	1.88							
*116503	M_1	3.10		1.68	1.42	2.21	2.21	a0.15	1.18	1.00
116507	M_1	3.27		1.68	1.59	2.05	2.12	<0.10?	1.06	0.97
*116503	M_2	3.40		1.74	1.66	2.64	a2.34	<0.10	1.05	1.13
116508	M_2	3.62		1.80	1.82	2.85	2.54	0.16	0.99	1.12
116509	M_2	3.64		1.97	1.67	2.60	2.54	<0.10	1.18	1.02
	$\bar{x}=$	3.55		1.84	1.72	2.70	2.47	<0.10	1.07	1.09
*116503	M_3	4.05		1.82	2.23	2.42	1.94	0.22	0.82	1.25
116510	M_3	—		—	2.34	—	1.94	—	—	—

* Associated teeth belonging to dentary, UCMP 116503. M_{1-3} length for UCMP 116503 is 10.40 mm.
a Approximate measurement.

Discussion. In most aspects of its dentition *Oxyprimus erikseni* strongly resembles *Protungulatum donnae*. The differences that do exist are important in assessing the question of what dental characters are primitive for condylarths, and the probable affinities of *O. erikseni*.

In contrast to *Protungulatum donnae, Oxyprimus erikseni* has a large P_4 metaconid, a longer P_4 relative to M_1, a relatively longer and wider talonid on M_{1-2}, a probably more closely appressed paraconid and metaconid on the lower molars, and a relatively shorter talonid on M_3. The DP^4 and upper molars described previously probably belong to *O. erikseni*, but this is by no means certain. Other than smaller size, they are quite similar to their counterparts in *P. donnae*. They do differ from *P. donnae* in having a slightly more developed cingular hypocone and more distinct postparaconule and premetaconule wings. However, the conules are not larger than in *P. donnae*. This is not true of at least one of the other species of *Oxyprimus* described by Van Valen (1978). In *O. galadrielae* the conules are distinctly enlarged relative to the condition seen in *P. donnae* (Van Valen, 1978, pl. 1, fig. 1).

I regard the above character states in *Oxyprimus erikseni* (and *O. galadrielae*) to be derived relative to their condition in *Protungulatum donnae*. Regarding an enlarged P_4 metaconid as being derived requires further explanation, because the presumed primitive state of this tooth is contested among various workers.

Although the primitive character state of the P_4^4 is a moot point among early mammal workers, I regard the "premolariform" condition as primitive for the Theria. A premolariform P_4 is here regarded as having a large protoconid, low anterior accessory cusp ("paraconid"), no metaconid, and a small trenchant heel or talonid. A premolariform P^4 would have a large paracone, a wide stylar shelf, a well-developed parastylar and less-developed metastylar region, no conules, no metacone, and a slightly lingually projecting protocone.

Lillegraven (1969) has quite convincingly shown that *Procerberus formicarum* has affinities with the palaeoryctids and not the leptictoids (see Sloan and Van Valen, 1965). Assuming, as has been done here, that the premolariform conditions of P_4^4 are primitive for eutherians, then the development of a "submolariform" (especially one possessing a large metacone and metaconid) P_4^4 was accomplished separately in a minimum of three lineages in Late Cretaceous eutherians. These lineages are the leptictoids (e.g., *Gypsonictops*) the palaeoryctids (e.g., *Procerberus*), and the condylarths (e.g., *Oxyprimus erikseni*). Until P^4's of *O. erikseni* are recognized it will not be known if this species has developed a P^4 metacone in *Procerberus* and *Gypsonictops*. However, if *O. galadrielae* and *O. putorius* are indicative, *O. erikseni* probably lacked an enlarged P^4 metacone.

In the preceding character states, which are taken to be derived, *Oxyprimus erikseni* (and probably the entire genus) resembles several groups of more advanced condylarths, even though in overall similarity it resembles *Protungulatum donnae*. The expansion (widening and lengthening) of the M_{1-2} talonids and the shortening of the M_3, both slightly developed in *O. erikseni*, are found in a great number of early Tertiary condylarths (excepting at least periptychids). However, lengthening of the P_4 and development of a large P_4 metaconid are not as widespread among condylarths. Some arctocyonids lengthen the P_4 (e.g., *Oxyclaenus* or the triisodontines) and some have a well-developed P_4 metaconid (e.g., *Ragnarok nordicum, Chriacus*, or *Thangorodrim*), but to my knowledge none shows these characters in combination. Also, some arctocyonids (other than *Protungulatum*) have a lingually placed paraconid (the primitive condition, based on comparisons with other early therians), but generally do not show fusion of the paraconid and metaconid. In this family, when the paraconid is undergoing reduction it moves sublingually.

In contrast to arctocyonids, many hyopsodontids (and some mioclaenids, in the sense of Van Valen, 1978) show the preceding derived character states seen in *Oxyprimus erikseni*, and also, as in this taxon, most show a low but distinct cingular hypocone and enlarged conules, as in *O. galadrielae*. It should be noted that phenacodontids show many of these trends; however, phenacodontids from very early in their phylogeny acquired more bulbous, low-crowned molars, larger size, and greater molarization of the premolars (particularly P^4 and later P^3_3, in contrast to most hyopsodontids).

In comparisons with various hyopsodontid taxa, *Oxyprimus* most closely resembles such genera as *Litomylus, Haplaletes, Haplomylus,* and *Hyopsodus*. However, in *Litomylus* the P_4 metaconid appears to be relatively more reduced. The reduction is even greater in *Aletodon*, and this in part led Gingerich (1977) to suggest affinities between the two genera.

Aside from the characters that indicate that *Oxyprimus* is more primitive (higher-crowned and less bulbous lower molars and smaller talonids), it also differs from most of the preceding taxa in its trigonid morphology. In his analysis of Paleocene hyopsodontids, Simpson (1937a) noted that paraconids were reduced or vestigial, median to subinternal, and not fusing to the metaconid in *Litomylus, Haplaletes, Haplomylus*, and other taxa (see also discussion of *Oxyacodon*). From the admittedly few specimens assigned to these genera which I have examined I agree with Simpson's assessment, except possibly for some specimens referred to *Haplaletes* (*H. disceptatrix*). In these specimens I am not convinced that the paraconid has not been lost through fusion with the metaconid. This seems likely to have been the case in various specimens assigned to *Hyopsodus* examined from the UCMP and YPM collections. In some individuals there is no trace of a paraconid, although the metaconid is enlarged in an anteroposterior direction. In others, what I take to be a paraconid is lingual and very closely appressed to the metaconid. The hyopsodontids (and mioclaenids) are greatly in need of reanalysis, but based on current knowledge in trigonid morphology *Oxyprimus* most resembles some individuals of *Hyopsodus* rather than *Haplomylus*, *Haplaletes*, or *Litomylus*.

Oxyprimus also resembles some but not all the taxa currently included in the Mioclaenidae (in the sense of Van Valen, 1978). It shows similarities to mioclaenids in the enlargement of upper molar conules and it shows similarities in P_4 morphology to specimens referred to species of *Litaletes*, *Ellipsodon*, and *Promioclaenus*, but not to species of *Mioclaenus* or *Choeroclaenus*. If the relatively enlarged cingular hypocone in *Oxyprimus erikseni* is assumed to indicate a derived trend, this species resembles hyopsodontids more than mioclaenids, in which this cusp is reduced or lost. The one character in which *Oxyprimus* more closely resembles mioclaenids is the lingual position and close appression of the paraconid and metaconid (although this is true to a lesser degree in *Oxyprimus* than in mioclaenids). However, the lingual position of the paraconid is a primitive therian character. Except for the slightly more appressed paraconid and metaconid (probably also at least in *Hyopsodus*), enlarged upper molar conules, and the resemblances in P_4 morphology with some mioclaenids (the latter two characters are also seen in hyopsodontids) *Oxyprimus* is most similar to hyopsodontids.

Clearly, review of the Hyopsodontidae and Mioclaenidae is needed to understand more fully the evolutionary trends in these two families and to unravel taxonomic problems. Based on the current understanding of these two families, *Oxyprimus* (at least *O. erikseni* and *O. galadrielae*) is tentatively considered to be the most primitive member of the Hyopsodontidae. It is less clear whether mioclaenids should be assigned familial status as was done by Van Valen (1978). A comparison of *Oxyprimus* to members of these two families does suggest separate origins (as advocated by Van Valen), but this would require some parallel evolution in the two families (at least as they are now constituted).

Family PERIPTYCHIDAE Cope, 1882d
Subfamily ANISONCHINAE Osborn and Earle, 1895

Comments. Material from the Hell Creek Formation and the lowermost portion of the Tullock Formation indicates the presence of two and possibly three periptychids. Two of the samples are referable to two species of *Mimatuta*, *M. morgoth* and *M. minuial* (Van Valen, 1978). The conclusion that these two species are aligned with the anisonchine periptychids was arrived at separately by Van Valen and myself, and thus strengthens allocation to this family. The third sample is composed of a dentary fragment preserving an M_3. This specimen is clearly not referable to either of the above species, but is very questionably placed in the Periptychidae.

The two new species are quite primitive and show that the morphology typical of the upper molars (and probably the lowers) of anisonchine (and possibly periptychine) periptychids evolved before the P^4 (and ?P_4) had become inflated. This trend is supported by the finding of two undescribed maxillae in the UCMP collections from the lower Nacimiento Formation, New Mexico (Puercan). In an analysis of the lower dentition of *Oxyacodon* it became clear that these two maxillae are the previously unknown upper dentitions of two of the species of this genus (*O. agapetillus* and *O. priscilla*). These maxillae are described and *Oxyacodon* is transferred to the Anisonchinae. J. Keith Rigby, Jr., and Leigh Van Valen (pers. comms.) have also separately determined that *Oxyacodon* is an anisonchine. However, Van Valen's (1978) and my views concerning *Oxyacodon* differ slightly. These differences are discussed under this genus.

Mimatuta, Van Valen, 1978
Mimatuta morgoth, Van Valen, 1978

(Figs. 60–63 & Tables 47–49)

Mimatuta morgoth Van Valen, 1978, p. 62

Type. UMVP 1560, right maxilla with M^2 (Van Valen, 1978, pl. 7, fig. 5).

Type locality. Harbicht Hill, Hell Creek Formation, Montana.

Referred specimens. 1 P^3; 1 P^4; 1$M^{2?}$; right maxilla with M^{2-3}, UCMP 116519; right mandible with P_1 alveolus, P_{2-4}, M_1, UCMP 116520; associated teeth for left mandible P_{1-4}, M_{2-3}, UCMP 116522; right mandible with M_2 and alveoli for M_3, UCMP 116526; left M_1 in bone, UCMP 116523 (may belong with UCMP 116522); right M_3 in bone, UCMP 116521 (may belong with UCMP 116520); 2 P_3's; 1 P_4; 2 M_1's; 2 M_2's from UCMP loc. V74111. 1 M_3 from UCMP loc. V74110. Dentary with M_2 fragment, UCMP loc. V73098.

Localities. UCMP locs. V73098, V74110, and V74111.

Distribution. Hell Creek Formation (Lancian) and lowermost Tullock Formation (premantuan), Montana.

Diagnosis. Van Valen, 1978, p.62 (but see comments regarding size under description of upper molars).

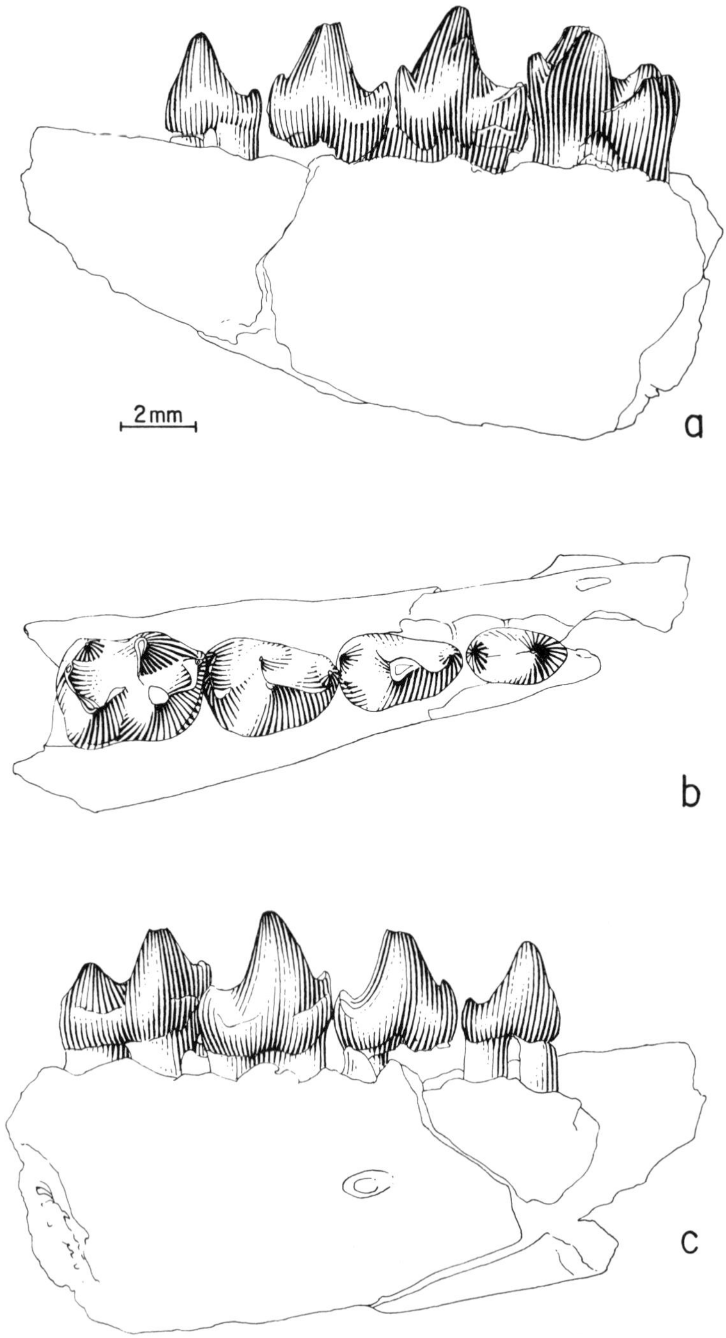

FIG. 60. *Mimatuta morgoth* Van Valen. UCMP 116520. Right dentary with P_{2-4}, M_1, loc. V74111: (a) lingual view; (b) occlusal view; (c) labial view. UCMP 116521, right M_3, loc. V74111 may belong with this dentary (see Fig. 62d, e, & f). (A. L.)

Description. Mandibles. Although *Mimatuta morgoth* is represented by more mandibular fragments and isolated teeth than *Oxyprimus erikseni*, none of the fragments is as complete as the one jaw, UCMP 116503, referable to the latter species. Two mandibles, UCMP 116520 and UCMP 116524, preserve portions of the jaw below P_{1-4} and M_1. Based on these two specimens the jaw of *M. morgoth* appears to be slightly deeper and more robust than *Protungulatum donnae.* As in *P. donnae*, there are two mental foramina, one below P_1 and another below the posterior root of P_3. Like *P. donnae* and *O. erikseni, M. morgoth* has four premolars, with P_1 having one root and the remainder having two. *P. donnae* has small diastemata between the canine and P_1, between P_1 and P_2, and often between P_2 and P_3. These diastemata, except between P_2 and P_3, are present in *M. morgoth.* UCMP 116524 preserves the posterior portion of the canine alveolus, which slopes posteriorly at about 45° from the vertical.

P_{1-4}. P_1 is preserved in association with P_2 in UCMP 116522. P_1 is the smallest premolar and has one large root, larger than either of the roots for P_2. The crown of P_1 has a large, vertically oriented central cusp and a small posterior basal cusp. There is no anterior basal cusp, but the tooth is slightly swollen on its anterior edge. The tip of the central cusp is missing but probably was not as tall as the same cusp on P_2.

P_2 has two roots, the posterior being slightly larger. There is a large central cusp which bears faint ridges on its anterior and posterior edges. The heel is longer than in P_1, but like P_1 has only one posterior basal cusp. Of the two P_2's present in the sample, one, UCMP 116522, has a small, anterior basal cuspule.

Six P_3's are present in the sample, four of which are associated with other teeth. The P_3's are larger than the P_2's, with the heel being the widest portion of the crown. The central cusp (protoconid) is tall and more robust than in P_2. A variable, faint ridge is present on the anterior edge of P_3 leading down to a small anterior basal cusp. The anterior basal cusp varies from being on the midline to lingual to the midline, with the second condition most common. There is a single, small cusp located at the posterior edge of the heel with a ridge running anteriorly from it to the posterior base of the central cusp. In UCMP 116520 the cusp on the heel is lingual to the midline and a second cuspule is located labial to it. None of the P_3's has a metaconid, but each of the unworn teeth bears a ridge running from the apex of the protoconid halfway down its posterolingual side. The anterior root on P_3 is smaller than the posterior root.

The P_4 is the largest and tallest of the premolar series. Like *Protungulatum donnae,* but unlike *Oxyprimus erikseni*, the P_4 is shorter than M_1 in *Mimatuta morgoth*. The P_4 of *M. morgoth* is more robust than that of *P. donnae.* This is best expressed in the more bulbous protoconid and expanded talonid. Although more bulbous, the protoconid is not relatively taller. The paraconid and metaconid are relatively and sometimes absolutely smaller, but not shorter than their counterparts in *P. donnae.* The P_4 talonid of *M. morgoth* has increased in relative width and length. In *P. donnae* the widest point of the P_4 is through the protoconid-metaconid. In *M. morgoth* the widest point has moved posterior and lies at about the midpoint of the talonid. In two P_4's of *M. margoth* (both worn) there is a ridge running along the midline of the talonid from the back of the protoconid to a thickened transverse ridge running along the back of the talonid. When unworn, a cusp was probably present at the juncture of the two ridges. In a third, unworn P_4 (UCMP 116520) the midline talonid ridge slants labially in its posterior aspect, making contact with the transverse posterior ridge labial to the midline. At this juncture there is a cusp in a similar position to a molar hypoconid. On the posterior transverse ridge lingual to the "hypoconid" there is a very small "hypoconulid" and larger "entoconid."

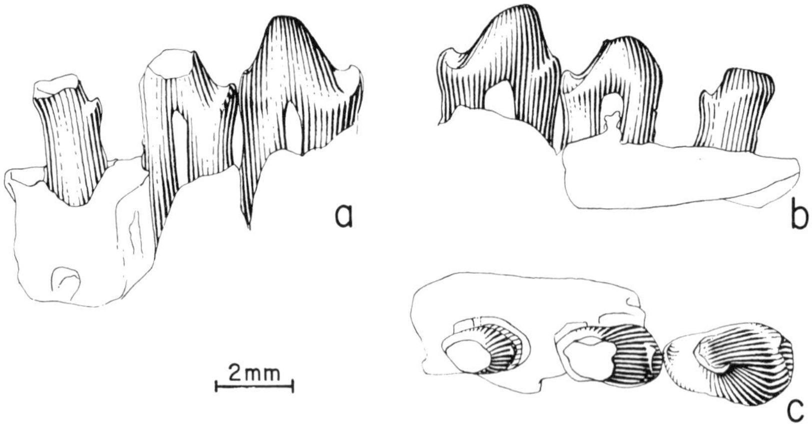

FIG. 61. *Mimatuta morgoth* Van Valen. Lower left P_{1-3}, UCMP 116522, loc. V74111: (a) labial view; (b) lingual view; (c) occlusal view. P_{1-2} are associated, P_3 (plus isolated P_4,M_{2-3}) is probably associated (all UCMP 116522), and an M_1 (UCMP 116523) is possibly associated. (A. L.)

Lower molars. Morphologically, the lower molars of *Mimatuta morgoth* differ from those of *Protungulatum donnae* in a slight expansion of the molar cusps and a labial shift of the paraconids. The molar talonids of *M. morgoth* are relatively wider than in *P. donnae* (i.e., the trigonid width/talonid width ratio is lower in *M. morgoth*). However, the molar talonids of the former species appear to be relatively shorter. The relative shortening of the M_3 talonid is a general trend among early condylarths. This trend is not true for M_{1-2} (especially M_2) and to my knowledge, among early condylarths only perptychids shorten M_{1-2} talonids. (This trend is considered further in the discussion section.)

P^{3-4}. Upper premolars are rare in the Hell's Hollow local fauna (UCMP locs. V74110 and V74111). Only one P^3 and a broken P^4, both from V74111, can be very tentatively assigned to *Mimatuta morgoth* on the basis of size. The P^3 (UCMP 116516) is very similar to P^3's of *Protungulatum donnae* and differs only in having a slightly lower paracone and better-developed protocone. The P^4 (UCMP 116517) also differs from the P^4's of *P. donnae* in these same two features. Although missing the parastylar region, UCMP 116517 is unworn and preserves sharp crests on the anterior and posterior faces of the paracone. The posterior crest has two distinct bulges near its lower end, but a metacone is not present. In *P. donnae* the postprotocrista is generally better developed. In UCMP 116517 the pre- and postprotocrista are more nearly equal in development, making the lingual portion of the tooth appear more nearly symmetrical when viewed occlusally. UCMP 116517 has narrow pre- and postcingula, as in some P^4's of *P. donnae*.

Upper molars. In the loc V74111 sample, upper molars referable to *Mimatuta morgoth* are poorly represented. A maxilla, UCMP 116519, preserves M^{2-3}, and UCMP 116518 may be an M^2 of this species. The M^2 in UCMP 116519 differs in several important features from M^2's of *P. donnae.* The M^2 is subtriangular in outline in the former species,

TABLE 47. Measurements of lower premolars of *Mimatuta morgoth* from UCMP loc. V74111.

UCMP Number	*Position*	*L*	*W*
[1]116522	P_1	1.90	1.15
[2]116520	P_2	2.48	1.44
[1]116522	P_2	2.74	1.61
[2]116520	P_3	3.29	1.94
[1]116522	P_3	3.33	2.08
[3]116524	P_3	3.28	2.13
[4]116525	P_3	3.03	1.90
116527	P_3	3.26	1.98
116528	P_3	3.33	2.11
	$\bar{x}=$	3.25	2.02
[1]116522	P_4	3.76	2.50
[2]116520	P_4	3.50	2.47
[3]116524	P_4	3.62	2.52
116529	$P_{4?}$	3.36	2.02
	$\bar{x}=$	3.56	2.38

[1] 116522 (P_{1-4}, M_{2-3}), probably also [1]116523 (M_1), M_{1-3}: length of 116522 plus 116523 equals 13.78 mm.
[2] 116520 (P_{2-4}, M_1), possibly including [2]116521 (M_3).
[3] 116525 (P_{3-4}).
[4] 116526 (P_3, P_4 trigonid).

TABLE 48. Measurements of lower molars of *Mimatuta morgoth* from UCMP loc. V74111, except UCMP 116534, from loc. V74110. See Table 47 for explanation of superscripts.

UCMP Number	*Position*	*L*	*L-TR*	*L-TA*	*W-TR*	*W-TA*	*P-M*	*L-TR/L-TA*	*W-TR/W-TA*
[2]116520	M_1	3.86	2.03	1.83	2.59	2.80	0.23	1.11	0.93
[1]116523	$M_{1?}$	4.27	2.44	1.83	a2.86	a3.12	0.49	1.33	0.92
116530	M_1	4.08	2.30	1.78	2.78	2.80	0.30	1.29	0.99
116531	M_1	4.03	2.31	1.72	a2.80	a2.80	0.27	1.34	a1.00
	$\bar{x}=$	4.06	2.27	1.79	2.76	2.88	0.32	1.27	0.96
[1]116522	M_2	4.51	2.42	2.09	a3.47	a3.49	0.43	1.16	0.99
116526	M_2	4.19	2.30	1.89	3.27	3.11	0.16	1.22	1.05
116532	M_2	4.43	2.40	2.03	3.41	3.23	0.20	1.18	1.06
116533	M_2	a3.97	a2.29	a1.68	3.25	a2.98	0.22	a1.36	a1.09
	$\bar{x}=$	4.28	2.35	1.92	3.35	3.20	0.25	1.23	1.05
[2]116521	M_3	4.85	2.15	2.70	2.80	2.53	0.14	0.80	1.11
[1]116522	M_3	5.15	2.14	3.01	a3.12	a2.63	0.44	0.71	1.19
116534	M_3	5.30	2.40	2.90	2.80	2.47	0.14	0.83	1.13
	$\bar{x}=$	5.10	2.23	2.87	2.91	2.54	0.24	0.78	1.14

a Approximate measurement.

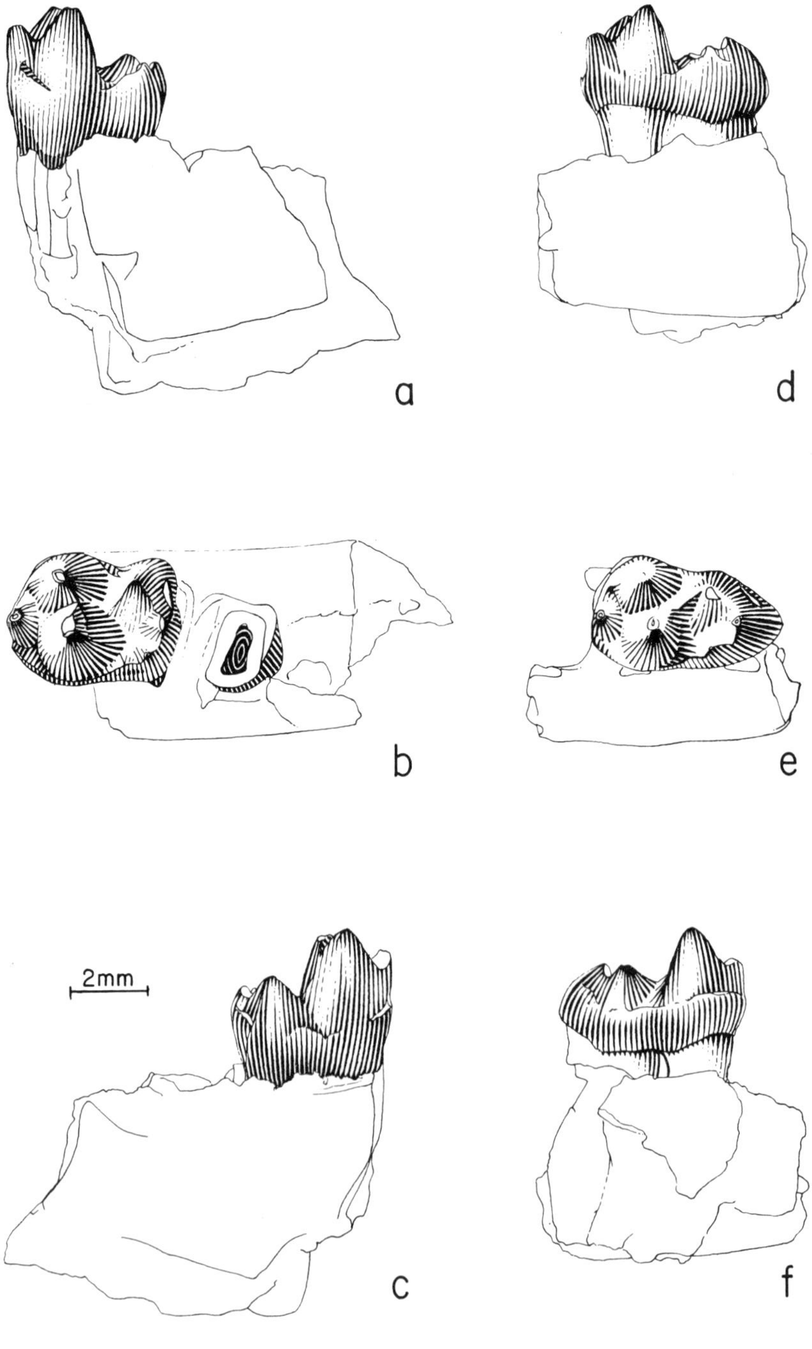

FIG. 62. *Mimatuta morgoth* Van Valen. Isolated lower molars, loc. V74111. UCMP 116526, right M_2: (a) lingual view; (b) occlusal view; (c) labial view. UCMP 116521, right M_3: (d) lingual view; (e) occlusal view; (f) labial view. UCMP 116521 may belong with UCMP 116520, right dentary (see Fig. 60). (A.L.)

as a result of the reduction in the bulbousness of the protocone and a slight decrease in the pre- and postcingula width. The protocone has also shifted labially (i.e., the A/W–A ratio is greater in *M. morgoth*), which has resulted in a closer apposition of the protocone and conules. Additionally, the postparaconule and premetaconule wings are more distinct and there is greater offset of the paracone and metacone, with the metacone being more lingual. UCMP 116518 is somewhat damaged, but in all preserved portions it agrees in general with the above description. However, this specimen does possess a small cusp on the postcingulum in the hypocone region. Based on its size UCMP 116518 is an M^2, but based on its high length/width ratios it could be an M^1.

The one definite M^2, UCMP 116519, is morphologically quite similar to the type of *M. morgoth*, UMVP 1560, also an M^2. However, the type is about 13 percent larger than UCMP 116519. Such a range in size is not uncommon in a single species, particularly since the two specimens come from localities differing slightly in age. Until a thorough comparison is made between the sample of *M. morgoth* from Harbicht Hill, which yielded the

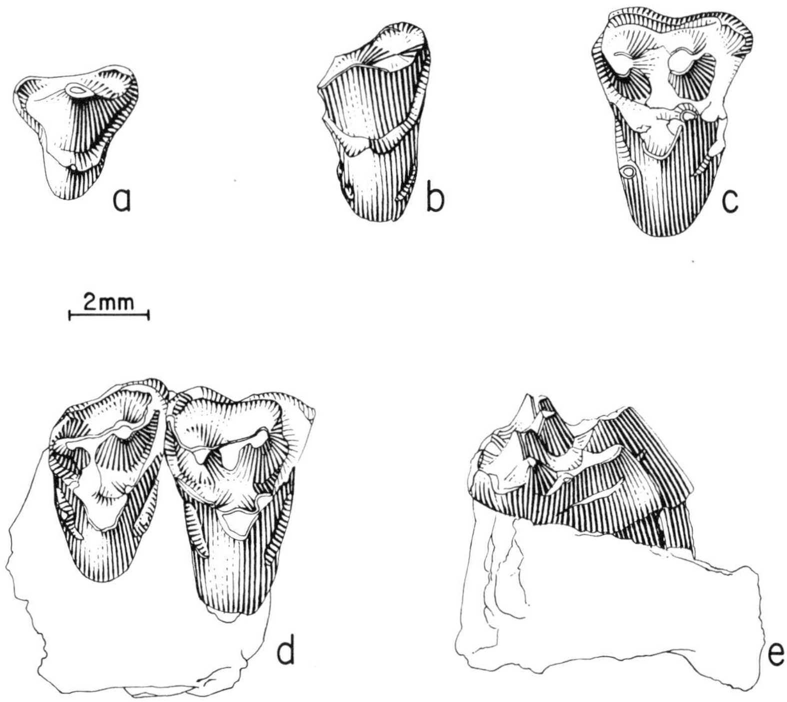

FIG. 63. *Mimatuta morgoth* Van Valen. Upper teeth, loc. V74111. UCMP 116516, left P^3: (a) occlusal view. UCMP 116517, left P^4: (b) occlusal view. UCMP 116518, right M^2: (c) occlusal view. UCMP 116519, right maxilla with M^{2-3} (d) occlusal view; (e) posterior view. (A. L.)

TABLE 49. Measurements of upper teeth of *Mimatuta morgoth* from UCMP loc. V74111.

UCMP Number	Position	*L*	*W–A*	*W–P*	*A*	*B*	*C*	*D*	*L*/*W–P*	*L*/*W–A*	*A*/*W–A*	*B*/*W–A*	*C*/*W–A*	*D*/*W–A*
116516	P^3	3.16	2.89	3.38	0.65	(no conules)		0.41						
116517	P^4	—	—	4.73	0.80	(no conules)		—	—	—	—	—	—	—
116518	$M^{2?}$	4.09	5.26	5.47	2.45	0.48	1.34	0.99	0.75	0.78	0.47	0.09	0.25	0.18
*116519	M^2	3.95	5.56	5.75	2.28	0.53	1.52	1.23	0.69	0.71	0.41	0.10	0.27	0.23
*116519	M^3	3.29	5.01	4.57	1.53	0.73	1.45	1.30	0.72	0.66	0.31	0.15	0.29	0.26

* M^{2-3} associated in maxilla, UCMP 116519.

type specimen, and the sample from the Hell's Hollow local fauna, it is assumed that the latter sample is referable to *M. morgoth.*

Discussion. Although no maxillae and mandibles of condylarths (or any other mammal) have been found in association at locs. V74110 and V74111, there are a number of size and morphological arguments for associating the upper and lower teeth at these localities. This is particularly true for *Mimatuta morgoth.* As noted previously, the M_{1-2} talonids of *M. morgoth* are relatively shorter than those of *Protungulatum donnae.* In fact, the former species has relatively the shortest M_{1-2} talonids of any of the condylarths from locs. V74110 and V74111. Of the upper molars from these two localities, only the three molars just described are characterized by a reduction in the length (anterior-posterior) across the protocone. This results in a distinctly subtriangular trigon, as opposed to the more subquadrate trigon in *P. donnae* and the other condylarths from locs. V74110 and V74111. The reduction in length of the M_{1-2} talonids and the subtriangular trigons of the upper molars strongly suggests that these lower and upper molars come from the same species.

Mimatuta morgoth, although still relatively primitive, shows definite trends toward a periptychid-type (particularly anisonchine) dentition. Based on this I agree with Van Valen (1978) that the species (and genus) is referable to the Anisonchinae.

Mimatuta cf. *M. morgoth* Van Valen, 1978

(Fig. 64 & Table 50)

A mandible from UCMP loc. V72211, lowermost Tullock Formation, Garfield County, is tentatively referred to this species. A right mandible, UCMP 116536, preserves P_4, M_{1-2}, and roots for M_3. The size and proportions of the teeth are most similar to *Mimatuta morgoth,* although there are some differences, such as the relatively longer talonid on M_2 of UCMP 116536 (see Table 50).

The greatest difference in the molars is in the paraconid. In UCMP 116536 the paraconids seem slightly enlarged and are not as labially shifted as in *Mimatuta morgoth.*

The P_4 of UCMP 116536 is almost identical to those of *Mimatuta morgoth* except for the larger metaconid in the former. However, the trigonid of the UCMP 116536 P_4 is worn fairly low by dental attrition and the seemingly large metaconid may not have been comparatively as large when the tooth was unworn. The paraconid on the P_4 of UCMP 116536 is also slightly larger than in P_4's of *M. morgoth* (recalling the condition in *Protungulatum donnae)* and the P_4 is slightly less inflated.

TABLE 50. Measurements of UCMP 116536, *Mimatuta* cf. *M. morgoth,* from UCMP loc. V72211.

Position	*L*	*W*	*L–TR*	*L–TA*	*W–TR*	*W–TA*	*P–M*	*L–TR* / *L–TA*	*W–TR* / *W–TA*
P_4	3.50	2.27							
M_1	3.64		2.10	1.54	2.63	2.76	0.14	1.36	0.95
M_2	4.08		2.14	1.94	3.19	3.23	<0.10	1.10	0.99

a Approximate measurement.

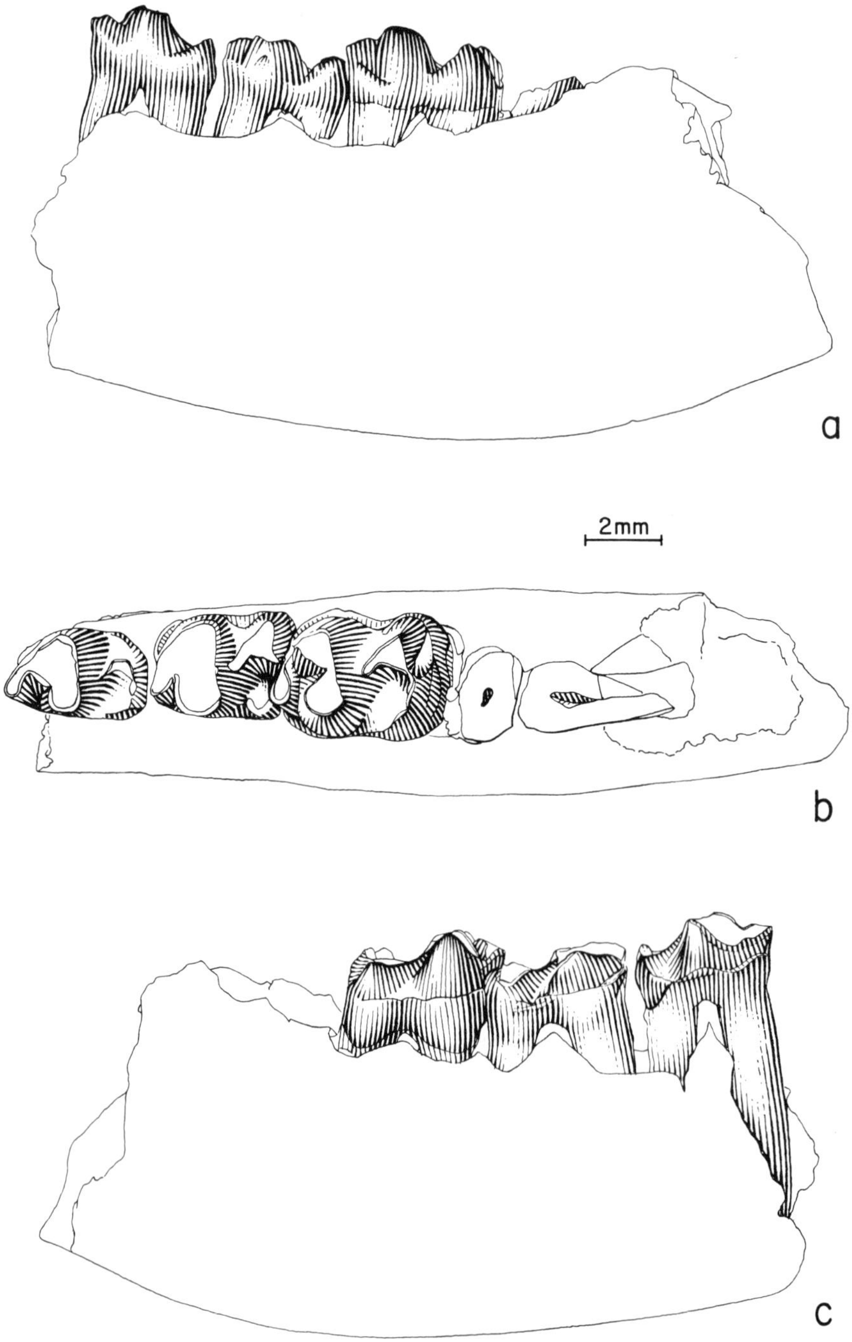

FIG. 64. *Mimatuta* cf. *M. morgoth* Van Valen. UCMP 116536. Right dentary with P_4, M_{1-2}, loc. V72211: (a) lingual view; (b) occlusal view; (c) labial view. (A. L.)

Mimatuta minuial Van Valen, 1978

(Figs. 65 & 66 & Table 51)

Mimatuta minuial Van Valen, 1978, p. 62

Type. PU 14211, left maxilla with P^3–M^3 (Van Valen, 1978, pl. 7, fig. 4).

Type locality. Mantua lentil, Polecat Bench Formation, Wyoming.

Referred specimens. Left maxilla with P^3 alveoli, P^4, M^{1-2}, LACM 112901; left M^2, UCMP 120410; and left M^2? fragment, UCMP 120411, from LACM loc. 3103 (=UCMP loc. V72210). Possibly right M_1, UCMP 120408, and left M_3, UCMP 120409, from UCMP loc. V77087.

Localities. See Referred specimens.

Distribution. Lowermost Tullock Formation, Montana, and Mantua lentil, Polecat Bench Formation, Wyoming (both mantuan).

Diagnosis. Van Valen, 1978, p. 62.

Comments. The provenance ot the maxilla, LACM 112901, is not known with certainty. The probable discoverer, H. Garbani, does not recall finding this specimen at LACM loc. 3103. However, there are several lines of evidence to suggest that the specimen does come from LACM loc. 3103. First, when the specimen was located in LACM collections, it was found with two mandibles (LACM 112902, *Ragnarok nordicum*, and LACM 112903, ?Periptychidae, gen. et sp. indet.), an M^3 of an insectivoran (LACM 112904), fragments of lower invertebrates, and seed and gastropod endocasts. H. Garbani agrees that the latter specimens come from loc. 3103. On a subsequent visit to the locality in 1975 with H. Garbani we were unable to find additional mammalian specimens. The bone pocket is a yellowish sandy silt and bears small rust-colored concretions. One of the mandibles (LACM 112902, *R. nordicum*) plus the endocasts are stained by ferric (red) iron, and they all have patches of dendritic pyrolusite. This is the type of preservation seen in the maxilla, LACM 112901, which suggests that the specimen may have come from this locality; this type of preservation is found at several localities within the Hell Creek Formation, but to date has not been found at other localities in the Tullock Formation, Garfield County. Second, in 1977 a complete left M^2, UCMP 120410, was recovered from LACM loc. 3103. UCMP 120410 is unworn, in contrast to the heavily worn M^2 in LACM 112901, but except for some differences in the stylar shelf the two molars are very similar and almost certainly belong to the same species. Third, although the "state of evolution" is weak circumstantial evidence, comparisons with other condylarths (particularly periptychids) from the Hell Creek and Tullock formations, Garfield County, suggest that LACM 112901 would have come from a locality very low in the Tullock Formation. LACM loc. 3103 is very low in the Tullock Formation (the locality sits immediately on the lowest Z coal) and thus fits within this interval. Also, the type of *Mimatuta minuial* described by Van Valen (1978), to which LACM 112901 is referred, comes from Mantua lentil, Wyoming. The Mantua lentil is regarded as being earliest Paleocene (mantuan) in age.

Description. Maxilla and upper dentition. A substantial portion of the maxillary bone is preserved dorsal to the teeth. A large infraorbital foramen is present above the alveoli for P^3. The portion of the maxilla exposed in the nasal cavity is relatively smooth and is slightly convex lingually.

Of the three alveoli for P^3, the labial two are smaller (the anterolabial alveolus still bears the root). The larger lingual alveolus extends lingually nearly as far as do the P^4, M^{1-2}, indicating that a moderately large protocone may have been present.

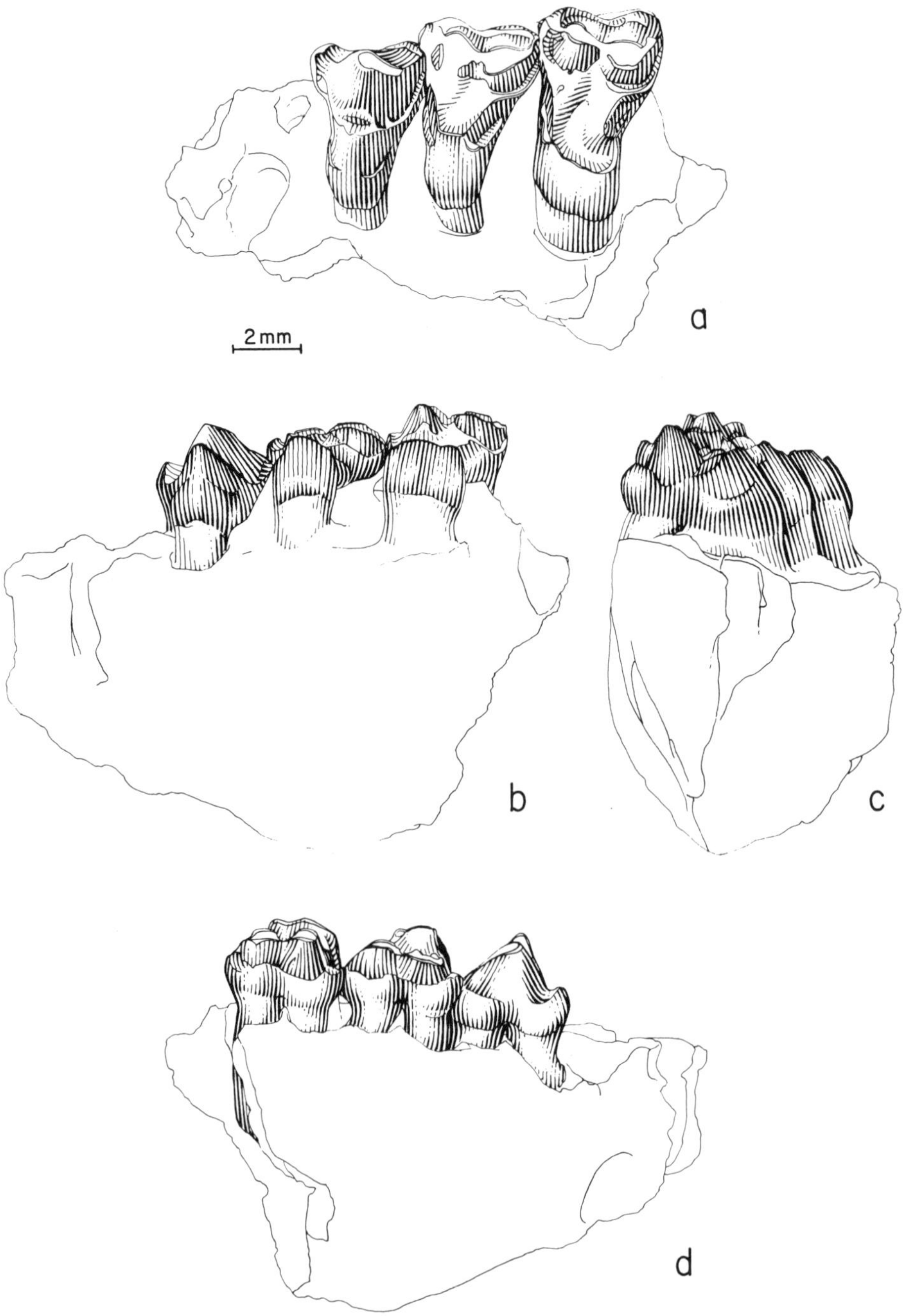

FIG. 65. *Mimatuta minuial* Van Valen. LACM 112901, left maxilla with P^4, M^{1-2}, LACM loc. 3103? (= UCMP loc. V72210): (a) occlusal view; (b) lingual view; (c) anterior view; (d) labial view. (A.L.)

In most respects the P^4 is quite similar to the P^4 of the *Protungulatum donnae* (and the P^4 fragment tentatively referred to *Mimatuta morgoth*). As in *P. donnae;* the prominent parastylar and metastylar regions form a line labial to the paracone; there is no metacone, but a ridge is present running from the paracone to the metastylar region; the protocone is well developed, and narrow pre- and postcingula are present (not always in *P. donnae*). The differences seen in *M. minuial* involve the shortening and slight inflation of the paracone and the labial shift of the protocone with an increase in the length of the lingual face of this cusp. The pre- and postprotocrista are heavily worn and there is no indication that conules were ever present. Because of the closer apposition of the paracone and protocone, the protofossa is reduced to a dorsally convex surface running between the two cusps. A minute ridge runs from the protocone to the base of the paracone near the anterior edge of the protofossa.

As was the case in *Mimatuta morgoth*, it is the upper molars that most clearly suggest the periptychid affinities of *M. minuial.* The molars are transversely wider than in *Protungulatum donnae*, the pre- and postcingula are narrower, the protocone is shifted labially and the lingual face of the protocone is oriented at a lower angle to the palate; the hypocone is long and cylindrical, with its apex shifted labially like the apex of the protocone; the conules are shifted labially, and the stylar shelf is narrower (at least in LACM 112901), particularly in the parastylar and metastylar regions. In LACM 112901 the crowns of the molars are heavily worn, especially in the protofossa and on the apices of the three major cusps. The apices of conules on both molars are worn, but enough of the conular wings is preserved to show that these cusps were very closely approximated to the para- and metacones. Determination of the position of the protocone apices, especially on M^2, is made difficult by the heavy wear. Fortunately, the isolated left M^2, UCMP 120410, is unworn. It shows that the protocone is more labial in its position than in *P. donnae*, but it may be only slightly more labial in position than in *M. morgoth.* A fragment of an upper molar, UCMP 120411 (probably an M^2), also shows the labial shift of the protocone. Both UCMP 120410 and UCMP 120411 also confirm that the conules are shifted labially relative to their position in *P. donnae* but approximately similar to their position in *M. morgoth.* As in *M. morgoth* but not *P. donnae*, *M. minuial* has slightly more distinct postparaconule and premetaconule wings. The only difference between the isolated M^2, UCMP 120410, and the M^2 preserved in LACM 112901 is in the stylar region. The metastylar regions in the two specimens are similar in size. However, in UCMP 120410 the parastylar region is slightly wider than the metastylar region, while the reverse is true in LACM 112901.

Lower molars. No lower molars have been recovered from LACM loc. 3103 that could belong to *Mimatuta minuial.* However, in 1977 two lower molars were recovered from a new locality, UCMP loc. V77087, that appears to be an extension of LACM loc. 3103. The two molars, an M_1 and an M_3, could belong to *M. morgoth*, but there are several minor features which suggest, rather, that they can tentatively be referred to *M. minuial.* The M_1, UCMP 120408, is within the observed ranges of three of the six measurements and both ratios of the four M_1's of *M. morgoth* given in Table 48. The three measurements of UCMP 120408 that fall outside the observed ranges are the talonid length, where it is less than the observed range, and the trigonid and talonid widths, where they are greater than the observed ranges (Table 51). In the ratio of trigonid width to talonid width UCMP 120408 is similar to M_1's of *M. morgoth.* However, in comparing total length to maximum width (i.e., the talonid width), UCMP 120408 is noticeably wider

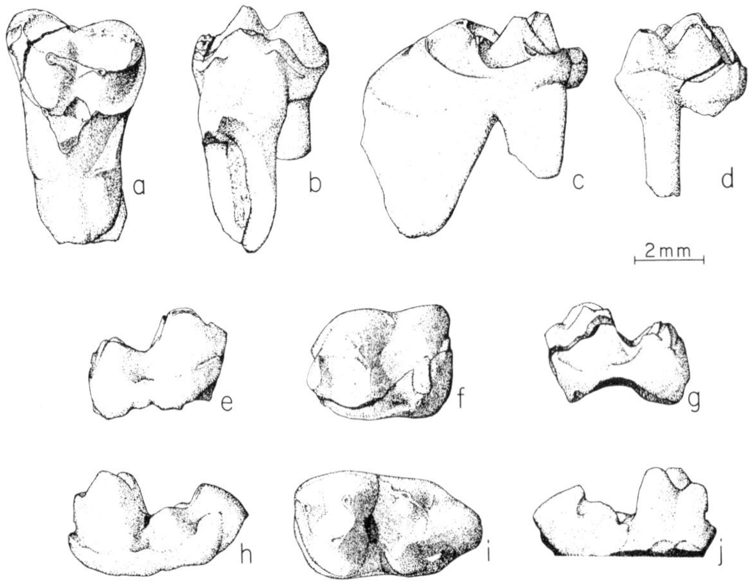

FIG. 66. *Mimatuta minuial* Van Valen. Isolated molars. UCMP 120410, left M^2, loc. V72210 (= LACM loc. 3103): (a) occlusal view; (b) lingual view; (c) posterior view; (d) labial view. UCMP 120408, right M_1, loc. V77087: (e), (f), & (g). UCMP 120409, left M_3, loc. V77087: (h), (i), & (j). Labial view: (e) & (h); occlusal view: (f) & (i); lingual view: (g) & (j).

relative to length. In the four M_1's of *M. morgoth* (Table 48) the total length ranges from 1.36 to 1.46 times the maximum width ($\bar{x} = 1.41$), while for UCMP 120408 the total length is only 1.23 times the maximum width. This increase in relative width is interpreted as a derived trend among periptychids and thus supports referral of the isolated M_1 to the slightly more derived *M. minuial* rather than to *M. morgoth*. There are also some minor morphological details that are interpreted as being more derived in UCMP 120408 relative to the M_1's of *M. morgoth*. In UCMP 120408 the talonid cusps are slightly more bulbous, are cusplike, and are more similar in size; a small but distinct entocristid connects the entoconid to the posterior side of the metaconid; and a very narrow lingual cingulum is present below the metaconid, extending from the paraconid to just below and anterior to the entoconid. Some of these characters are slightly developed in the M_1's of *M. morgoth*, but not in combination. The isolated M_3, UCMP 120409, is even more questionably referred to *M. minuial*. The M_3 is slightly larger than the three M_3's of *M. morgoth* (Table 48) in total length and in talonid length and width. Also, the isolated M_3 has a slightly wider talonid relative to the trigonid. The posterior of the trigonid is slightly worn on UCMP 120409, but a small entocristid may have been present. Other than these slight differences, the morphology of UCMP 120409 is very similar to that seen in the M_3's of *M. morgoth*.

TABLE 51. Measurements of teeth of *Mimatuta minuial*. All upper teeth are from LACM loc. 3103 (=UCMP loc. V72210), and both lower molars are from UCMP loc. V77087.

Specimen Number	Position	*L*	*W–A*	*W–P*	*A*	*B*	*C*	*D*	*L*/*W–P*	*L*/*W–A*	*A*/*W–A*	*B*/*W–A*	*C*/*W–A*	*D*/*W–A*
LACM 112901	P^4	3.44	4.48	4.87	2.20	(no conules)		0.39	0.71	0.77	0.49	(no conules)		0.09
LACM 112901	M^1	3.61	5.13	5.43	—	—	a1.12	0.96	0.66	0.70	—	—	a0.22	0.19
LACM 112901	M^2	3.60	5.61	6.01	—	—	a1.44	0.77	0.60	0.64	—	—	a0.26	0.14
UCMP 120410	M^2	3.95	6.40	6.25	3.05	0.64	1.50	1.25	0.63	0.62	0.48	0.10	0.23	0.20

Specimen Number	Position	*L*	*L–TR*	*L–TA*	*W–TR*	*W–TA*	*P–M*	*L–TR*/*L–TA*	*W–TR*/*W–TA*
UCMP 120408	M_1	3.90	2.25	1.70	3.05	3.20	a0.30	1.32	0.95
UCMP 120409	M_3	5.35	2.25	3.10	3.05	2.80	a0.15	0.73	1.09

a Approximate measurement.

Discussion. Many of the other early and middle Paleocene condylarths show some of the trends just described for *Mimatuta minuial* (and *M. morgoth*). However, only the periptychids, especially the anisonchines, share those characters which are interpreted as being derived in relationship to *Protungulatum donnae. M. minuial* and *M. morgoth* have been included in the Anisonchinae because they lack any character that could allocate them to the Periptychinae. This is done with the realization that they could just as likely be periptychines, or belong to a third, unrecognized subfamily of periptychids.

More important than correctly allocating this genus to one of the currently recognized subfamilies is the analysis of the character states of the premolars and molars, showing a possible way in which the peculiar periptychid dentition developed. These trends are considered below the description of two other anisonchines from the San Juan Basin, which also have a bearing on the development of the periptychid dentition.

Oxyacodon Osborn and Earle, 1895

Oxyacodon Osborn and Earle, 1895, p. 25

Genotypic species. Oxyacodon apiculatus Osborn and Earle, 1895.

Referred species. Oxyacodon agapetillus (Cope, 1884a) and *O. priscilla* Matthew, 1937.

Distribution. Lower Nacimiento Formation, New Mexico (Puercan).

Revised diagnosis. Small anisonchine periptychids, slightly inflated P^4; P^4 smaller or subequal to M^1; upper molars with moderately wide stylar shelf, hypocone lingual as in *Conacodon*, metacingulum continuous to metastylar region as in *Protungulatum donnae*, *Mimatuta morgoth*, and *M. minuial*, narrow precingulum with no protostyle; lower premolars and molars similar to *Conacodon cophater* except that premolars (only P_{3-4} compared) are slightly less inflated, paraconids are on midline but less reduced, labial cingulum present but not as well-developed, lingual cingulum mostly absent, molar cusp apices not closed, talonids not as reduced in length, hypoconulid on M_3 expanded posteriorly, and trigonids, especially M_3, less inflated.

Comments. Although not part of the faunas from the Hell Creek and lowermost Tullock formations of Garfield County, a palate and a maxilla from the lower Nacimiento Formation, New Mexico, add further information concerning early periptychid evolution. The two upper dentitions probably represent two distinct species of anisonchines that are intermediate, in a number of morphological features, between *Mimatuta* and the previously described Puercan genera *Conacodon, Anisonchus,* and *Hemithlaeus.* As a group the two species (represented by the palate and maxilla) resemble most closely *Conacodon cophater* and *C. entoconus,* based on both shared primitive and derived characters (which will be evaluated in the closing section on anisonchines).

During the earlier stages of this study it seemed reasonable to name a new species for each of these upper dentitions and place them in *Conacodon* or, possibly, to recognize a new genus. However, after examining specimens of *Oxyacodon* also from the lower Nacimiento Formation, it was concluded that this genus is an anisonchine periptychid and not a hyopsodontid (Matthew, 1937; Simpson, 1937a and 1945). The three species of *Oxyacodon* recognized since Simpson (1937a) are known only from lower dentitions. Therefore, it seems probable that the maxillae may belong to previously described species of *Oxyacodon*. This is especially reasonable since the upper and lower dentitions, when evaluated separately, are the most primitive ansonchines from the lower Nacimiento Formation and *Oxyacodon* is known from the locality at which the two upper dentitions were recovered (UCMP loc. V2811 = V3860). After the above comparisons had been made, separate personal communications from J. Keith Rigby, Jr., and Leigh Van Valen indicated that upper dentitions of *Oxyacodon* had also been recognized by these researchers. (The specimens described here have not been compared in detail with the material being studied by these individuals.)

The taxonomic status of *Oxyacodon* has had a somewhat checkered history. When named by Osborn and Earle (1895) *Oxyacodon apiculatus* was placed in Primates, *incertae sedis*. However, the authors did note that "The true molars in this genus resemble somewhat those of *Anisonchus*, but the structure of the last premolar is widely different" (p. 25). Matthew (1897) added a new species to the genus, *O. agapetillus (Anisonchus agapetillus* Cope, 1884a). He felt that "it [*Oxyacodon*] may belong to the Anisonchinae, but the molars do not show the closed basin of the heel peculiar to that group, and are wide instead of compressed" (p. 292). In 1937 Matthew referred two more species to *Oxyacodon, O. priscilla* and *O. turgidunculus (= Mioclaenus turgidunculus* Cope, 1888), thus recognizing four species of *Oxyacodon.* In addition, he placed *Oxyacodon* in the Hyopsodontidae in a new subfamily, Mioclaeninae (= Mioclaenidae, Osborn and Earle, 1895). Of the four species only *O. turgidunculus* was represented by upper dentitions. In 1937 Simpson erected a new mioclaenine genus, *Choeroclaenus*, for Matthew's *Oxyacodon turgidunculus* and removed the remaining three species of *Oxyacodon* to the Hyopsodontinae. The taxonomic status of *Oxyacodon* had not been reviewed since Simpson (1937a) and was still considered to be a hyopsodontine hyopsodontid with three included species, *O. apiculatus, O. agapetillus,* and *O. priscilla*, until Van Valen's brief review (1978) (see comments below).

It is not readily apparent why Matthew (1937) or Simpson (1937a) decided to place *Oxyacodon* in the Hyopsodontidae. As noted above, Matthew (1937) placed *Oxyacodon* in the Mioclaeninae, as he did all Paleocene hyopsodontids, leaving only *Hyopsodus*, the sole Eocene genus, in the Hyopsodontinae. However, Matthew did recognize the structurally primitive and intermediate position of the genus "between the Mioclaenidae, Anisonchidae, and Oxyclaenidae" (p. 206). It also seems reasonable that Matthew (1937) referred the four species of *Oxyacodon* recognized by him to the Hyopsodontidae because the one species known on upper dentitions, "*O.*" *turgidunculus*, was similar to the hyopsodontid *Mioclaenus*. Simpson (1937a) recognized the distinct features of "*O.*" *turgidunculus* by erecting a new genus, *Choeroclaenus*, for it. He modified Matthew's (1937) subfamilial assignments, and one of his changes was to retain *Choeroclaenus* in the Mioclaeninae while moving *Oxyacodon* to the Hyopsodontinae. Of the differences Simpson (1937a, p. 232) recognized between the lower dentitions of *Choeroclaenus* and *Oxyacodon*, two are considered as noteworthy here. In *Choeroclaenus*, the posterior premolars (especially P_4) are much more inflated than in *Oxyacodon* and the molar paraconids, although small as in *Oxyacodon*, are mostly internal and fusing to the metaconid. Simpson (1937a) also notes that the M_3 in *Choeroclaenus* has a "projecting hypoconulid." Although this is indeed the case, the M_3 hypoconulid in *Oxyacodon* is even more prominent and posteriorly projecting. As Matthew (1937) had done, Simpson (1937a) recognized the primitive features of *Oxyacodon*. However, he considered in much greater detail the intrafamilial relationships of the hyoposodontids. One of Simpson's (1937a) considerations was a possible structural sequence *Oxyacodon-Haplaletes-Hyopsodus*. At the same time he noted that there are no " 'cross specializations' " to argue against this sequence, he was careful to state that the probabilities of this being the exact line of descent are very low. More recent data on lower dentitions as well as that available to Simpson (1937a) suggest that *Haplaletes* is too derived to be a structural or direct ancestor of *Hyopsodus*. In *Hyopsodus* (e.g., *H. miticulus*) it can be seen that the paraconid tends to be only slightly shifted labially and is being lost by fusion to the metaconid. This is clearly not the case in *Oxyacodon* and is possibly not true for *Haplaletes*. In the earliest species of *Haplaletes*, *H. discreptatrix* (Simpson, 1935), the paraconid is reduced and median according to Simpson (1937a, p. 243). In examining casts of the type specimen, a lower jaw with P_{3-4}, M_{1-3}, only a medially positioned paralophid was obvious. This single character indicates that *Haplaletes* is already too derived to be a structural ancestor for *Hyopsodus*. The paraconid in *Oxyacodon*, although only slightly reduced, is distinctly median. This is regarded as a derived character state, and it seems unlikely that the paraconid would move lingually and fuse to the metaconid as in *Hyopsodus*. It is still possible, but not likely, that *Oxyacodon* is structurally ancestral to *Haplaletes*. A much more probable structural sequence (based only on lowers) is *Mimatuta morgoth-Oxyacodon-Conacodon*. *M. minuial* probably also fits in this sequence, but, unfortunately, the lower dentition is poorly known. The morphological features used in this sequence are considered in the closing section on early anisonchine relationships.

It was noted previously that my interpretations regarding *Oxyacodon* differ somewhat from those presented by Van Valen (1978). Rather than disperse these comments throughout the text, they will all be noted here. First, Van Valen (1978) considers *O. priscilla* to be a synonym of *O. agapetillus*. I have not compared the type specimens (both lower dentitions) of the two species; therefore, I cannot refute Van Valen's synonomy, based on the type specimens. However, I regard the maxillae described below to belong to two separate species. Their allocation to *O. agapetillus* and *O. priscilla* was based on size considerations. The large maxilla referred to *O. priscilla* could belong to a new species, but until I am convinced that this species is synonymous with *O. agapetillus* I will continue to use the name. Second, Van Valen (1978) feels that *Conacodon cophater* is probably better referred to *Oxyacodon*. Although I agree that there are a number of simlarities, I feel that it is best included with *Conacodon*. The reasons for this are made clear at the end of this section in a cladistic analysis including *Oxyacodon* and *Conacodon*. Third, although I am by no means certain, I doubt that Van Valen's new species *O. josephi* and *Escatepos campi*, which he indicated to be a junior synonym of *O. agapetillus*, should be included in the genus *Oxyacodon*. Both of these species lack the lingually expanded hypocone that is present in the two maxillae referred below to *O. agapetillus* and *O. priscilla*. Fourth, Van Valen (1978, fig. 6) suggested that *Oxyacodon* is ancestral to *Conacodon*. He clearly labeled his phylogeny as provisional and, based on his interpretation of these two genera, this may be a reasonable relationship. However, based on my interpretation of these two genera, *Oxyacodon* possesses some specializations in its upper dentition (see cladistic analysis) that would eliminate it from an ancestral position to most, if not all, other periptychids. In a more general sense, it is possible to arrange the character states seen in *Oxyacodon* and *Conacodon* in a morphocline, with the former taxon being primitive in most of its characters states.

Oxyacodon agapetillus (Cope, 1884a) Matthew, 1897

(Fig. 67 & Table 52)

Anisonchus agapetillus Cope, 1884a, p. 320
Oxyacodon agapetillus Matthew, 1897, p. 292

Lectotypes. AMNH 3557, "two fragments of the lower jaw, M_{1-2} r., M_2 l." (Matthew, 1937).

Type locality. "Probably Puerco formation [lower Nacimiento Formation], San Juan Basin, no exact locality recorded" (Matthew, 1937).

Referred specimen. UCMP 89690, palate with fragment of left P^3, P^4, M^{1-2}, lingual portion M^3 and posterior roots of right P^3, P^4, M^{1-3}.

Locality. UCMP loc. V2811, Mammelon Hill, lower Nacimiento Formation, New Mexico.

Distribution. Lower Nacimiento Formation, New Mexico (Puercan).

Revised diagnosis. Smallest species of *Oxyacodon*; parconule absent on M^{1-3}, mesostyle (crest) only on M^2, hypocone less lingually extended and relatively shorter than in *Oxyacodon priscilla*; M_{1-3} length equals 9.5 mm (measurements from Matthew, 1937, p. 208).

Comments. The association of the palate with the lowers is tentative pending analysis of the species of *Oxyacodon*. Allocation of the palate to the smallest species, *O. agapetillus*, is based on the good occlusion pattern of the palate with lower dentitions of *O. agapetillus* in the UCMP San Juan Basin collections.

Description. Only a small portion of the P^3 crown is preserved on the left side. Not enough of the crown is preserved to demonstrate whether a protocone was present. However, it is clear, from the portion preserved and the occurrence of a lingual root, that at least a lingual swelling, if not a small protocone, was present. A narrow postcingulum and stylar shelf and a "metacrista" are preserved on the left P^3 fragment.

Of the teeth preserved in the palate, P^4 shows the most striking differences from other periptychids considered in this study (other than *Mimatuta*). As in *Protungulatum donnae* and *Mimatuta*, cusps are not at all, or only slightly, inflated, although, as in at least *M. minuial* (and in periptychids), the paracone has been reduced in height. Also as in *Mimatuta*, the pre- and postprotocristae form narrow, almost continuous cingula to the stylar region. *Mimatuta* has narrow pre- and postcingula, while *Oxyacodon agapetillus* has only a narrow postcingulum on P^4. In contrast to *Mimatuta,* the pre- and postprotocristae in *O. agapetillus* are almost vertical as they descend the sides of the protocone. Further, as in other anisonchines, the parastyle and metastyle are in line with the paracone and not labial as in *Mimatuta*.

The upper molars clearly show the anisonchine affinities of *Oxyacodon agapetillus*, while retaining at least two primitive condylarth (or eutherian) characters. First, in contrast to other anisonchines but as seen in *Protungulatum donnae*, *Mimatuta morgoth*, and *M. minuial*, the stylar shelves on the molars of *O. agapetillus* are moderately wide with a distinct ectoflexus on M^{1-2}. Second, in *O. agapetillus*, as in *P. donnae* and *Mimatuta*, the postmetaconule wing forms a narrow metacingulum that runs labially to just below the metastylar region. In these three taxa the postcingulum terminates below the metacingulum at the level of the metaconule. The metacingulum and postmetaconule wing are more reduced in other anisonchines and the postcingulum forms a continuous shelf running labial to the metastylar region.

The premetaconule wing is well developed on molars of *Oxyacodon agapetillus*, as in *Conacodon cophater*. Unlike other Puercan anisonchines, in *O. agapetillus*, a small paraconule is not present, and its position is occupied by the labially positioned protocone. The protocone in *O. agapetillus* has three crests emanating from its apex. The first contacts the metacone and is the postprotocrista. The other two crests trend anterolabially from the protocone, but their homologies are less certain. The more posterior of the two crests runs from the apex of the protocone to the center of the paracone base. The second, more anterior crest descends along the anterior face of the protocone almost vertically and then trends anterolabially, terminating at the anterolingual corner of the paracone on M^1 but joining the parastylar region in M^{2-3}. Although variously developed, specimens of *C. cophater*, *C. entoconus, Hemithlaeus kowalevskianus*, and *Anisonchus gillianus* in the UCMP collections all have a paraconule, with a postparaconule wing and preparaconule wing that is a continuation of the preprotocrista. The position of the preprotocrista and the paraconular wings in these taxa suggests that the above two crests in *O. agapetillus* are modified homologues of the preprotocrista-preparaconule wing and the postparaconule wing.

The paracone and metacone are low and approximately the same size. The metacrista is well developed on M^{1-2} and the paracrista is also present on M^{1-3} but is slightly subdued. A low centrocrista is present between the paracone and metacone. In the "mesostylar" region on M^2 a crest runs labially from the centrocrista to the ectocingulum.

On M^1 the parastylar and metastylar lobes of the stylar shelf are subequal in size, with the metastylar lobe extending slightly more labially. On M^2 the parastylar lobe is more pronounced. On M^3 the parastylar lobe is large and the metastylar lobe is almost completely absent.

The hypocone varies its position on M^{1-3}. In going from M^1 through M^3 the hypocone becomes relatively more lingual to the protocone until M^3, where the hypocone and protocone are almost in direct lingual-labial alignment. Also, only on the M^3 are both the pre- and postcingulum continuous with the hypocone. Only the postcingulum connects with the hypocone on M^{1-2}. The position and relationships of the hypocone in *Oxyacodon agapetillus* are almost identical with that seen in *Conacodon cophater* except that the hypocone is longer in *C. cophater* (measured along the cusp from base to apex).

Discussion. Even more clearly than *Mimatuta morgoth* and *M. minuial*, *Oxyacodon agapetillus* possesses character states in its upper molars that are unambiguously interpreted as derived for anisonchines. The two most important shared derived characters that indicate anisonchine affinities are the development of a lingually placed hypocone and the accompanying labial shift of the protocone. In the two maxillae now tentatively referable to *Oxyacodon* (the second species, *O. priscilla*, is described below) the lingual placement of the hypocone and labial placement of the protocone are most similar to *Conacodon*. In *Conacodon* most other features of the upper dentition (e.g., more bulbous premolars, reduction of the stylar shelf, and continuous postcingulum from hypocone to metastylar region) are probably derived characters shared with other Puercan anisonchines. The lingual placement of the hypocone seen in *Conacodon* could be regarded as a shared derived character for *Conacodon* and *Oxyacodon*. Another equally possible interpretation is that the lingually positioned hypocone is a primitive condition in relationship to the more posterolingual placement of the hypocone seen in all periptychids (except *Haploconus*).

Disregarding for the moment other possible specializations, this scenario would begin with the protocone and hypocone (both with a low-angled lingual face) lingually positioned on a transversely widened molar (e.g., *Mimatuta*). Next, the protocone would shift labially, leaving a lingually positioned hypocone (e.g., *Oxyacodon, Conacodon*). Finally, with a trend toward size increase and "squaring" of the occlusal surface, the hypocone would shift posterolabially and a protostyle would develop (*Anisonchus*, *Hemithlaeus*).

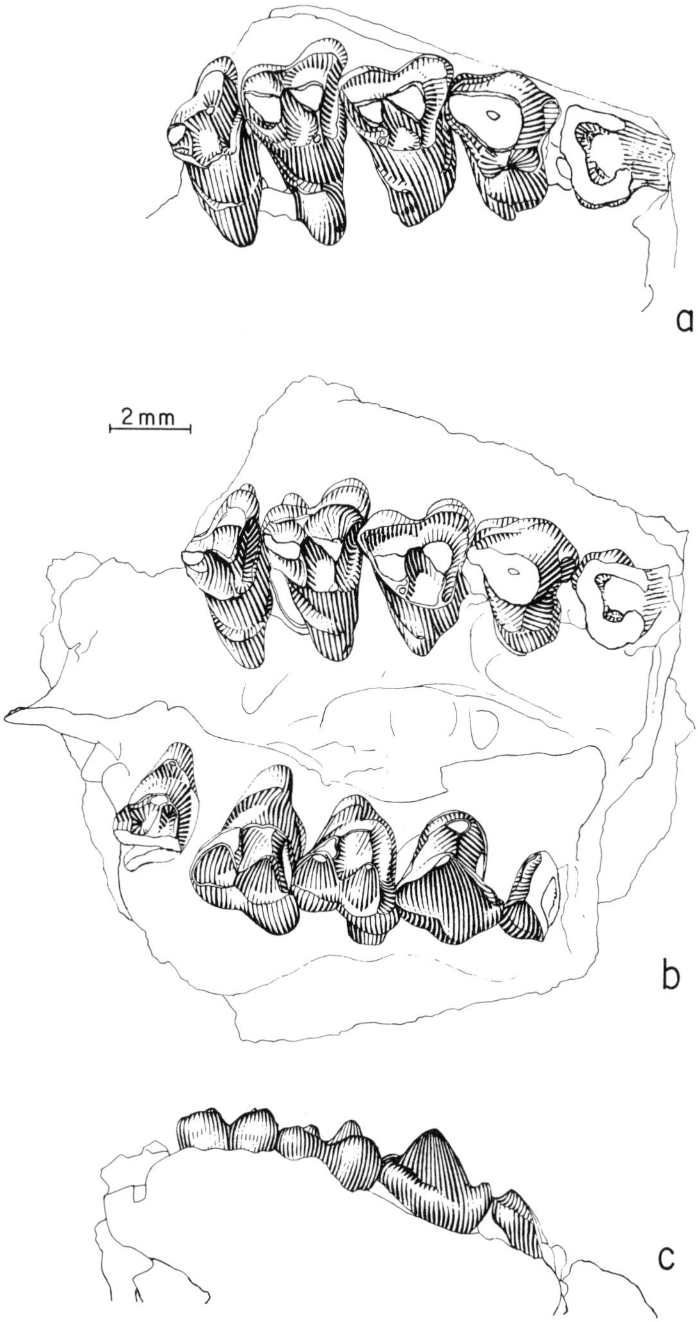

FIG. 67. *Oxyacodon agapetillus* (Cope). UCMP 89690, palate with fragment of left P^3, P^4, M^{1-2}, lingual portion of M^3 and posterior roots of right P^3, P^4, M^{1-3}, loc. V2811: (a) occlusal view of right side; (b) approximately occlusal view of palate, opposite sides slightly facing each other due to distortion along midline; (c) labial view of left side. (A. L.)

TABLE 52. Measurements of premolars and molars of *Oxyacodon agapetillus*, UCMP 89690, UCMP loc. V2811.

Position	L	W–A	W–P	A	B	C	D	L/W–P	L/W–A	A/W–A	B/W–A	C/W–A	D/W–A
LP^3	—	—	a2.80	—	(?no conules)		—	—	—	—	(?no conules)		—
LP^4	2.99	a3.20	4.08	1.41	(no conules)		*0	0.73	0.93	0.44	(no conules)		*0
RP^4	2.88	a3.27	4.07	1.39	(no conules)		*0	0.71	a0.88	a0.42	(no conules)		*0
LM^1	3.00	4.07	4.38	1.92	(no paraconule)		0.84	0.68	0.74	0.47	(no paraconule)		0.21
RM^1	2.98	4.01	4.33	1.89	(no paraconule)		a0.96	0.69	0.74	0.47	(no paraconule)		a0.24
LM^2	3.16	4.80	5.02	2.55	(no paraconule)		1.15	0.63	0.66	0.53	(no paraconule)		0.24
RM^2	3.14	4.83	5.03	2.62	(no paraconule)		1.00	0.62	0.65	0.54	(no paraconule)		0.21
LM^3	—	—	—	a2.20	(no paraconule)		—	—	—	—	(no paraconule)		—
RM^3	2.55	4.51	3.90	2.33	(no paraconule)		1.01	0.65	0.57	0.57	(no paraconule)		0.22

*There is no stylar shelf in the anterolabial corner of P^4. The small parastylar cusp is slightly lingual to the paracone when aligned for measurement (see text and Fig. 67).

a Approximate measurement.

The two contrasting possibilities, that is series of grades indicating a lingually placed hypocone may be primitive, or that this is a derived condition, must both be considered cautiously if mosaic evolution has occurred. However, the former of the two possibilities is favored, with reservations, here and in the section dealing with early anisonchine evolution.

Unlike the molars, the P^4 of *Oxyacodon agapetillus* shows no characters that can be considered as shared derived for anisonchines. The decrease in height of the paracone, the alignment of parastyle-paracone-metastyle, and the cusplike form of the para- and metastyle are found in other condylarths. These P^4 character states probably developed in parallel in the Condylarthra, but an analysis of these features is beyond the scope of this study. What does appear clear is that the otherwise primitive P^4 of *O. agapetillus* (and *O. priscilla* and *Mimatuta*) was retained in some early anisonchines, indicating that the typically anisonchine (and probably periptychid) molars developed before the P^4.

At least one character in the upper molars of *Oxyacodon agapetillus* appears to be an autapomorphy of the genus. This is the only anisonchine species in which the paraconule is completely absent (it is being reduced in *O. priscilla*) and what are probably the postparaconule wing and a combined preprotocrista-preparaconule wing emanate from the protocone. Since the upper dentition is known from only one individual, this character could be variable within the species.

The lower dentition has not been considered because topotypic material was not on hand during the study and it was not possible to evaluate Matthew's taxonomy of *Oxyacodon*.

Oxyacodon priscilla Matthew, 1937

(Fig. 68 & Table 53)

Oxyacodon priscilla Matthew, 1937, p. 207

Type. AMNH 3547a, right lower mandible with P_{2-4}, M_{1-3}. (figured by Matthew, 1897, fig. 10, under the name *Oxyacodon agapetillus*).

Type locality. Not given by Matthew (1937), but probably lower Nacimiento Formation, New Mexico.

*Referred specimens.*UCMP 36640, left maxilla with P^4, M^{1-3}.

Locality. UCMP loc. V3860 (=V2811), Eduardo Canyon (=Mammelon Hill), lower Nacimiento Formation, New Mexico.

Distribution. Lower Nacimiento Formation, New Mexico (Puercan).

Revised diagnosis. Larger than *Oxyacodon agapetillus* (but smaller than *Oxyacodon apiculatus?);* paraconule progressively being lost on M^{1-3}, mesostyle present on M^{1-3}, hypocone more lingually extended and relatively longer than in *Oxyacodon agapetillus*; M_{1-3} length equals 11.2 mm (measurements from Matthew, 1937, p. 207).

Comments. The left maxilla, UCMP 36640, is tentatively assigned to *O. priscilla* on the basis of its size.

A right mandible with P_{3-4}, M_{1-3} identified as *Oxyacodon priscilla* (by the author based on Matthew, 1937) was originally catalogued with the left maxilla under UCMP 36640. Although the maxilla is tentatively identified as *O. priscilla*, the mandible and maxilla probably do not belong to the same individual. The mandible (now catalogued as UCMP 116452) is from the opposite side (right), its preservation is slightly different, and it shows more dental wear than the maxilla.

Description. The P^4 of *Oxyacodon priscilla* is very similar to that of *O. agapetillus* described in the preceding pages. The few differences that do occur are relatively minor. In the P^4 of *O. priscilla* the paracone and protocone are slightly more inflated. In place of a small metastylar cusp, as in *O. agapetillus*, there is a small posterolabially oriented metastylar crest. Also, the postcingulum extends further around the lingual side of the protocone in *O. priscilla*.

The most labial portion of the metastylar region on M^2 and part of the parastylar region on M^3 are broken in UCMP 36640. The anterolabial root on M^3 appears to be in its correct relationship to the other molars, but the rest of the M^3 has been displaced posterolingually. However, it is clear, based on the portions preserved, that the stylar shelf is almost as wide and the ectoflexus almost as deep on M^{1-2} of *Oxyacodon priscilla* as in *O. agapetillus*. Also as in *O. agapetillus*, the parastylar region of M^3 is large while the metastylar region is almost absent. The major difference between the stylar shelves in the molars of *O. agapetillus* and *O. priscilla* is in the "mesostylar region" of the ectocingulum. As noted previously, only the M^2 of *O. agapetillus* has a small "mesostylar" crest running from the ectocingulum to the lowest point on the centrocrista. In contrast, all three molars of *O. priscilla* bear a cuspule on the "mesostylar region" of the ectocingulum. These cuspules are relatively larger than the single ridge on the M^2 of *O. agapetillus* and increase in size going from M^1 to M^3. Only the mesostyle on M^3 bears a small ridge running toward the centrocrista.

All three molars of *Oxyacodon priscilla* possess metaconules with well-defined pre-and postmetaconule wings. The paraconular areas on M^1 and M^3 are worn (and slightly damaged on M^3), but on M^2 this area is not worn and shows that a paraconule was not present. The pre- and postprotocristae slope gradually away from the apex of the protocone leading, respectively, toward the antero- and the posterolabial corner of the M^2. In turn the cristae merge, respectively, into the preparaconule and postmetaconule wings. However, unlike the premetaconule wing, which joins the postmetaconule wing and postprotocrista to form a metaconule, the postparaconule wing almost reaches the apex of the protocone. If correctly interpreted, the M^{1-3} show progressive loss of the paraconule by shifting the lingual end of the postparaconule wing closer to the apex of the protocone. Although slightly worn, it appears that the postparaconule wing reaches the apex of the protocone on M^3. This is the condition seen in all three molars of *O. agapetillus*. The only addition seen in *O. agapetillus* is the lowering and almost cingulum-like placement of the combined preprotocrista-preparaconule wing.

The only major difference between the molars of *Oxyacodon priscilla* and *O. agapetillus*, besides the former species' larger size, is in the hypocone. In the former species the base of the hypocone is more bulbous, especially on M^2, which expands this cusp more lingually. This recalls the condition in *Conacodon*. Also as in *Conacodon* but not in *O. agapetillus*, the lingual expansion of the hypocone increases this cusp's height measured tangentially along the cusp. Finally, unlike both *Conacodon* and *O. agapetillus*, the precingulum on the M^3 of *O. priscilla* merges with the base of the protocone before reaching the hypocone.

Discussion. In the preceding section dealing with *Oxyacodon agapetillus* it was noted that this species possesses an interesting combination of shared derived anisonchine (and ?periptychid) characters and primitive condylarth characters. This is also true of *O. priscilla*.

The peculiar way in which the paraconule is being lost in *Oxyacodon priscilla* and is finally lost in *O. agapetillus* suggests a specialized trend that would eliminate at least these

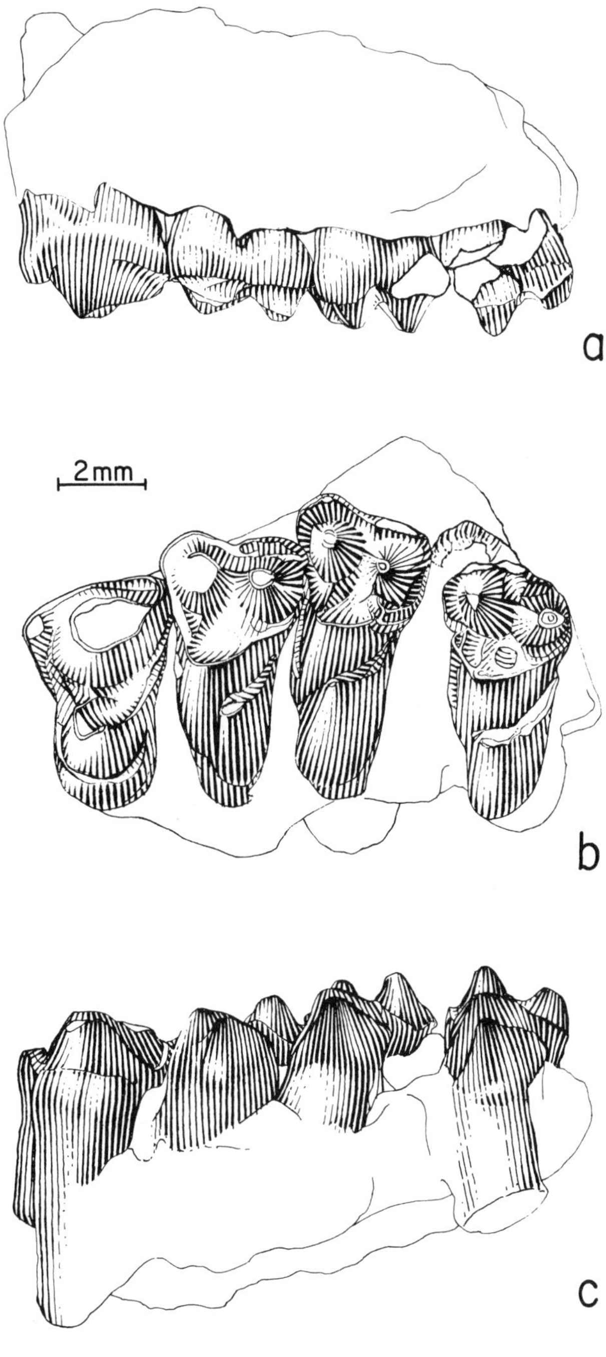

FIG. 68. *Oxyacodon priscilla* Matthew. UCMP 36640, left maxilla with P^4, M^{1-3} (parastyle of M^3 broken and tooth displaced posterolingually), loc. V3860 (= UCMP loc. V2811): (a) labial view; (b) occlusal view; (c) lingual view. (A.L.)

TABLE 53. Measurements of P^4, M^{1-3} of *Oxyacodon priscilla*, UCMP 36640, UCMP loc. V3860 (=UCMP loc. V2811).

Position	*L*	*W–A*	*W–P*	*A*	*B*	*C*	*D*	*L*/*W–P*	*L*/*W–A*	*A*/*W–A*	*B*/*W–A*	*C*/*W–A*	*D*/*W–A*
P^4	3.51	a4.55	5.02	1.82	(no conules)		*0	0.70	a0.77	0.52	(no conules)		*0
M^1	3.49	5.32	5.50	2.51	0.42	1.47	0.92	0.63	0.66	0.47	0.08	0.28	0.17
M^2	3.42	6.09	6.02	3.42	0.04	1.58	1.05	0.57	0.56	0.56	0.01	0.26	0.17
M^3	a3.55	a5.60	a5.27	2.78	(no paraconule)		a1.36	a0.67	a0.63	0.50	(no paraconule)		0.24

* There is no stylar shelf in the anterolabial corner of P^4.
a Approximate measurement.

two species of *Oxyacodon* from being either structural or patristic ancestors of later known anisonchines.

The *Conacodon*-like lingual placement of the hypocone was discussed earlier and was tentatively suggested as the primitive condition for at least Puercan anisonchines.

The "mesostyle" development in *Oxyacodon* and other anisonchines appears to be quite variable throughout the subfamily, although there may be a general trend to develop this cusp in more derived forms.

Thus, except for the loss of the paraconule, *Oxyacodon* possesses structurally the most primitive upper dentition among Puercan anisonchines.

As will be discussed briefly in the closing section on early trends in anisonchine evolution, the lower molars of *Oxyacodon* are also structurally quite primitive. This is probably one reason why *Oxyacodon* was not recognized as an anisonchine by Matthew (1937) and Simpson (1937a).

Family ?PERIPTYCHIDAE

gen. et sp. indet.

(Fig. 69 & Table 54)

In addition to the specimens of *Mimatuta minuial*, *Protungulatum engdahli*, and *Ragnarok nordicum*, LACM loc. 3103 (=UCMP loc. V72210) yielded a fourth condylarth, a left dentary fragment, LACM 112903, preserving an M_3.

The portion preserved indicates that the mandible was moderately deep. The anterior edge of the masseteric fossa is borderd by a well-delimited ridge oriented at about 50° to a line drawn through the alveoli. The ridge is broken along most of its surface, but even with this, it is quite robust.

The M_3 is moderately inflated, and the talonid and especially the trigonid appear to be closing by approximation of their cusp apices. The metaconid and protoconid are the same height, but the protoconid is more robust. The paraconid is only slightly smaller than the metaconid, but is not as tall and is shifted slightly labially. The M_3 is unworn and small crests join all the trigonid cusps.

The hypoconid is slightly larger and taller than the other talonid cusps. The hypoconulid and entoconid are approximately the same size, with the hypoconulid being taller. The hypoconulid is set somewhat posteriorly, but does not form a distinct lobe as in *Protungulatum donnae*. The talonid cusps are not quite equidistant, with the hypoconulid and entoconid more closely approximated. Low crests join all the cusps around the perimeter of the talonid. The cristid obliqua joins the base of the trigonid below the midpoint of the protolophid. A slightly lower entocristid joins the trigonid below the metaconid. A lingual cingulum also begins at the entoconid and becomes very narrow below the widest portion of the trigonid, merging with the trigonid below the paraconid. A

TABLE 54. Measurements of M_3, ?Periptychidae, gen. et sp. indet., LACM 112903, LACM loc. 3103 (=UCMP loc. V72210).

L	*L–TR*	*L–TA*	*W–TR*	*W–TA*	*P–M*	*L–TR* / *L–TA*	*W–TR* / *W–TA*
7.74	3.39	4.35	5.38	4.20	0.19	0.80	1.28

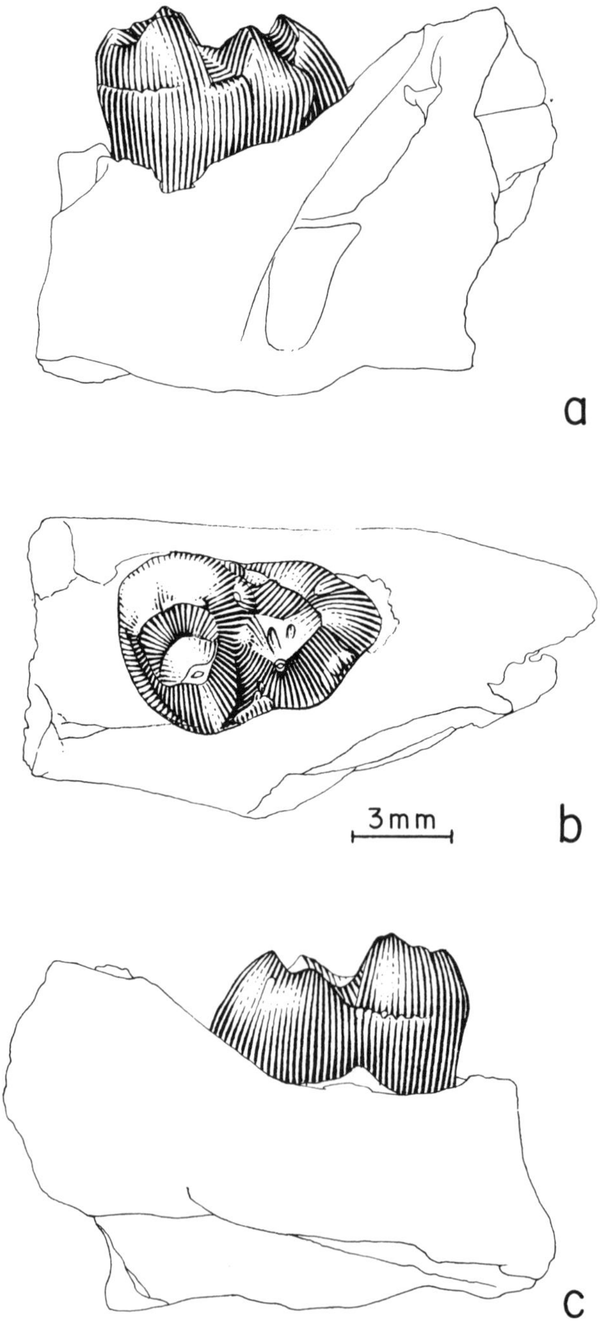

FIG. 69. ?Periptychidae, gen. et sp. indet. LACM 112903, left dentary with M_3, LACM loc. 3103 (= UCMP loc. V72210): (a) labial view; (b) occlusal view; (c) lingual view. (A.L.)

wider anterior and labial cingulum runs from the below the front of the paraconid to below the hypoflexid. This cingulum is interrupted at the widest portion of the protoconid. The hypoflexid forms a well-defined triangular area covered by numerous small crenulations and bumps. The labial cingulum is slightly expanded below the hypoflexid, forming a small ectostylid cuspule. The labial cingulum continues below the hypoconulid as a series of four small ridges and cuspules.

Discussion. The salient features, other than size, in this M_3 which distinguish it from *Protungulatum donnae* are (1) inflation of the crown; (2) presence of a metacristid; (3) the more labial position of the paraconid; (4) the "closing" of cusp apices; (5) relatively lower trigonid; (6) presence of an entocristid; (7) presence of narrow lingual cingulum; and (8) the hypoconulid not produced into a distinct posterior lobe. In contrast to *Protungulatum donnae*, these are all regarded as derived character states. Although not uniquely periptychid, the "closing" of the cusp apices, the position of the entocristid, and the type of inflation (i.e., nearer the base of the molar) strongly suggest affinities with this family. Such characters as the labially shifted although unreduced paraconid and the diminution of the hypoconulid lobe are not unique periptychid characters, but are found in this family. One feature, the presence of a metacristid, is not to my knowledge found in the periptychids. However, other condylarths, (e.g., *Protogonodon*) possess this presumably derived character.

Trends in Early Anisonchine Evolution

In the most recent comprehensive study of periptychids Matthew (1937) recognized the affinities of primitive members of this family with early members of other condylarth families. Although changes have been made in condylarth classification since Matthew (1937: e.g., the order Taligrada is no longer recognized), the recognition that families of this order share a close common ancestry has not been modified. If anything, this relationship has been strengthened by the discovery of Late Cretaceous condylarths, most notably *Protungulatum donnae*, which in almost every respect is a good structural and plausible patristic ancestor for later condylarths (Sloan and Van Valen, 1965). Material referable to the genus of anisonchine, *Mimatuta*, from Garfield County, Montana, was described in the preceding pages. The very primitive structure of this taxon's P^4, M^{1-3}, plus the description of the upper dentition of two primitive anisonchines (*Oxyacodon agapetillus* and *O. priscilla*) from the Puercan of New Mexico, suggests the mode of evolution of the anisonchine type of dentition from the sort of dental morphology seen in *P. donnae*.

Fig. 70 shows the trends in development of the upper dentition in anisonchines. Although Fig. 70 is a cladogram, in the sense that it indicates the branching sequence and utilizes shared derived characters to determine dichotomies, it is not intended as a definitive statement on anisonchine (or periptychid) affinities. A thorough revision of the Periptychidae is beyond the scope of this paper. Topotypic specimens, or in some cases representative samples of certain taxa (e.g., *Coriphagus*), were not on hand for this study. Therefore, Fig. 70 is intended to show only the trends in development of dental characters among Lancian through Puercan anisonchines. Finally, no attempt has been made to determine the most probable sister group of the anisonchines. Without a thorough analysis of the Periptychidae it is not possible to go beyond speculation, but it is doubted that the Periptychinae is the sister group of all anisonchines. Rather, periptychines are probably more closely related to a specific anisonchine genus, possibly *Hemithlaeus*.

Only features of the upper dentition were used to construct the cladogram in Fig. 70. The lower dentitions are considered below.

Table 55 lists the characters used in this analysis. Most of the data in Table 55 are self-explanatory and have been discussed on the preceding pages. Those character states which are considered to be derived are indicated by blackened rectangles in Fig. 70. A connection between blackened rectangles indicates that the character state for that group of genera is interpreted as being shared derived. In order to show more clearly how the characters in Table 55 were analyzed and to indicate where data are lacking, the array used in Fig. 70 was chosen. One shortcoming of this method is that it is difficult to show morphoclines. This has been partially overcome by using capital letters in some of the blackened rectangles. These letters indicate the approximate degree of change from the primitive condition within each of the morphoclines. Clearly, the way in which characters have been subdivided in Table 55 is somewhat arbitrary. A different grouping would yield a different number of characters; if character states have been correctly interpreted, however, the sequence of dichotomies would not change. In short, in this instance the number of characters recognized has little effect on the recognition of dichotomies.

The probable lower dentitions are known for all of the taxa considered in Table 55 and Fig. 70, but only poorly for *Mimatuta minuial*. In Table 56 characters of the premolar dentition (i.e., P_{3-4}) have been overly split to allow consideration of this part of the dentition. Unlike the lower molars, the P_{3-4} show a great amount of variation among Puercan anisonchines. The only possibly shared derived characters of the P_{3-4} in Puercan anisonchines (but to a lesser degree in *M. morgoth*) that suggest common ancestry are (1) the overall increase in the size of the P_{3-4} relative to the molars, and (2) the tendency to develop a slightly basined talonid on P_{3-4}. Beyond these two features, almost every species possesses a unique combination of characters in P_{3-4}. One important exception is the inflation of the P_{3-4}; interestingly, it is the three largest Puercan anisonchines, *Anisonchus gillianus*, *Hemithlaeus kowalevskianus*, and *Conacodon entoconus*, that show the greatest inflation of these teeth. It is also these three species that show the greatest "closing" of the molar talonids by approximation of the talonid cusp apices.

If the lower dentitions of these taxa are considered without regard for the upper dentition or for possible parallelisms,the cladogram in Fig. 71 results. The cladogram in Fig. 71 obviously shows a different branching sequence from that based on the upper dentition (Fig. 70). The cladogram in Fig. 71 has purposely been presented in this manner to point out what are regarded as important trends and parallelisms in early anisonchine (and periptychid) evolution.

First, it has been argued on the preceding pages and in the above discussion of upper dentitions that specialization of the premolars evolved after "typical" anisonchine molars were already developing. The P^4's of *Mimatuta morgoth*, *M. minuial*, *Oxyacodon agapetillus*, and *O. priscilla* suggest this trend. Thus, it is not at all odd that the premolars in the lower dentition show much greater variation within the subfamily than do the molars.

Second, as a corollary to the first trend, it appears that the inflation of the lower premolars is less a result of close phyletic relationship than of size-related features. This is probably also true for the "closing" of the molar talonids, which is most apparent in the larger species. Although not as clearly a result of inflation, the "closing" of the molars has probably occurred partly by inflating the molar bases. This seems to be supported by the fact that among the four most derived Puercan anisonchines (*Conacodon cophater*, *C. entoconus*, *Anisonchus gillianus*, and *Hemithlaeus kowalevskianus*) it is in the three largest forms (the three latter taxa) that the basal cingula are most muted.

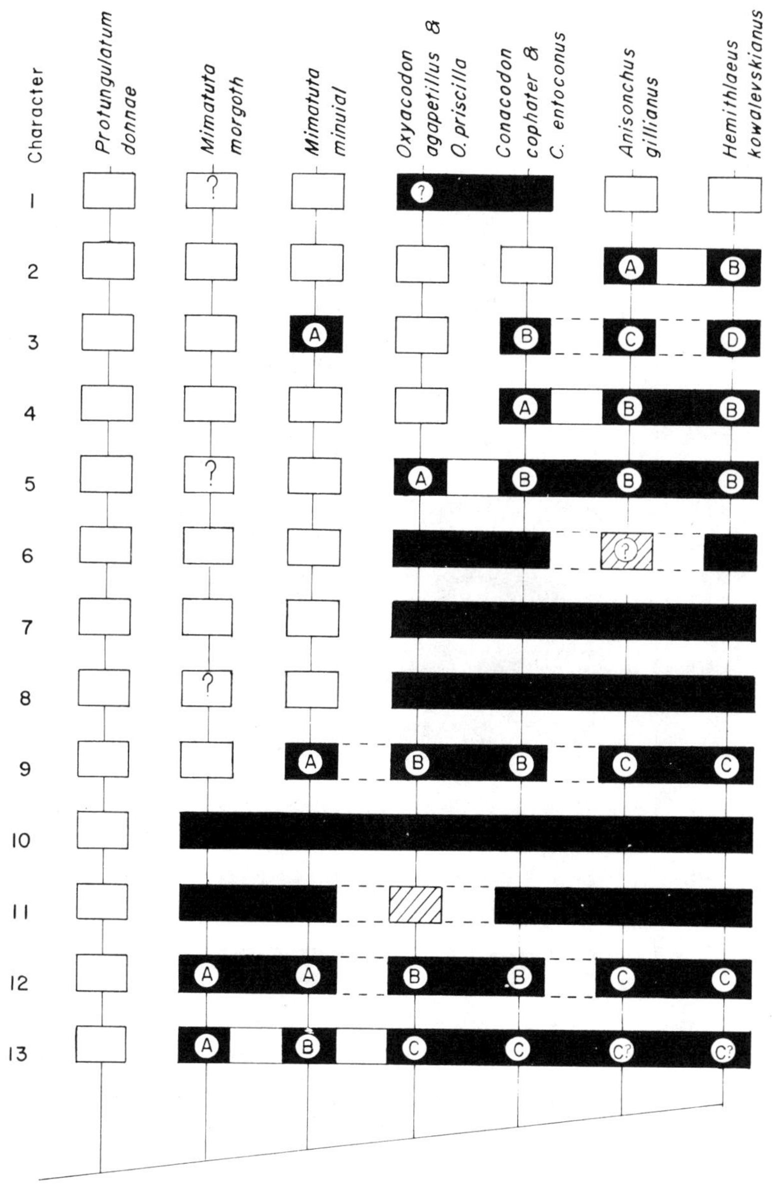

FIG. 70. Cladogram of *Protungulatum donnae* and Lancian through Puercan anisonchines, based on upper premolar and molar characters. Character states are represented as follows: Open rectangles = primitive. Black rectangles = derived (shared derived when connecting different taxa). Diagonal lines = derived differently from the general morphocline. Letters within black rectangles = level of morphocline for a particular character. Solid lines connecting black rectangles = a morphocline for a particular character. Dashed lines = possible but uncertain morphocline. Table 55 lists the characters used to construct the cladogram.

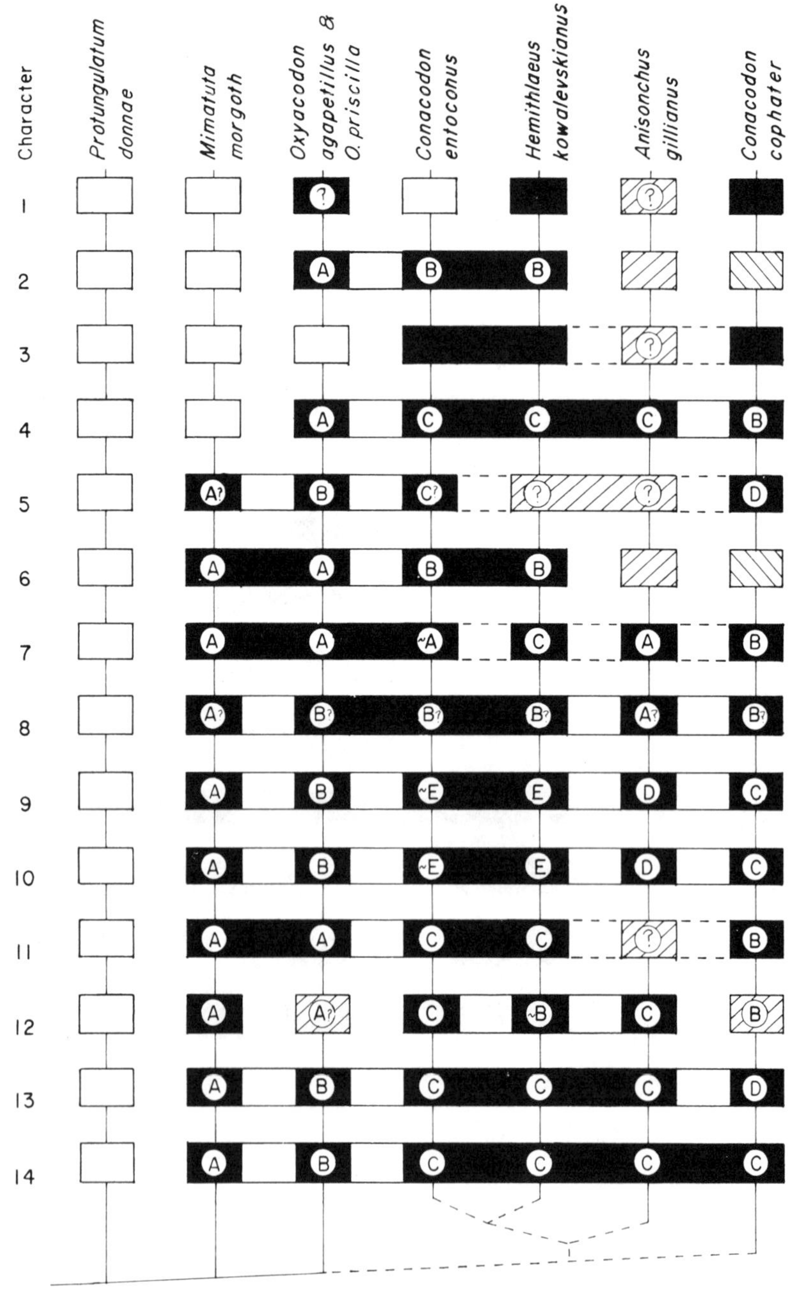

FIG. 71. Cladogram of *Protungulatum donnae* and Lancian through Puercan anisonchines, based on lower premolar and molar characters. See Fig. 70 for explanation. Table 56 lists the characters used to construct the cladogram.

TABLE 55. Character states of P^{3-4}, M^{1-3} in *Protungulatum donnae* and Lancian through Puercan anisonchines. Question marks indicate that the condition is not known. Arrows and dittos indicate similarity to the condition in preceding taxon or character.

Character	*Protungulatum donnae*	*Mimatuta morgoth*	*Mimatuta minuial*	*Oxyacodon agapetillus & O. priscilla*	*Conacodon cophater & C. entoconus*	*Anisonchus gillianus*	*Hemithlaeus kowalevskianus*
1. Protocone on P^3	present →	(ditto?)	present (Van Valen, 1978, pl. 7, fig.4)	?probably present, but very small	absent or very small	present →	(ditto)
2. Protostyle on molars	absent →	(ditto) →	(ditto) →	(ditto) →	(ditto)	absent or ?small	subequal to hypocone
3. Stylar shelf on molars	relatively wide, with distinct ectoflexus →	(ditto)	moderately wide, but parastylar region on M^2 can be reduced	similar to *Protungulatum*, with distinct ectoflexus	narrow but continuous; parastylar region M^2 slightly enlarged	narrow, with infrequent gaps	narrow, with frequent gaps
4. Post-cingulum & metacingulum on molars	both distinct & not connected; strong post metaconule wing forming metacingulum →	(ditto) →	(ditto) →	(ditto)	meta- & post-cingulum forming continuous shelf interrupted by slightly less prominent post-metaconule wing	cingula form continuous shelf not interrupted by postmetaconule wing	(ditto)
5. P^4 to M^1 size ratio	P^4 smaller	?	P^4 smaller	P^4 smaller to subequal	P^4 subequal to larger →	(ditto) →	(ditto)

Character	*Protungulatum donnae*	*Mimatuta morgoth*	*Mimatuta minuial*	*Oxyacodon agapetillus & O. priscilla*	*Conacodon cophater & C. entoconus*	*Anisonchus gillianus*	*Hemithlaeus kowalevskianus*
6. Mesostyle on molars	absent →	(ditto) →	(ditto)	small crest on M^2 of *O. agapetillus;* distinct on all molars of *O. priscilla*	occasional swellings on M^{1-2}, less so on M^3	?absent	swelling of ectocingulum through mesostyle & cusp "D" on at least M^{1-2}
7. Inflation of P^4 paracone & protocone	neither cusp swollen	similar to *P. donnae* →	(ditto)	slight swelling of both cusps	some swelling of both cusps, esp. protocone of *C. cophater* & paracone of *C. entoconus*	both cusps somewhat swollen →	(ditto)
8. Parastyle & metastyle on P^4	both labial to paracone	?same as in *P. donnae*	same as in *P. donnae*	styles shifted lingually & aligned with paracone →	(ditto) →	(ditto) →	(ditto)
9. Size & position of hypocone	small or absent, posterolingual to protocone →	(ditto)	distinct, posterolingual to hypocone	prominent structure, columnar to conical; more lingual than posterior →	(ditto)	tending to be more columnar; more posterolingual than *Oxyacodon-Conacodon* →	(ditto)

10. P⁴ paracone & protocone height	paracone much taller	paracone → taller, but difference lessened	(ditto) →	(ditto) →	(ditto) →	(ditto) →	(ditto)
11. Position of molar conules	relatively lingual	shifted → slightly labially	(ditto)	paraconule being lost; metaconule similar to *Mimatuta*	approximately as → in *Mimatuta*	(ditto) →	(ditto)
12. Shape of molars in occlusal view	transversely widened & subquadrate	transversely → widened & more triangular	(ditto)	transversely → widened, due to lingual hypocone & triangular	(ditto)	moderately → transverse, but cusp apices relatively labial & subtriangular	(ditto)
13. Position of protocone	protocone lingual; relatively high-angled lingual face	cusp apex slightly more labial; lower-angled lingual face	similar to *M. morgoth,* but lower-angled lingual face	cusp apex → shifted more labially, but angle of lingual face similar to *M. minuial*	(ditto)	similar to → *Oxyacodon-Conacodon*, but cusp base may be relatively more lingual	(ditto)

TABLE 56. Character states of P_{3-4}, M_{1-3} in *Protungulatum donnae* and Lancian through Puercan anisonchines. Notations are as in Table 55.

Character	*Protungulatum donnae*	*Mimatuta morgoth*	*Oxyacodon agapetillus & O. priscilla*	*Conacodon entoconus*	*Hemithlaeus kowalevskianus*	*Anisonchus gillianus*	*Conacodon cophater*
1. P_3 metaconid	small ridge →	(ditto)	?absent	small ridge	absent	small cusp or ridge?	?absent
2. P_3 anterior basal cusp	small cusp on midline →	(ditto)	very small or ?absent slightly lingual to midline	absent →	(ditto)	relatively large and lingual	low distinct cusp on midline
3. M_3 hypoconulid	well-developed, distinct posterior extension →	(ditto) →	(ditto)	distinct cusp, not extended posteriorly →	(ditto)	distinct cusp, slight posterior extension	similar to *C. entoconus* & *Hemithlaeus*
4. Placement of talonid cusps	entoconid & hypoconulid closer; talonid "open" →	(ditto)	all 3 cusps almost equidistant; talonid "open"	all 3 cusps equidistant; talonid "closing" →	(ditto) →	(ditto)	all 3 cusps equidistant; talonid relatively "open"
5. Molar cingula	narrow, discontinuous labial cingulum below protoconid & cristid obliqua	labial cingulum less distinct; some molars with slight lingual cingulum below metaconid	similar to *M. morgoth*, but lingual cingulum below paraconid more prominent	most similar to *C. cophater*, but cingula (esp. lingual) less prominent	portions of both cingula present, but lingual cingulum mostly absent →	(ditto)	both cingula best-developed of any species

6. P_4 paraconid	distinct but low & lingual	smaller but low & lingual	similar to *M. morgoth*, slight labial shift	absent	crest only	moderately large, just lingual to midline	distinct, low, almost on midline
7. P_4 talonid	trenchant, with small posterior cusp	not tren- → chant, with 2 indistinct cusps on pos- terior crest	(ditto)	basined, 2 cusps; small compared to trigonid	well-devel- oped & wide, usually with 3 cusps	similar to *C. entoconus*, but relatively larger	wide, well- developed basin with 2 cusps
8. P_3 heel (or talonid)	trenchant & small	trenchant, wider than *P. donnae*, with small cusp	?very small, *O. agapetillus; ?small. O. priscilla*	small, → cusplike heel	(ditto)	small heel, relatively as wide as in *M. morgoth*	wide, cusplike heel
9. P_4 protoconid	tall and not inflated ↓	tall and very slightly inflated ↓	relatively shorter and slightly more inflated ↓	moder- ately tall and well- inflated ↓	moderately short and well-inflated ↓	short, with moderate inflation ↓	similar to *Oxyacodon*, slightly more inflated ↓
10. P_3 protoconid	(ditto)	(ditto)	(ditto)	(ditto)	(ditto)	(ditto)	(ditto)
11. P_4 metaconid	small, dis- tinct, appressed to protoconid	smaller → but similar to *P. donnae*	(ditto)	small crest →	(ditto)	distinct cusp?, appressed to protoconid	small cusp, usually a thickening on a crest
12. Size and shape of P_{3-4} relative to molars	not inflated or elongated; smaller or subequal to M_1	slight inflation across talonid, not elongate; smaller or subequal to M_1	slightly more inflation than *M. morgoth;* P_4 longer than M_1 (at least 1 specimen)	inflated; P_4 slightly larger (esp. wider) than M_1	inflated; P_4 subequal to M_1	P_4 slightly larger than M_1 but not as in- flated as in *C. entoconus*	some inflation; P_4 distinctly larger and longer than M_1

Character	*Protungulatum donnae*	*Mimatuta morgoth*	*Oxyacodon agapetillus & O. priscilla*	*Conacodon entoconus*	*Hemithlaeus kowalevskianus*	*Anisonchus gillianus*	*Conacodon cophater*
13. M_{1-2} talonid size relative to trigonid	compared to other taxa, talonid is relatively narrow and long	slightly shorter & wider than in *P. donnae*	width similar to *M. morgoth,* but ? slightly shorter	slightly → shorter and wider than in *Oxyacodon*	(ditto) →	(ditto)	tends to have shortest & widest talonid, esp. M_2
14. Paraconid on molars	lingual and not reduced	close to midline & not reduced, not as closely appressed to trigonid as in following taxa	close to midline and slightly reduced	close to → midline and more reduction than in *Oxyacodon*, esp. M_3	(ditto) →	(ditto) →	(ditto)

Thus, unlike the upper dentition, there are no features in the lower dentition which can be unambiguously used to align the various genera or species. However, the taxa considered in this analysis do suggest a scenario for the development of the lower dentition of anisonchines (and periptychids). Two of these trends were mentioned above in the discussion of P_{3-4}, and they are repeated here. From *Protungulatum donnae* to *Mimatuta morgoth* to *Oxyacodon* to the other taxa, there was a general trend to increase the size of premolars relative to the molars and develop a basined talonid, especially on P_4. There was also a tendency to lower and slightly inflate the P_{3-4} protoconids. In the molars the paraconid quickly became medial in position (*M. morgoth*), and then was reduced in size and became more closely appressed to the other trigonid cusps. In *M. morgoth* there is already evidence of a trend to shorten the molar talonids relative to the trigonid. In later taxa the M_3 hypoconulid was reduced in prominence, the molars, especially below the trigonid, developed narrow lingual cingula, the talonid cusps became equidistant by increasing the talonid width, and in the larger forms the molar cusp apices appear to be "closing," probably because the molar bases were inflated.

The cladogram of the upper dentitions of Lancian through Puercan anisonchines (Table 55 and Fig. 70) is regarded as representing the most probable cladistic relationship of these species. The data on both upper and lower dentitions suggest that the molar patterns found in Tertiary periptychids began to develop before the enlarged, bulbous premolars evolved.

BIOSTRATIGRAPHY

INTRODUCTION

Various views have been held in the last hundred years concerning mammalian evolution during the Late Cretaceous. Although the reporting of "Paleocene-aspect" mammals from the latest Cretaceous of Montana by Sloan and Van Valen (1965) added important new data, some of these older views are still quite common. The general theme of most of these older views is that little, if any, mammalian evolution occurred during the latest Cretaceous of North America or it occurred elsewhere and that a major reorganization of the mammalian "community" occurred at the Cretaceous-Tertiary boundary through major extinctions, immigrations, and radiations.

Simpson (1929a, pp. 154–157) assailed parts of this viewpoint, particularly the thesis held by Marsh (e.g., 1892, pp. 250–251) that there is a great gap between Cretaceous and Tertiary forms. Simpson noted Osborn's opposing view (1893, p. 327) that the Laramie (essentially Lance Formation mammals at that time) and early Tertiary mammals are very similar to one another and that there is no "great gap" as postulated by Marsh. Simpson (1929a, p. 156) held that "the essential difference between the Lance and Paleocene faunas lies in the presence in the latter of very numerous new groups" (presumably referring to eutherians and some multituberculates). He felt there were three possible explanations: (1) the Paleocene forms arose from known Lancian forms or close relatives; (2) the ancestors of Paleocene forms were present in North America during "Lance times," but lived in a different facies from known Lance forms; or (3) most Paleocene mammals arose from forms living elsewhere which suddenly appeared in the record by migration.

Simpson said the first idea should be abandoned because the "true affinities of Lance mammals" were now known (for the time in which this was written, 1929) and the time span is too short between latest Cretaceous and earliest Paleocene mammals to allow the evolution of the latter from the former. Simpson also did not agree with the second model of facies differences, which he felt was implicit in Osborn's ideas but more explicitly given by Matthew (e.g., 1921, p. 221).

Without ignoring these authors' ideas, Simpson concluded: (1) that there is a distinct Cretaceous mammalian fauna, just as there are for the Paleocene and Eocene; (2) "that the Paleocene fauna was *not* [Simpson's italics] the culmination of any previous fauna in the same region"; but (3) that "the beginning of the Paleocene was here marked by migratory movements . . . involving the introduction of new groups" which evolved elsewhere.

In 1937 (pp. 352-354), Matthew responded to Simpson's arguments. The major arguments presented by these two authors are not given, but several comments by Matthew are important in relation to recent discoveries. Matthew suggested: (1) that "the main *ordinal* [Matthew's italics] differentiation of the Paleocene mammals must be referred back to the Cretaceous period . . ." (see Matthew's fig. 85, 1937); and (2) "The assumption that the Paleocene mammals must necessarily have been absent during the Lance from the whole of the 'large area' (western North America) through which it is found goes far beyond what the facts really warrant . . . ."

Recent work supports certain aspects of both of the models proposed by Simpson and Matthew. Clemens (1973; see also Jeletsky, 1962; Jeletsky and Clemens, 1965; and Lillegraven, 1969, for the Trochu local fauna) has shown that the Lance local fauna, Wyoming, is a correlative of at least part of the Maestrichtian, and thus by definition is Late Cretaceous, as was suggested by Simpson. The discovery and preliminary description of the Bug Creek Anthills mammals by Sloan and Van Valen (1965) showed that at least some of the Puercan species had closely related forms already living in western North America before the close of the Cretaceous. Matthew was wrong in his idea that the Lance and Puerco are similar in age and differing only in facies. However, he was correct in his suggestion of a pre-Tertiary radiation of mammals in North America. In addition, the two types of mammal "faunas" now known from the latest Cretaceous appear to differ in their paleoecology and environment of deposition (Estes and Berberian, 1970), as was anticipated by Matthew's model.

The mammals from the Hell Creek Formation and lower portion of the Tullock Formation of Garfield County, and the associated stratigraphic data, are discussed in the following pages. It will be shown that far from being a simple picture, mammalian evolution near the end of the Cretaceous was complex. This is particularly important when it is emphasized that eastern Montana is the *only* place where the continental Cretaceous-Tertiary transition is known in any detail. It is very likely that each new contribution to this period of mammalian evolution will bring added complexity. Precisely because this mammalian record is geographically limited, it is first necessary to discuss how the Cretaceous-Tertiary boundary has been estimated in Garfield County.

CRETACEOUS-TERTIARY BOUNDARY IN EASTERN MONTANA

In this study, as in many analogous studies, questions concerning lithologic, time-stratigraphic, and time boundaries become intertwined. In Garfield County the Hell Creek–Tullock formational contact can usually be established with relative ease and certainty as being at the base of the lowest, laterally persistent lignite (the Z coal). However, as discussed in the chapter on Stratigraphy, determining the formational boundary can be difficult in some instances.

Establishing where the Cretaceous-Tertiary (and of course Mesozoic-Cenozoic) boundary falls within this lithologic sequence is not as simple. When R. Brown (1952) popularized the concept of placing the Cretaceous-Tertiary boundary at the first coal above the highest dinosaurian remains he helped to stop the wrangling that had been going on concerning this boundary. For the time this was a very utilitarian solution, but as his "formula" clearly states (see chapter on Stratigraphy) it must be assumed that the close of the Cretaceous equals the extinction of the "last" dinosaur.

In this study I have followed tradition by placing the Hell Creek–Tullock contact at the base of the lowest lignite. The extinction of dinosaurs has also been used to mark the Cretaceous-Tertiary boundary. However, these two boundaries do not coincide; the latest dinosaurs occur approximately 6 to 20 ft. (1.8–6.1 m) below the lowest lignite. Therefore, the lowest lignite has not been used to mark the Cretaceous-Tertiary boundary because dinosaur extinction seems to have occurred before the deposition of this lignite. Also as discussed in the chapter on Stratigraphy, because biologic events should not be used to mark lithologic boundaries, dinosaur extinction has not been used in this study as part of the definition of the Hell Creek-Tullock formational contact. Using dinosaur extinction to mark the Cretaceous-Tertiary boundary remains the most operational definition for field studies; however, the practice can easily be questioned on theoretical grounds.

A specific problem in using this definition in eastern Montana is that dinosaur extinction cannot be traced with any certainty in lithologic sections. As noted above and in the chapter on Stratigraphy, the position of the stratigraphically highest dinosaur varies from section to section. This may represent a real biostratigraphic difference, an artifact of preservation and collection, or both. Whichever of these possibilities may be true, it is poor biostratigraphic practice to use extinctions to mark time-stratigraphic (and by extrapolation, time) boundaries.

The other problems with using dinosaur extinction to mark the Cretaceous-Tertiary boundary are more widely applicable. Dinosaur extinction may not have been synchronous worldwide, and even if it was synchronous within the limits of geologic resolution, it may not have coincided with the Cretaceous-Tertiary boundary as defined in European marine sections.

Both of the latter problems have been addressed in the past and have recently again been queried. Using admittedly flimsy evidence, Van Valen and Sloan (1977a) suggested that dinosaurian (as well as other terrestrial) extinctions may have occurred later in more southern areas. Butler and others (1977) presented magnetostratigraphic evidence to suggest that dinosaur extinction occurred later in San Juan Basin, New Mexico, than the terminal Cretaceous marine extinctions in Italy. Van Valen and Sloan (1977b) and Lindsay and others (1978) appear to have come to a consensus that dinosaurs persisted longer in more southern regions. Alvarez and Vann (1979), Fassett (1979), Lucas and Rigby (1979), and Lerbekmo and others (1979), variously drawing on lithostratigraphic, biostratigraphic, and paleomagnetic evidence, have argued that the magnetostratigraphic column was incorrectly correlated between the San Juan Basin and Italy. When recorrelated, particularly accounting for a hiatus of disputed length, dinosaur extinction in the San Juan Basin is found to have occurred at about the same time as the marine extinctions in Italy, according to these authors' comments. The apparent lack of definite Lancian through mantuan mammals from the San Juan Basin (although they may be present but have not been discovered) also suggests a recognizable hiatus in the San Juan Basin section.

The answer to whether dinosaur extinction was a synchronous event remains elusive. The evidence from eastern Montana cannot separately answer this question but, as will be discussed later, there is some suggestion for diachronous extinction in eastern Montana that may have wider implications.

In addition to dinosaur extinction, other paleontologic as well as non-paleontologic methods have been used to estimate the Cretaceous-Tertiary boundary in terrestrial sediments of western North America.

Radiometric dates are relatively few and when available (e.g., Folinsbee and others, 1965, 1970; Shafiqullah and others, 1968; Lerbekmo and others, 1979) their margins of error of a million or more years may completely encompass more finely scaled biological sequences. Radiometric dating is still the only method to establish geochronological dates; however, finer-scaled methods are needed for correlation.

Magnetostratigraphy has only recently been applied to the problem of determining the Cretaceous-Tertiary boundary in terrestrial sediments. The results of the first study of this sort conducted in San Juan Basin, New Mexico (Butler and others, 1977; Lindsay and others, 1978), have been challenged by a number of people, as was noted previously. A second study was conducted in the Red Deer River Valley, Alberta, by Lerbekmo and others (1979). One of the conclusions of this second study was that dinosaur extinction in Alberta occurred relatively close in time to the terminal Cretaceous foraminiferal extinctions in the Gubbio section, Italy. However, this conclusion as well as other aspects of this study have been challenged (Butler and Lindsay, 1980). Magnetostratigraphy may eventually prove to be a useful method for correlating events near the Cretaceous-Tertiary boundary within the terrestrial realm, but for now this is unfortunately not the case. R. Butler and E. Lindsay have been conducting magnetostratigraphic studies in eastern Montana and other areas of the western United States (written comm.), and this and similar studies may help to resolve some of the present controversy.

Vertebrates other than dinosaurs, particularly mammals, may eventually provide a relatively accurate means of correlation. The mammalian sequence discussed in this study is the most finely scaled sequence of its kind known at the Cretaceous-Tertiary boundary; however, without other such sequences this will remain a useful method of correlation only within eastern Montana. At present, only palynofloral studies offer a relatively finely scaled means of correlation over a large geographic area.

Palynofloral studies have not been conducted in conjunction with the present research in Garfield County. However, previous palynofloral studies in Garfield and McCone counties (Hall and Norton, 1967; Norton and Hall, 1969; Oltz, 1969; Tschudy, 1970) can be used to approximate the palynofloral change relative to dinosaur extinction that occurs about 10 ft. (3 m) below the Z coal. In Garfield County, the palynofloral change occurs above the last dinosaurs, but a more exact placement is difficult to establish because of differing sites and differing lithologic and time-stratigraphic interpretations. With the exception of Tschudy (1970), none of the palynofloral studies cited above presents detailed stratigraphic sections. All the authors follow R. Brown's (1952) formula of placing the Hell Creek–Tullock formational contact and the Cretaceous-Tertiary boundary at the base of the lowest coal (the Z) above the last dinosaurs. Although Norton and Hall (1969) do recognize a transition flora, these papers all implicitly or explicitly recognize a palynofloral change near the base of the lowest coal.

It cannot be determined with certainty whether these various palynofloral studies are recognizing the same Z bed that has been identified in this study. The one area that can be directly compared is in the vicinity of Brownie Butte. This butte lies on the eastern edge of the area mapped in Fig. 4. Both Norton and Hall (1969) and Tschudy (1970) have sampled near this butte. Norton and Hall (1969) indicate that they recognize a 3 ft. (0.9 m) lignite as the Z bed. As can be seen in Fig. 4, only the upper Z was definitely located in the vicinity of Brownie Butte. The lower Z was only inferred and in fact might be absent in this area. It seems highly likely that the Z bed of Norton and Hall (1969) equals the upper Z coal of the present study. There is a distance of approximately 20 to 25 ft. (6.1–7.6 m) between the upper and lower Z coals in the sections nearest Brownie Butte where both lignites are exposed. Adding these figures to the estimated average of 10 ft. (3 m) below the lower Z for the extinction of dinosaurs, a highly speculative range of 30 to 35 ft. (9.1–10.7 m) is obtained for the stratigraphic interval between dinosaur extinction and a palynofloral change.

At present it is not possible to determine with any certainty whether dinosaur extinction, palynofloral change, or neither most closely approximates the Cretaceous-Tertiary boundary as defined in marine sections in Europe. The only evidence that suggests that dinosaur extinction may be closest in time to the marine extinctions comes from the previously cited study of Lerbekmo and others (1979). It is interesting to note that in their study area in Alberta, dinosaur extinction also precedes palynofloral change. Russell and Singh (1978) previously reported that the opposite was true, but they have since found some of their stratigraphic placements of dinosaurs to be incorrect (D. A. Russell, written comm.).

Although encumbered by the variety of assumptions and potential problems discussed in the preceding pages, dinosaur extinction still remains the most utilitarian method for estimating the Cretaceous-Tertiary boundary in Montana. However, the decision as to where the Cretaceous-Tertiary boundary is in eastern Montana *cannot* be determined intraregionally but *must* be determined by correlation with the type areas in Europe, possibly through such techniques as magnetostratigraphy. It also must be emphasized that this is not simply a question of a legalistic definition. The more detailed fossil sequences that are gradually coming to light in areas such as eastern Montana require rigorous definition if this new material is to be of any use in dealing with biological events at the Cretaceous-Tertiary boundary.

HELL CREEK FAUNAL-FACIES MAMMALS

Before the discovery of moderately rich Hell Creek faunal-facies localities in Garfield County, large samples of typical latest Cretaceous mammals were known only from the Lance local fauna, Wyoming, and Trochu local fauna, Alberta. The localities in Garfield County are geographically intermediate, approximately 330 mi. (530 km) northwest of the region that yields the Lance local fauna and 420 mi. (680 km) southeast of the region that yields the Trochu local fauna.

Over 50 Hell Creek faunal-facies localities have been discovered in Garfield County. However, only one locality, Flat Creek 5 (loc. V73087) has produced a large enough screen-washing sample for comparison with the Trochu and Lance local faunas. The vertebrates from this locality, plus four other localities, constitute the Flat Creek local fauna (see the earlier description of localities for discussion). The Flat Creek local fauna includes all but five mammalian species that are known to occur in the Hell Creek faunal-facies, Hell Creek Formation of eastern Montana. Table 57 lists all the taxa of the Hell Creek faunal-facies that have been reported from the Hell Creek Formation, Montana. The Xs represent species that occur in the Flat Creek local fauna. The five 0s represent taxa that have been reported from other Hell Creek faunal-facies localities in Montana but have not been found in the Flat Creek local fauna. Mammals known from the Lance and Trochu local faunas and Hell Creek faunal-facies mammals occurring at Bug Creek faunal-facies localities are also given in Table 57. The columns in Table 57 (except the Bug Creek faunal-facies) are arranged in geographic sequence, with the more southerly Lance local fauna to the left. The table is also a listing of all typical latest Cretaceous mammals, except for *Meniscoessus conquistus*, which is known only from South Dakota and, possibly, the older Scabby Butte local fauna (Sloan and Russell, 1974).

TABLE 57. Species of typical latest Cretaceous mammals in the Lance (southernmost), Flat Creek (intermediate), and Trochu (northernmost) local faunas. 0s under the Flat Creek local fauna designate taxa known from other Hell Creek faunal-facies localities but not the Flat Creek local fauna. The fourth column lists Hell Creek faunal-facies mammals known from the Bug Creek Anthills local fauna (see Table 59). The fifth column lists Hell Creek faunal-facies mammals that persist into the Paleocene, are ancestral to Paleocene taxa, or share close common ancestry with a Paleocene species. The figures at the bottom are the number and relative frequencies of the three major groups of typical latest Cretaceous mammals in each local fauna. The two columns below the Flat Creek local fauna are with (+0) and without (−0) the five additional Hell Creek faunal-facies mammals.

	Lance local fauna	Flat Creek local fauna	Trochu local fauna	Bug Creek Anthills local fauna	Paleocene "relatives"
Mesodma hensleighi	X	X	X		✓
Mesodma formosa (& cf. *M. formosa*)	X	X	X		✓
Mesodma thompsoni	X	X	X		✓
?*Neoplagiaulax burgessi*, n. sp.		X			✓
?Neoplagiaulacidae gen. et sp. indet.		0			?✓
Cimolodon nitidus	X	X	X	✓	✓
Kimbetohia campi	X				✓
Ptilodontidae gen. et sp. indet.			X		?
Cimolomys gracilis	X	X	X	✓	
Cimolomys trochuus			X		
Meniscoessus robustus	X	X		✓	
Essonodon browni	X	X		✓	
Cimexomys minor	X	0		✓	✓
Paracimexomys priscus, n. gen.		X	X		
Alphadon marshi	X	X	X	✓	✓
Alphadon wilsoni	X	X	X	✓	
Alphadon lulli	X	X			
Alphadon rhaister	X	X	X		
Alphadon cf. *A. rhaister*	?X				
Glasbius intricatus	X				
Glasbius twitchelli, n. sp.		X		✓	
Pediomys elegans	X	X	X	✓	
Pediomys cooki (& cf. *P. cooki*)	X	X		✓	
Pediomys krejcii	X	X	X	✓	

Pediomys hatcheri	X	X	X	✓	
Pediomys florencae	X	X		✓	
?*Pediomys* cf. *P. florencae*		X			
Didelphodon vorax	X	X	X	✓	
Didelphodon padanicus	?X				
Gypsonictops hypoconus	X	0	X	✓	✓
Gypsonictops illuminatus (& cf. *G. illuminatus*)		X	X	?✓	✓
Cimolestes incisus	X	X		✓	✓
Cimolestes cerberoides (& cf. *C. cerberoides*)		0	X		✓
Cimolestes propalaeoryctes		X	X		✓
Cimolestes stirtoni	X	X			?✓
Cimolestes magnus	X	0	X	✓	✓
Batodon tenuis	X	X	X	✓	✓
Telacodon laevis	X				?

			−0		+0			
	No.	*(%)*	*No.*	*(%)*	*No.*	*(%)*	*No.*	*(%)*
Multituberculata	9	(32)	9	(35)	11	(35)	8	(38)
Metatheria	13	(46)	12	(46)	12	(39)	7	(33)
Eutheria	6	(21)	5	(19)	8	(26)	6	(29)
	28	(100)	26	(100)	31	(100)	21	(100)

The geographically intermediate position of the Flat Creek local fauna (and other similar localities) between those of Wyoming and Alberta is reflected in its taxonomic diversity and composition. Table 57 shows that thirteen species are present in all three local faunas from Wyoming, Montana, and Alberta. This number is raised to fifteen when taxa known from other Hell Creek faunal-facies localities are added to those from the Flat Creek local fauna. No species are in common between the Lance and Trochu local faunas that are not also found at Hell Creek faunal-species localities in Montana. However, four species are exclusively shared between the Hell Creek faunal-facies and the Trochu, and eight species are shared exclusively between the Hell Creek faunal-facies and the Lance.

Although it is intuitively obvious from Table 57 that the Lance, Trochu, and Flat Creek local faunas are taxonomically very similar, a numerical comparison can also be made. For this purpose, Simpson's (e.g., 1947, 1965) index 100 C/N_1 was employed (see Table 58 for definition of terms). Table 58 shows the results of these comparisons at the species level. The data in Table 57 and Table 59 were used for the computations. Each taxonomic listing in these figures was assumed to represent a separate species, even if unnamed. Taxa qualified by the term "cf." in the systematic descriptions are included with their respective species in these figures if it is believed they are conspecific; those qualified by "cf." or "aff." are listed separately if it is believed they may be distinct species.

The left-hand column in Table 58 is a pair-wise comparison of the Lance, Trochu, and Flat Creek local faunas and the Flat Creek and Bug Creek Anthills local faunas. The right column is a similar comparison, but all species reported from the Hell Creek faunal-facies are used, i.e., the Flat Creek local fauna plus five additional taxa. It is not certain whether the comparisons in the left- or right-hand column more accurately reflect the faunal resemblances among the various areas, but the results in the right-hand column are favored. The five species absent from the Flat Creek local fauna are rare members of the Hell Creek faunal-facies and their absence from this local fauna may be a result of collecting bias. However, there are unresolved problems regarding two of these taxa, *Gypsonictops* (various species) and *Cimexomys minor*. It is still not clear how many, and which species of *Gypsonictops* are present in the Hell Creek Formation. The recognition of *C. minor* at any given locality is dependent upon which tooth sites are represented. As noted in the systematic discussion of *Cimexomys minor*, I could only distinguish isolated P_4^4's and M^1's of this species from those of *Mesodma*. Thus, M_1's and M_2^2's of *C. minor* may have been present at a given locality but were not recognized as such.

However, both sets (columns) of pair-wise comparisons between the Lance, Flat Creek (or Hell Creek faunal-facies), and Trochu local faunas result in similar patterns. Only the comparison with the Bug Creek Anthills local fauna may be problematic.

TABLE 58. Faunal resemblances between various Lancian local faunas (l.f.) at the species level, using Simpson's (e.g., 1947, 1965) index 100 C/N$_1$. Data for the calculations are from Tables 57 and 59. C is the number of faunal units (in this case, species) in common between the two regions; N$_1$ is the number of faunal units (also species) in the smaller of the two samples. See text for further information.

	100 C/N$_1$		*100 C/N$_1$*	*Distance between the local faunas*
Lance l.f.-Flat Creek l.f.	77%	Lance l.f.-Hell Creek faunal-facies	82%	330 mi. (530 km)
Trochu l.f.-Flat Creek l.f.	76%	Trochu l.f.-Hell Creek faunal-facies	90%	420 mi. (675 km)
Lance l.f.-Trochu l.f.	71%	Lance l.f.-Trochu l.f.	71%	740 mi. (1190 km)
Flat Creek l.f.-Bug Creek Anthills l.f.	63%	Hill Creek faunal-facies-Bug Creek Anthills l.f.	79%	15 mi. (24 km)

Most geological evidence suggests that the Lance, Flat Creek, and Trochu local faunas existed in low-lying coastal environments, although sedimentation types and rates, distances to the midcontinental seaway, floras, etc., varied to some degree. The high values of faunal resemblance at the species level reflect this environmental homogeneity (as far as the mammals are concerned). To a lesser degree, the high values may reflect a close similarity in age, but this is less certain. It is also interesting to compare the degree of faunal resemblance and the distance between the areas. The distances between the areas are given in Table 58. They correspond almost inversely with the degree of faunal resemblance. Such a general inverse relationship is not unexpected. Simpson (e.g., 1965, p. 124) has previously noted this relationship. Simpson (1965, pp. 123–124) used an example from the late Paleocene to show that ecologically similar faunas from distant areas may resemble one another more than ecologically different faunas which are closer in geographic proximity. This explains why the ecologically distinct Bug Creek Anthills local fauna (Bug Creek faunal-facies) and Flat Creek local fauna (Hell Creek faunal-facies) do not show the greatest resemblance in Table 58, although they are only 15 mi. (24.1 km) apart.

Previous authors (e.g., Lillegraven, 1969) have been understandably cautious in trying to determine the relative ages of the Lance, Trochu, and Bug Creek local faunas. Lillegraven (1969) briefly discussed the possible age relationships of the Trochu and Lance local faunas and the mammals recovered by Sloan and Van Valen (1965) from the Hell Creek Formation, McCone County. Based on the multituberculates and metatherians (1969, pp. 33 and 50) he could find no significant difference in the ages of the Trochu and Lance local faunas. However, he agreed with Van Valen and Sloan's conclusion (1966, p. 265, fig. 1) "that the Bug Creek Anthills locality and those stratigraphically above it in the Hell Creek Formation are slightly younger than the upper part of the Edmonton Formation" (= Scollard Formation). Although this may still be a possibility, the stratigraphic evidence from the Hell Creek Formation, which will be discussed later, demonstrates that Hell Creek faunal-facies localities such as Flat Creek 5 (V73087) are coeval with (and possibly younger than) the majority of the Bug Creek faunal-facies localities. Thus if it can be shown that the Flat Creek, Trochu, and Lance local faunas are of similar age, then there is no basis for considering the majority of the Bug Creek faunal-facies localities to be younger.

As noted previously, the high degree of faunal resemblance at the species level between the Flat Creek, Trochu, and Lance local faunas may indicate a close similarity in age. However, the data in Table 58 are open to several interpretations. Following Simpson, it was suggested that greater faunal resemblance of ecologically similar local faunas was due to greater geographic proximity. It is also possible that greater faunal resemblance is due to similarity in age, and conversely, less resemblance is due to greater difference in age. Based on this criterion, the local faunas would be arranged (oldest to youngest): Lance, Flat Creek (plus the Bug Creek faunal-facies localities, as noted in the previous paragraph), Trochu. Although this interpretation is not favored, it remains a viable alternative.

One possible way to test this idea would be to have a well-sampled sequence of rich typical latest Cretaceous mammal sites. Unfortunately, such sequences are not known at the present time. Other techniques such as radiometric dating and paleomagnetic sequencing have not been on a fine enough scale to answer such questions. Comparisons with fossil organisms other than mammals are also possible. As an example, L. S. Russell (1975, p. 155) suggested that the apparent differences in eutherian abundance and diversity among the Lance, Hell Creek, and Scollard formations were not due to age differences; rather, the similarities in the reptilian faunas "suggest that the eutherian distribution indicates a southward incursion of these mammals, not yet completed in Lancian times." A closer analysis, below, favors Russell's interpretation. Lacking a better basis for correlation, a "simplistic dictum" recently phrased by Savage (1977, p. 431) is employed. "[L]acking evidence to the contrary, organisms appearing most alike are most closely related, and organisms which are alike lived during the same geochronological interval, no matter how extensive their geographic ranges." Based on this dictum, I consider the high degree of faunal resemblance (minimum in Table 58: = 71%) at the *species* level among the Lance, Trochu, and Flat Creek local faunas to be due in part to a close similarity in age.

There are a number of interesting and consistent patterns in numerical abundance and diversity among the Lance, Trochu, and Flat Creek local faunas. Relative abundances of the species from these local faunas have not been determined, except for species of *Alphadon* and *Pediomys* (Clemens, 1966, p. 85) and the genera (Estes and Berberian, 1970, table 1) in the Lance local fauna. However, it is possible to make qualitative comparisons based on the comments and data presented by Clemens (1964, 1966, 1973), Lillegraven (1969), and in this study.

As has been noted by several authors (e.g., Clemens, 1973; and Lillegraven, 1969), the Lance local fauna is dominated, both in numerical abundance and in taxonomic diversity, by marsupials and multituberculates. Clemens (1973, p. 94) estimated that the eutherians comprise only 10 to 15 percent of the numerical abundance in this local fauna. In terms of taxonomic diversity the marsupials comprise almost 50 percent (13 of 28 species-level taxa in Table 57) of the mammals in this local fauna.

In contrast, Lillegraven (1969, p. 50) found that in the Trochu local fauna eutherians rival the marsupials both in taxonomic diversity and numerical abundance. The relatively higher diversity can be seen in Table 57, where eutherians and marsupials each comprise about one-third of the taxa known from the Trochu, with multituberculates contributing slightly more than one-third.

The Flat Creek local fauna appears most closely to resemble the Lance local fauna in the diversity of the three mammalian groups (see Table 57). When the three species of eutherians known from other Hell Creek faunal-facies are added, eutherians comprise about one-quarter of the taxonomic diversity. However, this figure is quite deceptive. The Hell Creek faunal-facies localities, taken together, are essentially a mixture of the Lance and Trochu local faunas—which is not surprising, considering the geographically intermediate position of these localities. Thus, the Hell Creek faunal-facies localities share not only the increased relative diversity of the Trochu eutherians but also the diversity of the marsupials found in the Lance Formation.

The Flat Creek local fauna is also an admixture of the other two local faunas in terms of relative numerical abundances. Clemens (1966, table 16) found that *Pediomys* is by far the most common genus of marsupial in the Lance Formation. In contrast, Lillegraven (1969) found two species of *Alphadon*, *A. marshi* and *A. wilsoni*, to be numerically more abundant in the Trochu local fauna. Unlike the Trochu, the Flat Creek local fauna includes all the species of *Pediomys* Clemens recognized in the Lance; yet, as in the Trochu, *A. marshi* and *A. wilsoni* are more abundant. The Flat Creek local fauna is unique in its numerical abundance of *Glasbius twitchelli*. In this local fauna it is as abundant as either of the above species of *Alphadon*, while in the Lance the closely related species *G. intricatus* is rare, and the genus is totally absent from the Trochu. Finally, although the eutherians in the Hell Creek faunal-facies are as taxonomically diverse as in the Trochu, they are as numerically rare as in the Lance local fauna.

The multituberculates are a numerically abundant and taxonomically diverse group throughout all three local faunas. The neoplagiaulacid *Mesodma* is the most dominant genus, based on both of the above criteria. As Lillegraven (1969) noted, there are no significant differences among the multituberculates to suggest that the Lance and Trochu local faunas differ in age. The Flat Creek local fauna can be added to this list. There are some minor but conflicting variations in the multituberculates among these local faunas that suggest slight age differences; however, the variations may instead be due to small paleoecological changes.

The Lance local fauna includes a rare element, the ptilodontid *Kimbetohia campi*, the only known Cretaceous representative of this family. *Cimexomys minor* is also present. As noted earlier, it is sometimes difficult to recognize this species. However, if recognized occurrences are correct, this species has an odd distribution pattern. Besides being present in the Lance, it is known from both the Hell Creek and Bug Creek faunal-facies in Mc-

Cone County and only pre-mantuan localities in Garfield County, and it is absent from the Trochu. If this genus is as cladistically primitive as suggested previously, this may be a relict distributional pattern.

The Flat Creek local fauna includes the earliest record of a neoplagiaulacid of Paleocene aspect, ?*Neoplagiaulax burgessi*. This multituberculate has a high, arcuate P_4 blade previously known only in Tertiary representatives of the family. This local fauna also contains a P_4 of *Mesodma thompsoni* that is almost indistinguishable in size from *Mesodma ambigua* of the earliest Paleocene, Mantua lentil, Wyoming. However, two other similarly large P_4's of *M. thompsoni* were recovered from localities (locs. V72208 and V74116) that are believed to be almost 100 ft. (30.5 m) lower in the section. Also, a questionable neoplagiaulacid with a high, arcuate blade was recovered from one of these stratigraphically low localities (loc. V72208).

The Trochu local fauna is odd in that it lacks the ubiquitous species *Meniscoessus robustus*. It also includes the unique species *Cimolomys trochuus*. If Lillegraven (1969) is correct in his identification and assignment of this species to the Cimolomyidae, this is the first representative of this family that is known to have lost the small, peglike P_3.

One final trend among the Lance, Hell Creek, and Trochu local fauna is noteworthy—however, it should first be said that both the recognition and interpretation of this trend are very speculative. Lillegraven (1969, p. 29) noted that the Trochu representatives of *Cimolodon nitidus* "are consistently at the higher end of the Lance scale" of this species or exceed it. He suggested that "Bergmann's rule" might apply in this case. He further added that this indirectly suggests a cooler climate to the north. However, equating "Bergmann's rule" with a cooler northern climate has been challenged (Lillegraven, pers. comm.). McNab (1971) found that most latitudinally widespread mammals in North America do not follow "Bergmann's rule." The mammals that do follow this trend are usually carnivores and granivores. McNab suggested that latitudinal changes in body size of the carnivores are affected by changes in the prey size, due to distributional patterns of the prey species or competition with other predators utilizing the same prey. He found that among predators only the smallest species normally conform to "Bergmann's rule" and then only beyond the distribution of the larger species. This was considered to be an example of character displacement.

It might be possible to test McNab's thesis on Tertiary mammals; however, this would be exceedingly difficult to do for latest Cretaceous mammals, which can only be placed in very broad feeding niches. Even if this is done, relating predators to prey species would be even more speculative, and, further, most of the predators were probably insectivores. Nevertheless, there may be a trend to increase body size from the Lance to the Trochu local fauna in some forms.

The sizes of the twenty-six mammalian species (or closely related species) that occur in two or three of the local faunas were qualitatively compared. Six of the twenty-six species are too poorly represented to make a comparison. Unfortunately, four of these are palaeoryctids which may on occasion have taken smaller mammalian prey. Of the remaining twenty, six are of equal size throughout their range (*Cimolomys gracilis*, *Meniscoessus robustus*, *Cimexomys minor*, *Pediomys hatcheri*, *Didelphodon vorax*, and *Batodon tenuis*), three increase in size toward the south (*Mesodma hensleighi*, *M. formosa*, and *Cimolestes stirtoni*), seven increase in size toward the north (*Cimolodon nitidus*, *Paracimexomys priscus*, *Alphadon wilsoni*, *A. lulli*, *Glasbius intricatus–G. twitchelli*, *Pediomys florencae*, and *Gypsonictops hypoconus–G. illuminatus*), and four have other size-distribution patterns. There is no distinct taxonomic correspondence with these size distributions.

If the oversimplified assumption is made that the size distribution is random relative to latitude, given the three areas (Lance, Flat Creek, and Trochu) for comparison, there are thirteen possible size-distribution patterns. Since the patterns listed above were determined qualitatively and some taxa occur in only two of the three local faunas, a statistical comparison is not warranted. However, intuitively it seems odd that one-third (7 of 20) of the species show a pattern of size increase toward the north and another one-third appear to be unaffected by latitude. (Given the above assumptions, only one or two mammals should show each of these trends.) Even if these two trends are real, particularly that of size increase toward the north, there are myriad possibilities, both biological and geological, to explain them.

BUG CREEK FAUNAL-FACIES, PRE-MANTUAN, AND MANTUAN MAMMALS

Three local faunas, Bug Creek Anthills, Bug Creek West, and Harbicht Hill, which Sloan and Van Valen described in 1965, showed striking differences from previously described Lancian local faunas (Simpson, 1927a; Clemens, 1964, 1966; Lillegraven, 1969). In addition to typical latest Cretaceous mammals, these local faunas included a number of mammals of Paleocene aspect.

Sloan and Van Valen (1965; Van Valen and Sloan, 1977a; Van Valen, 1978) have demonstrated that a turnover in species occurred in their stratigraphic sequence—Bug Creek Anthills, Bug Creek West, Harbicht Hill—by immigration or *in situ* evolution of the Paleocene-aspect mammals and progressive reduction in diversity and numbers of typical latest Cretaceous mammals. Localities to the west in Garfield County discussed in this section lend support to the conclusions of these two authors regarding trends *within* the Bug Creek faunal-facies.

Fig. 72 summarizes the stratigraphic data given earlier in this study, plus the stratigraphic data from McCone County (Sloan and Van Valen, 1965; Sloan, 1976). Table 59 is an updated faunal list of Bug Creek faunal-facies, pre-mantuan, and mantuan localities from McCone and Garfield counties, arranged in their presumed stratigraphic and biostratigraphic sequence. Sloan and Van Valen (various papers) show (and this can also be seen in Table 59) that the Bug Creek faunal-facies was undergoing a rapid species turnover. This contrasts with the stasis that is seen in stratigraphic sequences containing typical latest Cretaceous mammals.

As can be seen in Table 59, species diversity appears to decline going up-section. Van Valen and Sloan (1977a, fig. 4) note that this decline may be somewhat spurious. Bug Creek Anthills has yielded a tremendously larger amount of specimens than the other localities. The sample includes some species (e.g., *Essonodon browni*) that are even rare at rich Hell Creek faunal-facies localities. The lower representation of Hell Creek faunal-facies mammals at the stratigraphically higher localities may be in part due to smaller sample sizes. However, even if Bug Creek Anthills is excluded there still remains a progressive reduction of Hell Creek faunal-facies mammals within the Bug Creek faunal-facies sequence on into the sequence of pre-mantuan and mantuan localities. The very small number of taxa known from the two mantuan localities (V72210 and V77087) is almost certainly spurious. Vertebrates, in addition to mammals, are rare at these two localities.

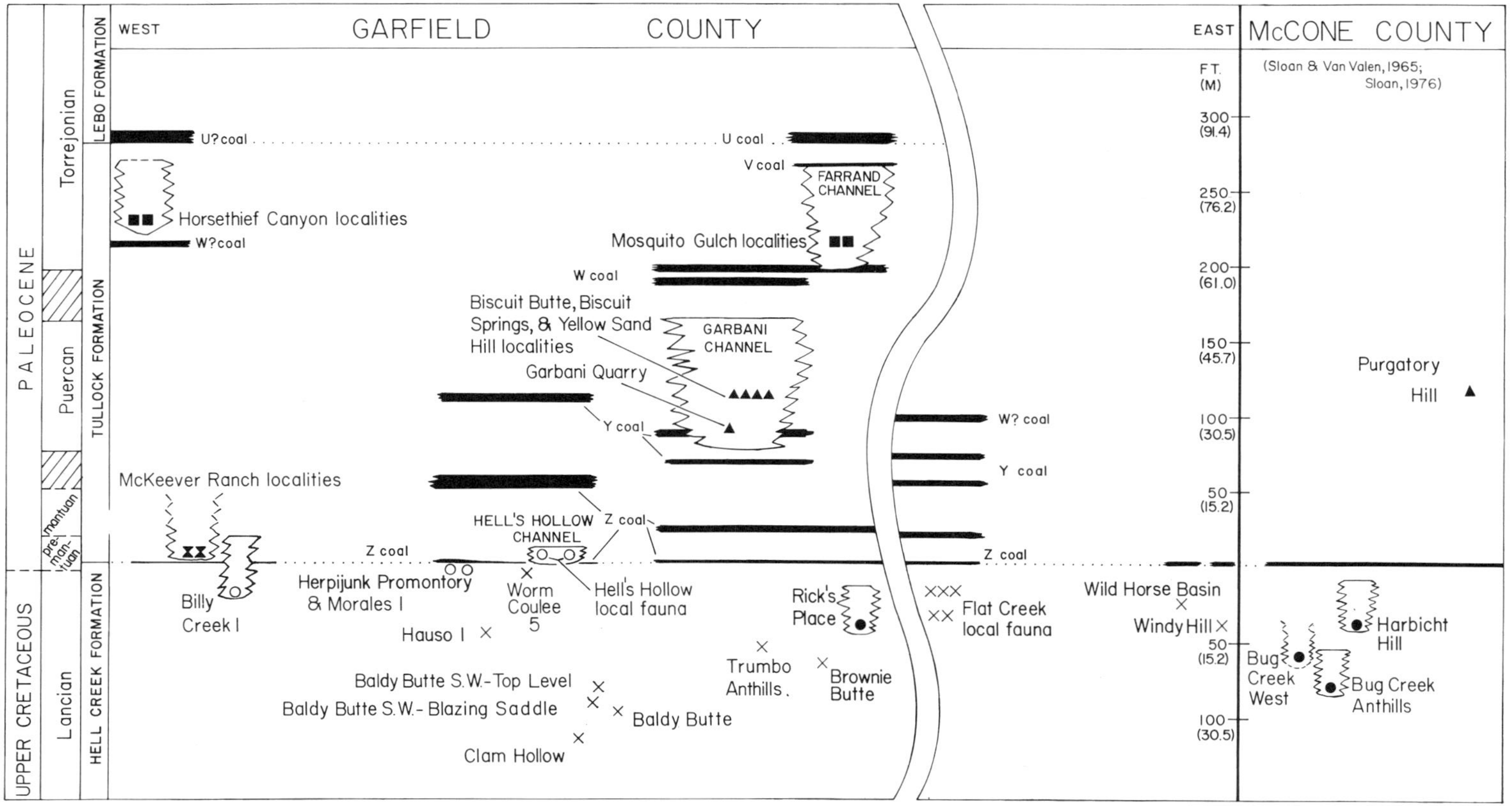

FIG. 72. Schematic summary of stratigraphic and biostratigraphic relationships discussed in this chapter and the chapter on Stratigraphy, Physical Geology, and Localities: × = Hell Creek faunal-facies localities; ● = Bug Creek faunal-facies localities; ○ = pre-mantuan localities; ⧗ = manutan localities; ▲ = Puercan localities; ■ = Torrejonian localities. Localities are shown in very approximate east-west position: approximately 60 mi. (97 km) separate the Horsethief Canyon localities and Purgatory Hill; approximately 30 mi. (48 km) separate Rick's Place and the Flat Creek local fauna.

TABLE 59. Species-level taxa in Lancian Bug Creek faunal-facies, pre-mantuan, and mantuan local faunas (or groups of localities) from Garfield and McCone counties. The local faunas and localities are arranged in biostratigraphic and stratigraphic order with the lowest to the left. Question marks indicate questionable occurrences or, in the case of *Gypsonictops*, the uncertainty as to which species is (are) present, and "cf.'s" indicate that specimens at a given locality are tentatively referred to the particular taxon. The sixth column lists Hell Creek faunal-facies mammals found in Lancian Bug Creek faunal-facies or younger local faunas. The seventh column lists Lancian mammals (some also shown in Table 57) that persist into the Paleocene, are ancestral to Paleocene taxa, or share ancestry with a Paleocene species. Column eight gives the taxonomic references used in this table: (1) Sloan and Van Valen, 1965, table 1; (2) Estes and Berberian, 1970, faunal list, pp. 7–8; (3) Clemens, unpublished data; (4) Novacek and Clemens, 1977; (5) present study; (6) Van Valen, 1978; (7) UCMP collections; (8) Van Valen and Sloan, 1965.

	Bug Creek Anthills local fauna	Bug Creek West local fauna	Harbicht Hill local fauna	Hell's Hollow local fauna	McKeever Ranch (V72210 & V77087)	Hell Creek faunal-facies	Paleocene "relatives"	References
Cimolomys gracilis	X					✓		1
Essonodon browni	X					✓		1
Alphadon marshi	X					✓	✓	1
Alphadon wilsoni	X					✓		2
Didelphodon vorax	X					✓		1
Gypsonictops hypoconus	?X					✓	✓	1
Cimolestes incisus	X					✓	✓	1
Cimolestes magnus	X					✓	✓	2
Batodon tenuis	X					✓	✓	3
Mesodma sp.	X	cf.					✓?	4
Cimolodon nitidus	X	X				✓	✓	1
Stygimys kuszmauli	X	X					✓	1
Pediomys elegans	X	X				✓		1
Pediomys krejcii	X	?				✓		2
Pediomys hatcheri	X	X				✓		1
Pediomys cooki	X	?				✓		2
Pediomys florencae	X	?				✓		2
Glasbius twitchelli, n. sp.	X	X				✓		5

Gypsonictops illuminatus	?X	?X				✓	✓	2
Meniscoessus robustus	X	X	X			✓		1
Catopsalis joyneri	X	X	X				✓	1
Cimexomys minor	X	X	X	X		✓	✓	1
Procerberus formicarum	X	X	X	X			✓	1
Protungulatum donnae	X	X	X	cf.			✓	1,6
Protungulatum gorgun		X	X				✓?	6
Mimatuta morgoth		X	X	X			✓	6
Cimexomys sp.			X				✓	1
Eucosmodontidae (undesc. gen. et sp.)			X				?	1
Marsupalia (undescribed)			X				?	7
Purgatorius ceratops			X				✓	8
Mesodma garfieldensis, n. sp.			cf.	X			✓?	5
Stygimys aff. *S. kuszmauli*			cf.	X			✓?	5
Oxyprimus erikseni			X	X			✓	6
Ragnarok harbichti			X	?			✓?	6
Ragnarok engdahli, n. sp.				X			—	5
Cimexomys hausoi, n. sp.				X			—	5
Acheronodon garbani, n. gen. et n. sp.				X			—	5
Catopsalis cf. *C. foliatus*				X			—	5
Peradectes cf. *P. pusillus*				X			—	3,5
Protungulatum mckeeveri, n. sp.					X		—	5
Ragnarok nordicum					X		—	6
Mimatuta minuial					X		—	6
?Periptychidae gen. et sp. indet.					X		—	5
Number of species-level taxa at each locality	24	14–17	15	12–13	4			

Bug Creek faunal-facies, pre-mantuan, and mantuan localities occur in relatively large, sandy channels, in which channel cappings are often discernible. UCMP locs. V77128 and V77129 discovered in paludal conditions in 1977, Garfield County, appear to be exceptions (Clemens, pers. comm.). The channels appear to be the first of several cycles of alternating channel-fill deposits and coals that are found in the overlying Tullock Formation.

All the localities known from the channel-fill deposits occur in the lower half of the deposits and most occur very near the base. This includes localities from both Garfield and McCone counties. Using the Hell Creek–Tullock formational contact (the base of the lowest lignite), the relative stratigraphic positions of the localities have been established (see chapter on Stratigraphy and Table 1). In all but one instance, the stratigraphic and biostratigraphic placement of the Bug Creek faunal-facies localities (including those from McCone County), pre-mantuan localities, and mantuan localities are in accord. The one exception is the pair of localities McKeever Ranch 1 (V72210) and McKeever Ranch, Harley's Point (V77087). As will be discussed below, the biostratigraphic evidence indicates that these localities are mantuan in age. However, as discussed in the chapter on Stratigraphy, these localities are closer to the underlying Z coal than the Hell's Hollow localities (V74110 and V74111) which have been determined to be pre-mantuan on faunal evidence (see also Fig. 72). Although discussed in the chapter on Stratigraphy, it must be emphasized again that the stratigraphically low position of a locality within a channel sequence represents the near-maximum age for the locality. The channel capping, if known, represents the highest possible stratigraphic position of the locality and thus the minimum age. The biostratigraphic evidence suggests that the channel capping for the deposit that includes the McKeever Ranch localities should be stratigraphically higher than the channel capping for the Hell's Hollow channel deposit. Further, the McKeever Ranch deposit may have cut more deeply into the underlying sediment. Unfortunately, the capping for McKeever Ranch is not known, so its stratigraphically higher position remains only a prediction.

With the vagaries and uncertainties that surround the use of the coals as stratigraphic markers (particularly the Z coal), I find it rather remarkable that the biostratigraphic and stratigraphic correlations among the localities in Garfield and McCone counties appear to be in accord. It will not be at all surprising if a future discovery uncovers incongruities between the two sorts of correlations. In such instances, as with the possible example of McKeever Ranch, I regard the biostratigraphic correlations as being the more accurate.

In the following discussion of the biostratigraphy of the localities from Garfield County, two sorts of correlations have been considered. The first is a comparison of the level of evolution of taxa between localities. This incorporates the morphological analyses considered in the chapter on Systematics and is separate from stratigraphic considerations. The second sort of comparison is based on range data (i.e., presence/absence, first appearances, and disappearances) of the taxa. Both types are useful for the correlation of localities within Garfield and McCone counties. However, the first type is less subject to the vagaries of sample size. The rates of species turnover (both from speciation and immigration) are high enough so that even poorly sampled localities such as the McKeever Ranch localities can be placed in relative biostratigraphic position with confidence. The second type of correlation relies in part on negative evidence. For some localities, such as those in the Hell's Hollow channel deposit, the sample sizes subjectively appear to be large enough that most absences of taxa, such as almost all Hell Creek faunal-facies mammals in Hell's Hollow, are probably not an artifact of collecting or preservation.

Table 59 shows that the Harbicht Hill and Hell's Hollow local faunas may have about half of their respective species in common. Close comparisons have not been made among all the material known from these two local faunas, so there may be even more taxa in common. Sloan and Van Valen (1965) reported an undescribed eucosmodontid as occurring at Harbicht Hill. A new microcosmodontine eucosmodontid, *Acheronodon garbani*, is present at Hell's Hollow, but it is not known if this is the form represented at Harbicht Hill. These authors (1965) did not report the occurrence of marsupials at Harbicht Hill. A UCMP party recovered a small marsupial dentary with two very worn molars from this locality. It could not be determined with certainty whether the specimen belonged to a small species of *Alphadon*, *Pediomys*, or *Peradectes*. Even if it does belong to the latter genus, it is clearly a rare taxon, in contrast to its abundant representation at Hell's Hollow.

A number of taxa in the Hell's Hollow local fauna suggest a younger age than Harbicht Hill. Sloan and Van Valen (1965) report that *Catopsalis joyneri* is present at Harbicht Hill. This genus is rare at Hell's Hollow, but available evidence suggests the presence of a larger animal, closer in size to *C. foliatus* but a new species, according to M. Middleton (pers. comm.). *Peradectes*, known only from the Tertiary, is a common element at Hell's Hollow. As noted above, if present at Harbicht Hill this taxon is extremely rare. Both Harbicht Hill and Hell's Hollow include two species of *Cimexomys*. One species, *C. minor*, is present in both local faunas. It was not reported from Harbicht Hill by Sloan and Van Valen (1965), but Sloan (unpub. data) indicates that it is present in this local fauna. This is also the only Hell Creek faunal-facies species definitely known from Hell's Hollow, but it also ranges into the Puercan (Van Valen and Sloan, 1965). *C. hausoi* from Hell's Hollow and an undescribed new species from Harbicht Hill are not the same, according to Sloan (pers. comm.), but it is not known whether *C. hausoi* is a more advanced form. *Ragnarok engdahli*, the smallest representative of this genus, is known only from Hell's Hollow. As indicated in the description of this poorly known new species, it is possible but unlikely that this form could belong to *R. harbichti* or, more possibly, *R. nordicum*. The specimens referred to *R. engdahli* were also compared to material that Van Valen (1978) named as a new species of *Protungulatum*, *P. gorgun*. There are similarities, particularly in size, but the proportions of the lower molars clearly indicate that the Hell's Hollow material is most like *Ragnarok*. The questionable occurrence of *R. harbichti* at Hell's Hollow shown in Table 59 is based on the large, worn M^3 that was only questionably referred to *R. engdahli* because of its large size. Thus it is only a supposition that *R. harbichti* is present at Hell's Hollow. The tooth could belong to *R. engdahli* or *R. nordicum*. Only one taxon, *Purgatorius ceratops*, reported from Harbicht Hill but not Hell's Hollow, suggests a younger age for Harbicht Hill. However, subsequent work at Harbicht Hill has not uncovered additional material beyond the original lower molar reported by Van Valen and Sloan (1965), and use of this species for biostratigraphic correlation remains suspect, at least for now.

The presence of at least one Hell Creek faunal-facies mammal, *Meniscoessus robustus* (= *borealis*), and dinosaurs at Harbicht Hill but not at Hell's Hollow further suggest that the latter local fauna is younger. Sloan (1970, fig. 6) also reported the presence of *Gypsonictops* and *Cimolestes* at Harbicht Hill; however, since species designations were not given these taxa are not shown in Table 59. Neither of these genera is represented in the Hell's Hollow local fauna, although *Cimolestes* is known from the Puercan of Montana and New Mexico.

The lack of dinosaurs (except for two probably reworked hadrosaurian teeth) and of all but one Hell Creek faunal-facies mammal (*Cimexomys minor*) in the Hell's Hollow local fauna is negative evidence that must be used with some caution. However, when compared with data presented by Sloan and Van Valen (various papers) these absences do not appear to be due to preservational or sampling biases. The evidence for the decrease in the number and diversity of typical latest Cretaceous mammals *within* the Bug Creek faunal-facies sequence of McCone County presented by these authors was noted earlier. They also present similar evidence to show that dinosaurs underwent a similar decline. Although dinosaurs are present at Harbicht Hill, the relative abundance of *Triceratops* is reduced to 10 percent of what it was at Bug Creek Anthills (Van Valen and Sloan, 1977a). The total absence of dinosaurs and the reduction to only one Hell Creek faunal-facies mammal (*C. minor*) in the Hell's Hollow local fauna appear to be a continuation of the trends observed by Sloan and Van Valen. In addition to being younger than the Harbicht Hill local fauna, Hell's Hollow local fauna is considered to be earliest Paleocene in age because it lacks dinosaurs. It is important to emphasize that this determination has *not* been made because the Hell's Hollow local fauna comes from localities in the lowermost

Tullock Formation. Evidence based on the mammals from the Hell's Hollow local fauna does not strongly support or contradict an assignment of this local fauna to the earliest Paleocene. As noted previously, the abundance of *Peradectes* and the presence of a new larger species of *Catopsalis* in the Hell's Hollow local fauna are data indicating that this local fauna is younger than Harbicht Hill. These are the only mammals that tend to corroborate an earliest Paleocene age for the Hell's Hollow local fauna.

Two other localities, Morales 1 (V77128) and Herpijunk Promontory (V77129), which may be very similar faunally to the Hell's Hollow local fauna, occur just below the lowest lignite and are thus in the Hell Creek Formation. Both localities lack dinosaurs and thus are considered to be earliest Paleocene in age. Herpijunk Promontory (V77129) has produced only one identifiable mammalian specimen, a dentary fragment of *Ragnarok harbichti*. However, preliminary discoveries from Morales 1, which occurs in a lateral continuation of Herpijunk Promontory, suggest the faunal similarity with the Hell's Hollow local fauna (W. A. Clemens, per. comm.).

Based on comparisons with previously recognized Lancian (but Bug Creek faunal-facies) and Puercan local faunas, the Hell's Hollow local fauna appears to be transitional between these two North American land-mammal ages. Van Valen (1978) has proposed the "Mantuan" land-mammal age, which is bracketed by Lancian and Puercan. As noted in the chapter on Techniques of Study, I agree with Van Valen that Mantua lentil represents a period of mammal evolution (and time) that is not sampled in the local faunas of the type Puercan from San Juan Basin, New Mexico. However, I do not feel the naming of a new land-mammal age is warranted. Mantua lentil could easily be included within the Puercan, but a complete discussion of this topic is beyond the intended scope of the present study. Therefore "mantuan" is used informally (indicated by the lower-case first letter). Similarly, the Hell's Hollow local fauna is informally considered to be pre-mantuan because it is faunally older than mantuan localities in Wyoming and Montana but younger than the Lancian Harbicht Hill local fauna.

A small sample of mammals from the McKeever Ranch localities (V72210 and V77087) appear to be most similar to the Mantua lentil, Wyoming, and thus are referred to the mantuan.

Since only four taxa, all condylarths, are known from the McKeever Ranch localities (see Table 59), the assignment to the mantuan rests entirely on joint occurrences between Mantua lentil and McKeever and on the relatively more derived evolutionary state of the condylarths at McKeever than that of the Hell's Hollow local fauna.

The two taxa common to Mantua lentil and McKeever Ranch are *Mimatuta minuial* and *Ragnarok nordicum*. Morphologically, *M. minuial* is more derived than *M. morgoth* known from the Hell's Hollow, Harbicht Hill, and Bug Creek West local faunas. Similarly, *R. nordicum* is more derived than *R. harbichti* from the pre-mantuan localities (questionable occurrence in the Hell's Hollow local fauna) and Harbicht Hill. *R. engdahli* from the Hell's Hollow local fauna seems to be more derived than *R. harbichti* but questionably less derived than the larger *R. nordicum*.

Of the two remaining condylarths known from McKeever Ranch localities, one, *Protungulatum mckeeveri*, is slightly more advanced than *P. donnae*. However, *P. mckeeveri* is known only from the type dentary. The final specimen, an M_3 in a dentary, is questionably referred to the Periptychidae. Even if this referral is incorrect, the tooth is distinctly more derived than any M_3's known from the Hell's Hollow local fauna. With only four species known from the McKeever Ranch localities, assignment to the mantuan is tentative. However, I am still quite convinced that this assignment is correct.

Two other localities, Billy Creek 1 (V72211) and Rick's Place (V73098), have each produced a single dentary fragment that most closely resembles the dentary of *Mimatuta morgoth*. These two localities cannot be placed biostratigraphically. However, stratigraphic placement suggests that Rick's Place may be similar to Bug Creek West or Harbicht Hill, while Billy Creek 1 may be similar to Hell's Hollow or McKeever Ranch (see Fig. 72).

In summary, the Bug Creek faunal-facies includes three distinct biostratigraphic levels, from the oldest to the youngest—the Bug Creek Anthills, Bug Creek West, and Harbicht Hill local faunas of Sloan and Van Valen (1965). Except possibly for Rick's Place (V73098), no Bug Creek faunal-facies localities are known from Garfield County. Based on stratigraphic correlations with the Hell Creek faunal-facies localities and on the presence of dinosaurs, all known Bug Creek faunal-facies localities are latest Cretaceous in age and are assignable to the Lancian. The Hell's Hollow local fauna is considered to be earliest Paleocene because of the lack of dinosaurs and is assignable to the informal "premantuan" level. Morales 1 (V77128) and Herpijunk Promontory (V77129) may be similar in age to the Hell's Hollow local fauna. Finally, the McKeever Ranch localities are considered to be earliest Paleocene and are assignable to the mantuan.

MAMMAL EVOLUTION NEAR THE CRETACEOUS-TERTIARY BOUNDARY AND COMMENTS ON RECENT MODELS OF DINOSAUR EXTINCTION

The model of latest Cretaceous mammal evolution that has resulted from the present study agrees in some aspects with the model of Van Valen and Sloan (1977a; Sloan, 1970, 1976), but the evidence from the Hell Creek faunal-facies mammals in Garfield County adds increased complexity. These results also have implications for various models of dinosaur extinction.

The major emphasis in Garfield County has been on the upper approximately 150 ft. (45.7 m) of the Hell Creek Formation and the Tullock Formation. The upper portion of the Hell Creek Formation is dominated by gray siltstones. Although a good stratigraphic sequence of Hell Creek faunal-facies localities is not known, all indications suggest that mammals of this faunal-facies were preserved in the sediments of small, sluggish, possibly ephemeral streams and less frequently in overbank and paludal sediments. These mammals frequented, or more probably lived in, the vicinity surrounding these sites of preservation. The lower, sandier portions of the Hell Creek Formation have not been as thoroughly examined, so it is not possible to extrapolate the preceding scenario to this part of the formation. However, the evidence, particularly from the upper portion of the formation, indicates stasis within this faunal-facies. There was either no or little speciation. There are a few exceptions, such as the appearance of ?*Neoplagiaulax burgessi*. Also, palaeoryctids may have radiated early in the Lancian and begun to spread southward during this time (Lillegraven, 1969; L. S. Russell, 1975).

There is also no indication of a general decrease in numerical abundance or taxonomic diversity of mammals within the Hell Creek faunal-facies. The Flat Creek local fauna is only 17 ft. (5.2 m) below the base of the Tullock Formation (the base of the lower Z coal) and yet the diversity in this local fauna is as great as in the Lance and Trochu local faunas.

Dinosaurs (mostly teeth) are also common. Other Hell Creek faunal-facies localities, although none as rich as Flat Creek 5 (loc. V73087), have been discovered within about 10 ft. (3.0 m) of the formational contact. Dinosaurs are common at these sites. It does not appear that dinosaurs decreased in relative abundance at the Hell Creek faunal-facies mammal localities. However, the evidence from surrounding flood-plain deposits is less certain. Van Valen and Sloan (1977a, p. 44) note that dinosaur remains became increasingly scarce up-section, at least in the Bug Creek area. Dinosaur remains are also less common in certain parts of the upper Hell Creek Formation in Garfield County. However, in other places they are very common. For example, parts of what appear to be seven *Triceratops* individuals were found within approximately 0.25 sq. mi. (0.40 sq. km.), about 35 ft. (10.7 m) below the base of the Tullock. Another portion of a *Triceratops* skull was found about 11 ft. (3.4 m) below the formational contact. The specimen of *Tyrannosaurus rex* collected by the Los Angeles County Museum in the late 1960s was within 60 ft. (18.3 m) of the Tullock Formation, and teeth of this animal have been found within 25 ft. (7.6 m) of the boundary. Clearly, a general survey is needed to establish whether dinosaur remains decrease in relative abundance in the flood-plain deposits of the upper Hell Creek Formation.

Finally, at between 6 and 10 ft. (1.8–3.0 m) below the base of the lower Z coal, the Hell Creek faunal-facies probably ceased to exist as a distinct assemblage. A great many of the mammals either continued to exist or at least left Paleocene descendants. However, all but one marsupial and all the dinosaurs became extinct, at least in eastern Montana. As noted, this level is considered to mark the Cretaceous-Tertiary boundary in eastern Montana. Slaughter (1978) has noted the occurrence of a *Pediomys*-like tooth from the Eocene of Texas, suggesting that pediomyids may have survived into the Tertiary along the Texas coast.

The evidence from the Hell Creek faunal-facies does not negate the model proposed by Sloan and Van Valen (1965 and other papers). The data from Garfield County suggest that there was a rapid speciation of Paleocene-aspect mammals and a related decrease in typical latest Cretaceous mammals and dinosaurs going upward in the Hell Creek Formation and into the lower Tullock Formation. However, these trends were *restricted* to Bug Creek faunal-facies localities. During much of the deposition of the upper Hell Creek Formation after Paleocene-aspect mammals first appear at Bug Creek Anthills, two assemblages, the Hell Creek faunal-facies and Bug Creek faunal-facies, coexisted. Any models must account for these two coeval assemblages.

Until the upper 6 to 10 ft. (1.8–3.0 m) of the Hell Creek Formation are reached, the sedimentological evidence strongly indicates the presence of these two faunal-facies. The inception of larger, sandier channels (or at least their return, if the sandier, lower portions of the Hell Creek Formation are analogous) first seen at Bug Creek Anthills undoubtedly disrupted existing habitats as well as created new habitats. Paleocene-aspect mammals (i.e., *Protungulatum*, *Procerberus*, *Catopsalis*, and *Stygimys*) also make their first appearance. Causal relationships are at best speculative, but the evidence from Garfield and McCone counties does suggest that these invaders (plus changes in the physical environment and probably the flora) had an impact on the preexisting animals, as has been advocated by Van Valen and Sloan (1977a). As noted by these authors, the invaders are numerous, although not taxonomically diverse when they first occur at Bug Creek Anthills. It seems very likely that the habitat disruption which accompanied the inception of the larger, sandier streams allowed colonization by these invaders. This is a reasonable supposition because at least to the upper 6 to 10 ft. (1.8–3.0 m) of the Hell Creek Formation the Paleocene-aspect mammals are restricted to these larger, sandier channels. In this same interval the other environments of deposition (mostly smaller, siltier streams, but also paludal or quiet-water conditions) have yielded only typical latest Cretaceous mammals.

Once the assemblage that is termed in this study the Bug Creek faunal-facies was established, relatively rapid changes occurred. Fig. 4 in Van Valen and Sloan (1977a) shows these changes for the mammals and dinosaurs. Table 59 also shows the mammalian changes. New, closely related Paleocene-aspect mammals appeared probably through local (but allopatric) speciation. At the same time, the numbers of typical latest Cretaceous (or Hell Creek faunal-facies) mammals and dinosaurs were decreased. As noted earlier, the evidence from Garfield County supports Van Valen and Sloan's (1977a) contention that the decrease is real and not an artifact of sampling. It must be emphasized again that these trends were restricted to the Bug Creek faunal-facies. Van Valen and Sloan (1977a) are not sure whether the Hell Creek and Bug Creek faunal-facies (which they term the *Triceratops* and *Protungulatum* communities, respectively) "were always physically separated by habitat differences. . . ." The data from Garfield County coupled with that from McCone County definitely point to such a separation until the Hell Creek faunal-facies disappears. Therefore, when they note that "[w]hile the *Protungulatum* community gradually expanded, the *Triceratops* community gradually contracted," this statement must be restricted to trends *within* the Bug Creek faunal-facies. From what can be seen of their distribution, the larger, sandier channel-fill deposits of the Bug Creek faunal-facies constitute a low percentage of the total sediment in the upper Hell Creek Formation.

Although the effect upon the Hell Creek faunal-facies of the invading and radiating Paleocene-aspect mammals apparently was minimal (other than the original disruption it caused) until possibly the very close of the Cretaceous, their effect in the Bug Creek faunal-facies appears to have been profound. Van Valen and Sloan (1977a) suggest that the extinction of the dinosaurs and some other vertebrates (presumably including typical latest Cretaceous mammals) was through diffuse competition, particularly for food sources. For the mammals this supposition is very reasonable, but it is less so for dinosaurs. It remains very doubtful that such invaders as *Protungulatum* or even the more rodentlike *Catopsalis*, with their low-crowned cheek-teeth, would be competing for the same food as *Triceratops* and *Anatosaurus*, with their batteries of "grinding" teeth. I agree with Van Valen and Sloan (1977a) that body size, by itself, may not necessarily be related to resource use. However, in the one example they cite (Webb, 1969) for possible replacement of large (ungulate) by small (rodent) mammals, it is at least reasonable that the potential competitors ate the same food. They point out that competition may have been with the young or baby dinosaurs. This remains plausible, particularly as an extension of the model proposed by Pond (1977).

Van Valen and Sloan (1977a) use paleobotanical evidence to suggest increasingly cooler winter temperatures, although they clearly note that the continued presence of such aquatic animals as crocodiles, turtles, and champsosaurs argues against severe winters. The subtropical forests also gave way to more temperate forests. As noted in Archibald and Clemens (1980), it would be of interest to the above thesis to determine whether these floral changes occurred first along the banks of the larger Bug Creek faunal-facies streams.

Some sort of competition and replacement, rather than total exclusion, seems to be indicated by the Bug Creek faunal-facies sequences of Garfield and McCone counties. The fact that typical latest Cretaceous mammals and dinosaurs are common at Bug Creek

Anthills and only decrease up-section argues against exclusion. However, other factors such as climatic and floral changes may have also played a role. Competition, diffuse or otherwise, remains a very plausible but speculative explanation for the trends within the Bug Creek faunal-facies. Dinosaurs and at least some of the typical latest Cretaceous mammals may have become locally extinct within the Bug Creek faunal-facies before their demise in the Hell Creek faunal-facies. Unfortunately, this is still not well documented. If later research shows dinosaur extinction was diachronous within the Hell Creek Formation, it will have wider implications. It would be very reasonable to assume that if dinosaur extinction was diachronous in one small region, it was also so on a worldwide scale.

Plausible explanations can be put forward for the extinction of dinosaurs and some of the typical latest Cretaceous mammals within the Bug Creek faunal-facies sequence, but these explanations are not as readily applicable to the disappearances within the Hell Creek faunal-facies. Indications at present are that the Hell Creek faunal-facies (mammals and dinosaurs) disappear rather suddenly at about 6 to 10 ft. (1.8–3.0 m) below the Hell Creek–Tullock formational contact. It is known that the Hell Creek faunal-facies persisted relatively unchanged until at least 17 ft. (5.2 m) below the formational contact, based on the Flat Creek local fauna. Typical latest Cretaceous mammals and dinosaurs do occur above the 17 ft. (5.2 m) level, but the material, particularly for the mammals, is very scrappy. The only circumstantial evidence suggesting that the Hell Creek faunal-facies may have continued relatively unchanged above this level is that the sediments in which this faunal-facies are found continue to be prevalent until at least the base of the lowest lignite is reached. No typical latest Cretaceous mammals or dinosaurs are known from the 6–10 ft. (1.8–3.0 m) interval below the formational contact, but as just noted, this interval shows no marked lithologic change from underlying sediments. This 6–10 ft. (1.8–3.0 m) interval is now being more closely examined (W.A. Clemens, pers. comm.). Lower vertebrates (excluding dinosaurs) and two mammal-producing localities, Morales 1 (V77128) and Herpijunk Promontory (V77129), have been found in this interval. The former locality has produced only one mammal, a dentary fragment, described in this study as *Ragnarok harbichti*. The second locality has produced a number of mammalian specimens that suggest affinities with the Hell's Hollow local fauna (Clemens, pers. comm.). By the definition used in this study, these localities are Paleocene in age and may be pre-mantuan. Of more interest is the fact that these two localities occur in paludal or overbank sediments and are thus the only known exceptions to the rule of finding Paleocene-aspect mammals in large, sandy channel-fill deposits in the Hell Creek Formation. It is possible that the Paleocene-aspect mammals were expanding beyond the environs of the large channels in the uppermost Hell Creek Formation. Thus, they may have finally had an impact on the Hell Creek faunal-facies. Without further data, this supposition remains a wild speculation. Even with the demise of the Hell Creek faunal-facies assemblage, many of the typical latest Cretaceous mammals (with the notable exception of most of the marsupials) either continued or left Paleocene descendants.

The evidence from eastern Montana is quite clear in showing that near the end of the Cretaceous at least two distinct environments containing mammals were present. In the Hell Creek faunal-facies, the mammals were essentially in evolutionary stasis, while in the Bug Creek faunal-facies the speciation rates of mammals increased dramatically. Van Valen (1978) has noted that origination rates for the latter mammals suddenly increased by several orders of magnitude and then gradually declined.

The origin and evolution of several Paleocene mammalian faunas are now being studied by various paleontologists. Van Valen (1978) has published his views on early condylarth evolution. Based on this work, the data I have presented here, and helpful discussions with W. A. Clemens and M. Middleton, it is clear that the tremendous radiation of mammals was well under way before the end of the Cretaceous. Although some taxa, particularly the marsupials, were drastically affected by events at the Cretaceous-Tertiary boundary, most were not. The reorganization of the mammals that had commenced earlier continued through this transition. The magnitude of this transition was no greater than is seen between other Tertiary land mammal ages. Whether these events can be extended worldwide remains to be seen. Similar sequences are not known, but it would be naive to assume that other regions are necessarily similar or that eastern Montana represents the center of origin for the great radiation of mammals. Even a cursory look at the biotas of South America and Australia shows differences, the most obvious of which is that the marsupials there probably did not suffer the fate that befell them at the end of the Cretaceous in North America.

Van Valen and Sloan (1977a) may be correct in stating that within the Bug Creek faunal-facies dinosaurs lost in competition with the rapidly expanding Paleocene-aspect mammals and that they were also affected by climatic deterioration. Within the Hell Creek faunal-facies, this combination of events cannot as easily be suggested as causal factors. When expanded worldwide, these explanations become tremendously less likely, especially when there are no other adequate sequences on which to test them. However, the evidence from eastern Montana does not support cataclysmic events for the extinction of dinosaurs (Archibald and Clemens, 1980; Van Valen and Sloan, 1977a). The possibility that such an event (e.g., Russell and Tucker, 1971) set the stage cannot be ruled out, but the diachronous extinction of various vertebrates within the Hell Creek and Bug Creek faunal-facies argues against it.

Another recent model, Bakker (1977), includes latest Cretaceous extinctions as one of six or seven cycles of diversification and extinction. Insofar as the extinction of large (> 10 kg) terrestrial tetrapods is concerned, Bakker's model involves the chaining of several assumptions.

First, he considers speciation rates to be higher in continental areas during times of "geographic-topographic diversity" because of the increased number of barriers. Speciation rates are lower in continental basins or continental margins where barriers are fewer. As long as barriers exist in the more continental areas, speciation is higher and immigration of new species into the lower-lying areas keeps pace with local extinction rates. When the topographic diversity decreases, speciation decreases and extinction rates in low-lying areas are higher than the immigration rates of new taxa. This may lead to, or at least contribute to, extinctions in these low-lying areas.

Second, according to Bakker's interpretation of the "Haug Effect" (Johnson, 1971) during maximum transgression, orogeny and uplift are also at a maximum. During this time, geographic-topographic diversity will be high, and thus speciation rates in continental areas will, through immigration, offset extinction rates in the low-lying areas. According to Bakker, during regression, orogeny and geographic-topographic diversity are minimized. Thus, speciation rates and immigrations from continental areas to low-lying areas are decreased, and extinction rates in the low-lying areas are relatively increased. This may lead to reduced diversity and contribute to the total extinction of large terrestrial tetrapods. According to Bakker, the microfauna may not react in the same way. He logically concludes that "[t]he magnitude of the geographical barrier necessary for speciation varies with size and mobility of the animal; for dinosaurs and elephants barriers must be of considerable size and duration; for mice a much smaller feature is sufficient."

He goes on to comment on how topographic features will affect speciation rates for large and small tetrapods differently. I do not disagree with this assessment, but other sorts of barriers (e.g., ecologic) besides geographic-topographic features can affect speciation. A case in point is the latest Cretaceous of North America, which can serve not only as a general exception but also a specific exception in regard to dinosaur extinction.

As previously noted, the typical latest Cretaceous Lance, Flat Creek, and Trochu local faunas are very similar in mammalian species composition, yet almost 750 mi. (1200 km) separate the two furthermost local faunas (Lance and Trochu). These similarities, plus physical geology, suggest that this north-south-trending coastal plain was very homogeneous. Even small topographic barriers do not appear to have impeded these small mammals. When the Bug Creek faunal-facies appeared in the upper Hell Creek Formation, the introduction of large streams brought physical as well as habitat changes to the region. However, there is no suggestion of great changes in topography. The effects on the typical latest Cretaceous mammals *within* the Bug Creek faunal-facies sequences were dramatic, but the effects on the dinosaurian fauna were equally dramatic. This is a clear example of the way other than topographic fetures can serve as barriers to macrofauna as well as to microfauna. It cannot be shown that in this case these ecological barriers affected the speciation rates of dinosaurs and possibly the typical latest Cretaceous mammals. Nevertheless, this example shows that it is too simplistic to assume that only topographic features can serve as barriers for speciation of large or small tetrapods.

In another section, Bakker noted that no special mechanism is required to cause the extinction of the "last survivors." He felt that the extinction of the last dinosaurs could have come about through a variety of agents (which he listed). I agree with Bakker that there is a real decrease in taxonomic diversity of dinosaurs going from Campanian through Maestrichtian within western North America (but see D. A. Russell, 1975). However, even in the upper part of the Hell Creek Formation, within the Hell Creek faunal-facies there may be as many as fifteen species (estimates vary) of dinosaurs that fill the medium-to-large herbivore and carnivore niches. To treat these forms simply as the "last survivors" is an oversimplification. This is even more the case when the total worldwide extinction of dinosaurs is considered. Unfortunately, the record of this event (or events—for it may well have been diachronous) is not well known in other areas of the world. I agree with Bakker that various agents probably caused the extinction of dinosuars in different regions, but for this to have happened on a worldwide scale in a relatively short span of geologic time is more than a coincidence. Bakker's view that mass extinctions of tetrapods are due to the speciation rates falling below the extinction rates is at least plausible. Whether it can be applied as a universal causation for all or most mass tetrapod extinctions remains to be documented. It certainly may have been a contributing factor to dinosaur extinction, particularly if the example of latest Cretaceous, western North American dinosaurs is used, but it is clearly not the only or possibly even the major factor.

Finally, Bakker (1977) appealed to the "Haug Effect" (Johnson, 1971), which correlates maximum orogeny with maximum transgression, as a method to increase geographic-topographic diversity. The "Haug Effect" is by no means a widely accepted theory (see, e.g., Stille, 1936), and thus it is not clear whether maximum orogenies and maximum transgressions occurred at the same time, particularly on a worldwide scale. As used in Bakker's model, these two maxima would have to have occurred worldwide over what was geologically a very short period of time. Jeletzky (1978) presented data demonstrating that there is almost no evidence for eustatic oscillations during the

Cretaceous and cited other works (i.e., Yanshin, 1973) to indicate that this was probably also the case for the Phanerozoic as a whole. He did not deny that such eustatic changes occurred, but he felt that they were of minor amplitude and are very seldom preserved in the record. However, he did indicate that one such global event may have been the Maestrichtian regression. Even if the "Haug Effect" is assumed to be the correct model, it would have to be demonstrated that topographic diversity would be maximized during an orogeny when base level is high, versus a regressive phase when base level is lowered and renewed down-cutting ensues.

From the example of the Hell Creek Formation mentioned earlier, it is apparent that barriers other than topography could serve to isolate or to exclude organisms, even large tetrapods. If such barriers are minimized, speciation rates may decrease, as Bakker believes, and this may be a factor, but not the only factor, leading to mass extinctions.

SUMMARY

Three topics were considered in this study: a description of the geology of the Hell Creek (Upper Cretaceous and lowermost Paleocene) and Tullock (lower and middle Paleocene) formations, particularly in the valley of Hell Creek; a systematic description of the mammals recovered from the Hell Creek Formation and lowest portion of the Tullock Formation; and a comparison of the mammals with similarly aged local faunas in other areas.

Microvertebrate sites were surface-collected and the richer of these were sampled using the underwater screening techinique. Only one locality, the Garbani Quarry, produced enough associated teeth to warrant both hand-quarrying and underwater screening.

Brief descriptions were given of the formations subjacent (Bearpaw Shale, Fox Hills Sandstone) and superjacent (the Lebo Formation) the the Hell Creek and Tullock formations. The latter formations were mapped in detail in the valley of Hell Creek. The base of the lowest lignite (the Z) in the Tullock Formation is taken as the contact between the Tullock and underlying Hell Creek Formation. The presence or absence of dinosaurs is not included in this definition of the Hell Creek–Tullock contact because biologic events should play no role in determining a lithologic boundary. Dinosaur extinction, which occurs about 6 to 10 ft. (1.8–3.0 m) below the lowest lignite is used to approximate the Cretaceous-Tertiary boundary. Thus, this uppermost interval of the Hell Creek Formation is considered to be lowermost Paleocene. The Hell Creek and Tullock formations are largely fluviatile in origin.

Mammal-producing localities have been found throughout much of northern Garfield County. However, discussion concentrated on four areas of study. The first and most important was the Hell Creek–Snow Creek area. The others were the Billy Creek–McKeever, Flat Creek, and Horsethief Canyon areas.

Two distinct latest Cretaceous faunal-facies, both of Lancian age, were coeval during much of the deposition of the Hell Creek Formation. These faunal-facies are separated primarily by differences in environments of deposition and by differences in the mammalian components of the faunas. Two earliest Paleocene mammal levels, the premantuan and mantuan, were present near the Hell Creek–Tullock boundary. Both of these informal levels are older than the presently recognized Puercan. Puercan and Torrejonian localities were also present in the Tullock Formation, but the mammals and biostratigraphy of these localities will be described elsewhere by W. A. Clemens.

The Lancian, Hell Creek faunal-facies localities are found throughout the stratigraphic extent of the Hell Creek Formation, except for the upper 6 to 10 ft. (1.8–3.0 m), and occur in small, poorly sorted, silty channel-fill deposits or in paludal sediments. These localities, like those from the Scollard Formation, Alberta, and Lance Formation, Wyoming, are often referred to as having typical latest Cretaceous mammals because of the abundance of ptilodontoid mulituberculates and marsupials. The richest locality of the faunal-facies, Flat Creek 5 (loc. V73087), is only 17 ft. (5.2 m) below the base of the Tullock Formation, and the vertebrates from it and four other localities are included in the Flat Creek local faunal. Dinosuars are common elements in this faunal-facies.

The Lancian, Bug Creek faunal-facies localities are found in the upper 80 ft. (24.4 m) of the Hell Creek Formation and occur in large, sandy, channel-fill deposits. This faunal-facies is only questionably known from Garfield County but is well known from (oldest to youngest) the Bug Creek Anthills, Bug Creek West, and Harbicht Hill local faunas of McCone County. These localities include mammals of Paleocene aspect, particularly condylarths and taeniolabidoid multituberculates. The taxonomic diversity and numerical abundance of typical latest Cretaceous mammals and dinosaurs decrease at successively younger localities *within* this faunal-facies (Van Valen and Sloan, various papers).

The Paleocene, pre-mantuan localities occur on both sides very close to the Hell Creek–Tullock contact. The richest and best-known pre-mantuan locality, Worm Coulee 1 (V74111), is 7 ft. (2.1 m) above the formational contact. The mammals from this locality and one other locality (V74110) constitute the Hell's Hollow local fauna. These localities occur in a moderately large, sandy, channel-fill deposit. Two other possibly pre-mantuan localities occur in overbank-like deposits just below the Hell Creek-Tullock formational contact. One locality (V77129) has produced only one condylarth, but the other has produced a number of taxa which are now under study by W. A. Clemens.

The Paleocene mantuan is represented by two localities (V72210 and V77087) that occur in a sandy, channel-fill deposit immediately overlying the Hell Creek–Tullock contact.

The multituberculates *Mesodma thompsoni* and *M. hensleighi* are present in the Hell Creek faunal-facies. *M. formosa* (cf. *M. formosa*) is tentatively recognized. The recognition of the latter species is complicated by the presence in the Flat Creek local fauna of a new neoplagiaulacid ?*Neoplagiaulax burgessi*, known only from five isolated symmetrical P_4 blades. The P_4's of *M. thompsoni* are very large and further suggest the phyletic relationship of *M. thompsoni* with the earliest Paleocene species, *M. ambigua.* Another new species of *Mesodma, M. garfieldensis*, is present in the Hell's Hollow local fauna. However, only isolated teeth (except associated P_{3-4}) are known and there are ambiguities which suggest that two species may be present.

Another isolated P_4 blade is high and arcuate and cannot be referred to *Mesodma* or ?*Neoplagiaulax burgessi*. It is questionably placed in the Neoplagiaulacidae as an indeterminate genus and species.

Isolated P_4's of *Cimolodon nitidus* fall into two size classes, as in the Lance local fauna (Clemens, 1964). One of the blades is very high and arcuate and strongly resembles the Paleocene species *Anconodon gidleyi.*

Three species of taeniolabidoids are present in the Hell's Hollow local fauna. The eucosmodontoids are represented by the eucosmodontine *Stygimys* aff. *S. kuszmauli* and a new microcosmodontine, *Acheronodon garbani*. The specimens of *Stygimys* from the Hell's Hollow local fauna were compared to *S. gratus* (Mantua lentil) and specimens from Harbicht Hill. Based on this comparison, it appears that *S. gratus* and *S. kuszmauli* (Bug Creek Anthills) are distinct. However, the specimens from Harbicht Hill, assigned to *S. gratus* by Sloan and Van Valen (1965), and those from Hell's Hollow may represent one, or possibly two new species. The new microcosmodontine, *A. garbani*, is represented by a single, minute P_4. *Acheronodon* shares characters with the other microcosmodontine genera, *Microcosmodon* and *Pentacosmodon*, and in fact could be congeneric with *Microcosmodon*. The presence of *Acheronodon* in the earliest Paleocene constitutes a range extension for the subfamily and supports Holtzman and Wolberg's (1977) suggestion of a long, separate evolutionary history for eucosmodontines and microcosmodontines. The taeniolabidid, *Catopsalis*, is rare in the Hell's Hollow local fauna, but because of its large size is best, but only tentatively, referred to *C. foliatus* from the Paleocene rather than the smaller Bug Creek Anthills species, *C. joyneri*. M. Middleton (pers. comm.) believes that it belongs to a new species he is describing from the lower Paleocene of Colorado.

The Cimolomyidae, including *Cimolomys*, *Meniscoessus*, and, questionably, *Essonodon*, are removed from the suborder Taeniolabidoidea (Sloan and Van Valen, 1965) because the enamel on the lower incisor of *M. robustus* is not restricted to a ventrolabial band as in taeniolabidoids. *C.gracilis* is relatively rare in the Hell Creek faunal-facies. *M. borealis* is synonymized with *M. robustus*, based on a comparison of the type M^2 of *M. borealis* with M^2's from the Lance local fauna and Hell Creek faunal-facies. *M. conquistus* is restricted to the type specimen, as suggested by Clemens (1964). The rostral region of a poorly preserved skull *M. robustus* is described. The skull preserves large incisive foramina. It is possible that large palatal vacuities are present, but the skull fragment and other palatal fragments give conflicting information. A reconstruction of the rostral region of *M. robustus* was attempted.

Among the sample of isolated P^4's referred to *Meniscoessus robustus*, three display a morphology atypical for this taxon. These three P^4's lack anteroexternal and posterior basal cusps. These P^4's resemble a worn and broken P^4 in a maxilla of *Essonodon browni* that also preserves M^{1-2}. However, the P^4 in the maxilla is so damaged that close comparisons are not possible. Thus, the three isolated P^4's are questionably retained within *M. robustus*. The P^4 and all molars of *Essonodon* are now known, although the lower molars are represented by only isolated M_1's and M_2's. *Essonodon* has a number of peculiar and sometimes unique features such as a very tall P^4, transversely elongated molar cusps, anteroposteriorly short M^2_2's relative to M^1_1's, and probably very posteriorly positioned palatal vacuities. The affinities of this genus are still uncertain, but it is better considered to be a questionable cimolomyid rather than a ptilodontid.

The genus *Cimexomys* is revised to include only *C. minor* (the genotype), *C. antiquus*, *C. hausoi* (a new species from the Hell's Hollow local fauna), and an undescribed species (Sloan and Van Valen, 1965). The new species, *C. hausoi*, differs from *C. minor* in its larger size and greater number of molar cusps and P_4 serrations.

A new genus, *Paracimexomys*, is recognized. It includes *P. priscus* (the genotype), *P. magister*, probably *"Cimexomys" judithae*, and possibly *"C." magnus*. *Paracimexomys* differs from *Cimexomys* in its lower M^1_1 cusp formula, divergence of the external and medial cusp rows of M^1, and "waisting" of the molars.

Paracimexomys and *Cimexomys* are more primitive than ptilodontoids in their lower cusp count, shorter M^1 internal cusp row (50% of the M^1 length), and lower number of P_4 serrations. These genera are removed from the Ptilodontoidea and placed in suborder and family *incertae sedis*.

Paracimexomys, *Cimexomys*, and the cimolomyids were compared cladistically with other multituberculates. The results suggest that these taxa should not be included in the Ptilodontoidea or Taeniolabidoidea. The characters these three groups share with each other and the members of the two suborders are almost all primitive retentions and thus are not indicative of close phyletic relationships.

All the marsupials that Clemens (1966, 1973) recognized in the Lance local fauna are at least tentatively recognized in the Hell Creek faunal-facies except for *Alphadon* cf. *rhaister* and *Glasbius intricatus*. The latter genus is represented in the Hell Creek Formation by a new species, *G. twitchelli*, that is larger and almost totally lacks stylar cusp C. *A. marshi*, *A. wilsoni*, and *G. twitchelli* are the most common marsupials in the Hell Creek faunal-facies. Numerous isolated teeth from the Hell's Hollow local fauna are referred to *Peradectes* cf. *P. pusillus*. It is suggested that the type of *Stagodon validus* (a dentary fragment) from the Lance local fauna may belong to *Didelphodon padanicus*, previously known by one specimen from the Hell Creek Formation, South Dakota. A large edentulous dentary from the Flat Creek local fauna is questionably referred to ?*Pediomys* cf. *P. florencae* but may be the first Lancian record of *Eodelphis*.

Gypsonictops and the palaeoryctids are very rare (except *Procerberus formicarum*) elements in the Hell Creek Formation. However, all the eutherians recognized by Lillegraven (1969) in the Trochu local fauna are at least tentatively recognized in the Hell Creek faunal-facies of Montana. Research by various people has shown that most of these taxa are not related to lipotyphlous insectivores. Accordingly, they are placed in Eutheria, *incertae sedis*, and are referred to informally as proteutherians, after Butler's (1972) usage.

Condylarths are diverse and relatively common in the Bug Creek faunal-facies, Hell Creek Formation, McCone County. They continue to be diverse at the pre-mantuan and mantuan localities in Garfield County. The pre-mantuan condylarths are represented by the oxyclaenine arctocyonid *Protungulatum* cf. *P. donnae*, the loxolophine arctocyonids *Ragnarok harbichti*, and *R. engdahli* (n. sp.), the ?hyopsondontid *Oxyprimus erikseni*, and the anisonchine periptychid *Mimatuta morgoth*. Those from the mantuan are represented by the oxyclaenine arctocyonid *P. mckeeveri* (n. sp.), the loxolophine arctocyonid *R. nordicum*, the anisonchine periptychid *M. minuial*, and a ?periptychid.

A maxillary fragment and a palate from the Puercan (San Juan Basin, New Mexico) were described and are refered to *Oxyacodon agapetillus* and *O. priscilla*. [J. K. Rigby, Jr., and L. Van Valen (pers. comms.) have also found upper dentitions which they believe to belong to *Oxyacodon*.] Both specimens show characters that are probably uniquely derived (e.g., tendency to reduce the paraconule), but in most features they are structurally more primitive than *Conacodon cophater* and *C. entoconus*. The genus *Oxyacodon* is transferred to the Anisonchinae based on the characters seen in the above specimens, plus those seen in the lower dentition. This referral was also done separately by Van Valen (1978) and Rigby (pers. comm).

A cladistic analysis was done separately for upper and lower dentitions of various Lancian, pre-mantuan, mantuan, and Puercan anisonchines. One of the most interesting trends is the development of the bulbous P^4. It appears that the typical anisonchine upper molar began to develop before the P^4 had been modified to any great extent. This in part explains why anisonchine lower molars are relatively morphologically consistent, while the lower premolars (mostly P_{3-4}) are quite variable.

The determination of the Cretaceous-Tertiary boundary in the terrestrial sediments of eastern Montana should be done by correlation to the type sections in Europe. Magnetostratigraphy may eventually become the best method for accomplishing this; until that time, dinosaur extinction remains the most utilitarian method for approximating the boundary in eastern Montana. This event occurs on the average at 10 ft. (3.0 m) below the Hell Creek–Tullock formational contact.

The typical latest Cretaceous (Lancian) local faunas from Lance, Flat Creek (Hell Creek faunal-facies), and Trochu were compared. The faunal resemblances (using Simpson's index of100 C/N_1) among these local faunas are very high at the species level. Even the two most distant local faunas, Lance and Trochu, are 71 percent similar at the species level. The Flat Creek local fauna is an admixture of the other two local faunas in species representation, species diversity, and relative abundances. There is little suggestion of an age difference between any of these three local faunas and the Bug Creek faunal-facies of McCone County. There is some indication that one-third of the species in common between the Lance, Hell Creek and Trochu local faunas increase in size toward the north. However, the reality of the trend is suspect and even if true, no likely explanations could be determined.

Based on biostratigraphic criteria, the Hell's Hollow local fauna is most similar to Sloan and Van Valen's (1965) Harbicht Hill local fauna, but this faunal comparison suggests that the Hell's Hollow local fauna is slightly younger. Unlike Harbicht Hill, Hell's Hollow has numerous specimens of *Peradectes* and a larger species of *Catopsalis* resembling the Puercan species *C. foliatus*, and it lacks dinosaurs. The only species present at Harbicht Hill that suggests that this local fauna is younger is a single lower molar of the primate *Purgatorius*. The Hell's Hollow local fauna is considered to be earliest Paleocene because it lacks dinosaurs.

The McKeever Ranch localities (V72210 and V77087) have yielded only four taxa of condylarths. However, two of these taxa, *Mimatuta minuial* and *Ragnarok nordicum*, are known from Mantua lentil, Wyoming, and thus the McKeever Ranch localities are considered to be mantuan in age. The Hell's Hollow local fauna (and locs. V77128 and V77129) appears to be biostratigraphically bracketed by the Lancian Harbicht Hill local fauna and the mantuan McKeever localities, and thus is informally considered to be pre-mantuan.

The inception (or return) of large channels within the upper 100 ft. (30.5 m) of the Hell Creek Formation disturbed the already existing habitat. Typical latest Cretaceous mammals and dinosaurs occupied these disturbed areas. However, Paleocene-aspect mammals (probably as well as other new elements, possibly including flora) also invaded with these habitat changes. The evidence from Garfield County supports Van Valen and Sloan's thesis (e.g., 1977a) that the diversity and relative abundance of the typical latest Cretaceous mammals and dinosaurs decreased up-section *within* this new environment, the Bug Creek faunal-facies.

Within this sequence the Paleocene-aspect mammals replaced the typical latest Cretaceous mammals and dinosaurs through speciation and immigration. Whether competition, diffuse or otherwise, was involved is uncertain, but the manner of replacement suggests some form of competition. Competition between the two groups of mammals is more easily explained than between the Paleocene-aspect mammals and the dinosaurs. Competition (for food sources) between small and large organisms is reasonable, but it is doubtful whether adult dinosaurs and the invading mammals utilized the same sources. It remains possible that such competition did occur between juvenile or baby dinosaurs and the invading mammals. Whatever the cause(s), dinosaurs and typical latest Cretaceous mammals may have become locally extinct *within* the Bug Creek faunal-facies before their total extinction elsewhere in the upper Hell Creek Formation. The evidence for this conclusion (which differs from the model proposed by Van Valen and Sloan, 1977a) comes from the Flat Creek local fauna and other localities that occur to within 6 to 10 ft. (1.8–3.0 m) of the base of the Tullock Formation. At these localities, termed the Hell Creek faunal-facies, typical latest Cretaceous mammals and dinosaurs are common, and Paleocene-aspect mammals are absent.

The Hell Creek faunal-facies localities represent smaller, siltier, possibly ephemeral streams (and paludal and overbank conditions) that were present before and continued after the inception of the Bug Creek faunal-facies. At least during the deposition of the approximately upper 100 ft. (30.5 m) of the Hell Creek Formation, these two faunal-facies coexisted. The Bug Creek faunal-facies showed rapid turnover (speciation, immigration, and local extinctions), as was documented by Van Valen and Sloan (1977a). In contrast, the Hell Creek faunal-facies is characterized by evolutionary stasis.

The presence of this second, coeval faunal-facies, not previously well documented, is more difficult to explain in terms of the model of dinosaur (and typical latest Cretaceous mammal) extinction proposed by Van Valen and Sloan (1977a).

In the upper 6 to 10 ft. (1.8–3.0 m) of the Hell Creek Formation, the assemblage of vertebrates that constitutes the Hell Creek faunal-facies ceased to exist. This probably marks the final extinction of dinosaurs and some typical latest Cretaceous mammals (notably the majority of the marsupials) in eastern Montana.

Paleocene-aspect mammals may have finally had an impact upon the Hell Creek faunal-facies. Within this 6 to 10 ft. (1.8–3.0 m) interval, various non-dinosaurian lower vertebrates and as yet unstudied earliest Paleocene (pre-mantuan) mammals have been recovered from paludal or overbank deposits. Although it remains wild speculation, this occurrence may indicate that Paleocene-aspect mammals expanded beyond the margins of the larger channels as a result of the demise of the Hell Creek faunal-facies or that they may have contributed to this demise.

The change-over in the mammalian fauna at the end of the Cretaceous does not appear to have been drastic. When current studies on earliest Paleocene local faunas by various researchers are complete, it is probable that the evidence will show that the Lancian–pre-mantuan–mantuan–Puercan turnover was similar in magnitude to that seen among other land mammal ages. The change-over to a more Paleocene-like mammalian fauna had already begun well before the close of the Cretaceous, at least in Montana.

The data from Garfield County supports, or at least does not contradict, some of the portions of the extinction models proposed by Van Valen and Sloan (1977a), as discussed above. However the model of Van Valen and Sloan (1977a) does not easily explain the coexistence of the Bug Creek and Hell Creek faunal-facies. Catastrophism theories (e.g., Russell and Tucker, 1971) cannot be ruled out, but the evidence from eastern Montana does not support them.

The evidence from Garfield County does suggest a decrease in the taxonomic diversity of dinosaurs going from the Campanian to Maestrichtian which supports Bakker's (1977) view that extinction rates may have become greater than speciation rates. Although, as Bakker (1977) states, topographic-geographic barriers do affect large and small vertebrates differently, data from eastern Montana show that other barriers, in this case ecological, probably affected dinosaurs and typical latest Cretaceous mammals in a similar fashion. Bakker feels that a variety of agents may have caused the extinction of the last surviving dinosaurs. Although various agents were probably involved, the Hell Creek faunal-facies indicates that dinosaurs enjoyed at least moderate diversity and abundance and were not simply the "last survivors."

Bakker utilizes the "Haug Effect" to expand his model of increasing extinction rates/decreasing speciation rates to a worldwide scale to explain the extinction of various tetrapod dynasties. The "Haug Effect" is not accepted by all workers. Eustatic changes in sea level, as is at least required by Bakker's use of the "Haug Effect," have been shown to be of minor amplitude. However, the Maestrichtian regression may have been eustatic (Jeletsky, 1978). Even if the "Haug Effect" is accepted, it is not clear that topographic-geographic diversity would be maximized during the presumed joint maxima of transgression and orogeny. Without a global mechanism such as the "Haug Effect," Bakker's model of tetrapod extinction may have been a contributing factor but was not a worldwide agent of extinction.

Appendices

APPENDIX 1.
Measurements of molars of *Mesodma hensleighi* from various Lancian localities.

		LENGTH					WIDTH				
	Locality	*N*	$\bar{x}$	*s*	*CV*	*OR*	*N*	$\bar{x}$	*s*	*CV*	*OR*
M^1	V5620	7	2.44	—	—	2.35–2.60	7	1.14	—	—	1.10–1.20
	V5711	2	2.58	—	—	2.55–2.60	2	1.25	—	—	1.20–1.30
	Trochu	3	2.20	—	—	2.10–2.30	9	1.02	—	—	1.00–1.10
	V73087	1	—	—	—	2.17	2	1.09	—	—	1.08–1.10
M_1	V5620	8	1.86	—	—	1.70–1.95	8	0.90	—	—	0.80–1.00
	V5711	2	1.95	—	—	1.95	2	0.98	—	—	0.95–1.00
	Trochu	5	1.80	—	—	1.70–1.95	6	0.78	—	—	0.75–0.85
	V73087	12	1.87	0.076	4.04	1.73–1.97	12	0.88	0.084	9.58	0.72–0.99
M^2	V5620	—	—	—	—	—	—	—	—	—	—
	V5711	—	—	—	—	—	—	—	—	—	—
	Trochu	9	1.09	—	—	1.00–1.15	9	1.08	—	—	1.00–1.15
	V73087	15	1.14	0.053	4.69	1.04–1.22	15	1.11	0.057	5.12	0.97–1.17
M_2	V5620	1	—	—	—	1.10	1	—	—	—	0.95
	V5711	—	—	—	—	—	—	—	—	—	—
	Trochu	2	1.02	—	—	0.95–1.10	2	0.85	—	—	0.80–0.90
	V73087	10	1.12	0.072	6.45	1.05–1.23	10	0.93	0.066	7.15	0.86–1.07

APPENDIX 2.
Measurements of molars of *Mesodma formosa* from various Lancian localities.

		LENGTH					WIDTH				
	Locality	*N*	$\bar{x}$	*s*	*CV*	*OR*	*N*	$\bar{x}$	*s*	*CV*	*OR*
	V5620	18	2.88	0.125	4.35	2.65 – 3.05	18	1.33	0.049	3.70	1.25 – 1.40
M^1	V5711	9	3.14	—	—	3.00 – 3.25	9	1.41	—	—	1.35 – 1.45
	Trochu	2	3.13	—	—	3.10 – 3.15	7	1.36	—	—	1.25 – 1.40
	V73087	1	—	—	—	2.53	4	1.27	—	—	1.21 – 1.32
	V5620	50	2.33	0.126	5.40	2.10 – 2.55	50	1.03	0.076	7.40	0.90 – 1.20
M_1	V5711	14	2.36	0.025	6.72	2.05 – 2.50	13	1.08	0.170	9.89	0.95 – 1.35
	Trochu	3	2.30	—	—	2.10 – 2.50	11	1.07	—	—	0.90 – 1.15
	V73087	—	—	—	—	—	—	—	—	—	—
	V5620	27	1.51	0.117	7.77	1.30 – 1.70	27	1.44	0.082	5.72	1.30 – 1.60
M^2	V5711	2	1.70	—	—	1.65 – 1.75	2	1.63	—	—	1.60 – 1.65
	Trochu	17	1.44	0.100	6.94	1.30 – 1.60	17	1.37	0.090	6.93	1.25 – 1.50
	V73087	9	1.42	—	—	1.21 – 1.69	9	1.39	—	—	1.23 – 1.46
	V5620	6	1.35	—	—	1.25 – 1.50	6	1.16	—	—	1.10 – 1.25
M_2	V5711	2	1.45	—	—	1.40 – 1.50	2	1.28	—	—	1.25 – 1.30
	Trochu	16	1.41	0.140	9.79	1.25 – 1.60	17	1.18	—	—	1.00 – 1.40
	V73087	—	—	—	—	—	—	—	—	—	—

APPENDIX 3.
Measurements of molars of *Mesodma thompsoni* from various Lancian localities.

		LENGTH					WIDTH				
	Locality	*N*	$\bar{x}$	*s*	*CV*	*OR*	*N*	$\bar{x}$	*s*	*CV*	*OR*
M^1	V5620	22	3.27	0.080	2.44	3.15 – 3.45	22	1.43	0.053	3.70	1.35 – 1.55
	V5711	15	3.53	0.094	2.67	3.40 – 3.70	15	1.54	0.082	5.33	1.40 – 1.65
	Trochu	—	—	—	—	—	4	1.51	—	—	1.45 – 1.60
	V73087	1	—	—	—	3.27	4	1.47	—	—	1.41 – 1.55
M_1	V5620	1	—	—	—	2.65	1	—	—	—	1.05
	V5711	7	2.74	—	—	2.65 – 2.90	7	1.21	—	—	1.15 – 1.30
	Trochu	1	—	—	—	2.85	1	—	—	—	1.25
	V73087	—	—	—	—	—	1	—	—	—	1.21
M^2	V5620	2	2.00	—	—	1.85 – 2.15	2	1.70	—	—	1.60 – 1.80
	V5711	1	—	—	—	2.15	1	—	—	—	1.75
	Trochu	2	1.87	—	—	1.85 – 1.90	2	1.72	—	—	1.65 – 1.80
	V73087	—	—	—	—	—	—	—	—	—	—
M_2	V5620	1	—	—	—	1.75	1	—	—	—	1.35
	V5711	3	2.03	—	—	1.95 – 2.15	3	1.63	—	—	1.50 – 1.60
	Trochu	2	1.95	—	—	1.90 2.00	2	1.55	—	—	1.50 – 1.60
	V73087	—	—	—	—	—	—	—	—	—	—

APPENDIX 4.
Measurements of associated upper molars of *Protungulatum donnae*, Bug Creek Anthills (UCMP loc. V70201), McCone County, Montana.

	M^1			M^2			M^3		
	N	$\bar{x}$	*OR*	*N*	$\bar{x}$	*OR*	*N*	$\bar{x}$	*OR*
L	3	3.83	3.66–3.95	6	4.18	3.84–4.37	2	3.32	3.28–3.26
W-A	3	4.86	4.72–5.03	6	6.13	5.83–6.47	2	5.54	5.47–5.60
W-P	3	5.21	5.09–5.39	6	6.22	5.88–6.51	2	4.82	4.56–5.07
A	3	1.70	1.59–1.79	6	2.19	2.09–2.27	2	1.65	1.63–1.66
B	3	0.62	0.57–0.69	6	0.72	0.57–0.85	2	0.64	0.62–0.65
C	3	1.54	1.50–1.56	6	1.76	1.59–1.85	2	1.59	1.50–1.67
D	3	1.00	0.97–1.06	6	1.46	1.31–1.82	2	1.67	1.66–1.68
L/W-P	3	0.73	0.71–0.73	6	0.67	0.65–0.70	2	0.69	0.66–0.72
L/W-A	3	0.79	0.76–0.82	6	0.68	0.64–0.72	2	0.60	0.60
A/W-A	3	0.35	0.34–0.37	6	0.36	0.35–0.37	2	0.30	0.29–0.30
B/W-A	3	0.13	0.12–0.14	6	0.12	0.10–0.14	2	0.12	0.11–0.12
C/W-A	3	0.32	0.31–0.33	6	0.29	0.25–0.31	2	0.29	0.27–0.30
D/W-A	3	0.21	0.20–0.21	6	0.24	0.21–0.28	2	0.30	0.30

APPENDIX 5.

Measurements of associated lower premolars and molars of *Protungulatum donnae*, Bug Creek Anthills (UCMP locs. V65127 and V70201), McCone County, Montana. Only one specimen, UCMP 100896, preserves M_{1-3} in association, and the length of these molars is 12.95 mm.

	P_2		P_3			P_4		
	N	*OR*	*N*	$\bar{x}$	*OR*	*N*	$\bar{x}$	*OR*
L	1	1.87	2	3.04	2.95–3.12	2	3.48	3.45–3.51
W	1	1.20	2	1.67	1.67	2	2.12	2.09–2.14

	M_1			M_2					M_3		
	N	$\bar{x}$	*OR*	*N*	$\bar{x}$	*s*	*CV*	*OR*	*N*	$\bar{x}$	*OR*
L	6	3.68	3.43–3.89	9	3.97	0.182	4.58	3.74–4.23	5	4.72	4.08–5.07
L-TR	6	1.97	1.82–2.06	9	2.17	0.085	3.91	2.08–2.29	5	1.95	1.45–2.20
L-TA	6	1.71	1.47–1.85	9	1.80	0.120	6.68	1.62–1.94	6	2.81	2.56–3.05
W-TR	6	2.62	2.30–2.77	9	3.20	0.254	7.94	2.79–3.64	5	3.04	2.87–3.18
W-TA	6	2.62	2.47–2.87	9	2.93	0.256	8.75	2.48–3.32	6	2.38	2.19–2.47
P-M	5	0.16	0.11–0.20	9	0.10	—	—	<0.10–0.16	5	<0.10	<0.10–0.20
L-TR/L-TA	6	1.16	1.02–1.33	9	1.21	0.067	5.54	1.09–1.31	5	0.71	0.55–0.80
W-TR/W-TA	6	1.00	0.93–1.04	9	1.10	0.041	3.73	1.05–1.13	5	1.28	1.25–1.31

Literature Cited

Alvarez, W., and D. W. Vann.

1979. Comment on "Biostratigraphy and magnetostratigraphy of Paleocene terrestrial deposits, San Juan Basin, New Mexico." Geology 7:66–67.

Ameghino, F.

1890. Los plagiaulacideos argentinos y sus relaciones zoológicas, geológicas y geográficas. Bol. Instituto Geográfico Argentino, 9:143–201.

American Commission on Stratigraphic Nomenclature.

1961. Code of stratigraphic nomenclature. Am. Assoc. Petroleum Geol. Bull. 45:645–665.

Archibald, J. D.

1977. Ectotympanic bone and internal carotid circulation of eutherians in reference to anthropoid origins. Jour. Human Evol. 6:609–622.

1979a. Mammalia. Pp. 278–280 *in* McGraw-Hill 1979 Yearbook of Science and Technology. D. N. Lapedes, ed. New York: McGraw-Hill.

1979b. Oldest known eutherian stapes and a marsupial petrosal bone from the Late Cretaceous of North America. Nature 281:669–670.

Archibald, J. D., and W. A. Clemens.

1977. The beginnings of the Age of Mammals (abs.). Jour. Paleont. 51 (Suppl. to No. 2):1.

1980. The beginnings of the Age of Mammals in Montana. *in* Catastrophes in Earth History: The New Uniformitarianism, W. A. Berggren and J. A. Van Couvering, eds. Princeton: Princeton University Press (in press).

Archibald, J. D., and J. H. Hutchinson.

1979. Revision of the genus *Palatobaena* (Testudines, Baenidae), with the description of a new species. Postilla 177:1–19.

Bakker, R. T.

1977. Tetrapod mass extinctions—a model of the regulation of speciation rates and immigration by cycles of topographic diversity. Pp. 439–468 *in* Patterns of Evolution, as Illustrated by the Fossil Record. Developments in Paleontology and Stratigraphy No. 5, A. Hallam, ed. Amsterdam, Oxford, New York: Elsevier.

Brown, B.

1907. The Hell Creek beds of the Upper Cretaceous of Montana. Am. Mus. Nat. Hist., Bull. 23:823–845.

Brown, R.

1939. Fossil plants from the Colgate Member of the Fox Hills Sandstone and adjacent strata. U.S. Geol. Survey Prof. Pap. 189-I:239–275.

1952. Tertiary strata in eastern Montana and western North and South Dakota. Billings Geol. Soc. Guidebk. 3:89–92.

Bryson, R. P.

1951. The Coalwood coal field, Powder River County, Montana. U.S. Geol. Survey Bull. 973B:23–106.

Butler, P. M.
1956. The skull of *Ictops* and the classification of the Insectivora. Zool. Soc. London, Proc. 126:453–481.
1972. The problem of insectivore classification. Pp. 253–265 *in* Studies in Vertebrate Evolution, K. A. Joysey and T. S. Kemp, eds. New York: Winchester Press.
Butler, R. F., and E. H. Lindsay.
1980. Magnetostratigraphy, biostratigraphy, and geochronology of Cretaceous-Tertiary boundary sediments, Red Deer Valley (a comment on Lerbekmo and others, 1979). Nature 248:375.
Butler, R. F., E. H. Lindsay, and L. L. Jacobs.
1977. Magnetostratigraphy of the Cretaceous-Tertiary boundary in the San Juan Basin, New Mexico. Nature 267:318–323.
Calvert, W. R.
1912. Geology of certain lignite fields in eastern Montana. Contributions to Economic Geology (short papers and preliminary reports). U.S. Geol. Survey Bull. 471:187–210.
Clemens, W. A.
1964. Fossil mammals of the type Lance Formation, Wyoming: Part I. Introduction and Multituberculata. Univ. Calif. Publ. Geol. Sci. 48:1–105.
1966. Fossil mammals of the type Lance Formation, Wyoming: Part II. Marsupalia. Univ. Calif. Publ. Geol. Sci. 62:1–122.
1968. A mandible of *Didelphodon vorax* (Marsupalia, Mammalia). Los Angeles Co. Mus. Contri. Sci. 133:1–11.
1973. Fossil mammals of the type Lance Formation, Wyoming: Part III. Eutheria and Summary. Univ. Calif. Publ. Geol. Sci. 94:1–102.
1974. *Purgatorius*, an early paromomyid Primate (Mammalia). Science 184:903–905.
1977. North American Puercan Didelphidae (Marsupalia, Mammalia). (ms).
1979. Marsupalia. Pp. 192–220 *in* Mesozoic Mammals: The First Two-thirds of Mammalian History, J. A. Lillegraven, Z. Kielan-Jaworowska, W. A. Clemens, eds. Berkeley: Univ. of Calif. Press.
Clemens, W. A., and J. D. Archibald.
1980. Evolution of terrestrial faunas during the Cretaceous-Tertiary transition. Pp. 67–74 *in* Ecosystémes Continentaux du Mésozoique. Mémoire de la Société Géologique de France, N.S. 59.
Clemens, W. A., R. C. Fox, A. Sahni, R. E. Sloan, and L. Van Valen.
1975. Cretaceous Mammalia. (ms).
Clemens, W. A., and Z. Kielan-Jaworowska.
1979. Multituberculata. Pp. 99–149 *in* Mesozoic Mammals: The First Two-thirds of Mammalian History, J. A. Lillegraven, Z. Kielan-Jaworowska, W. A. Clemens, eds. Berkeley: Univ. of Calif. Press.
Clemens, W. A., J. A. Lillegraven, E. H. Lindsay, and G. G. Simpson.
1979. Where, when, and what—a survey of known Mesozoic mammal distribution. Pp. 7–58 *in* Mesozoic Mammals: The First Two-thirds of Mammalian History, J. A. Lillegraven, Z. Kielan-Jaworowska, W. A. Clemens, eds. Berkeley: Univ. of Calif. Press.
Clemens, W. A., and L. G. Marshall.
1976. American and European Marsupalia. Pp. 1–114 *in* Fossilium catalogus, I. Animalia, Pars 123, F. Westphal, ed. The Hague: W. Junk.
Clemens, W. A., and L. S. Russell.
1965. Mammalian fossils from the upper Edmonton Formation. Pp. 32–40 *in* Vertebrate Paleontology in Alberta: Univ. of Alberta Bull. (Geol. 2).
Collier, A. J., and M. Knechtel.
1939. The coal resources of McCone County, Montana. U.S. Geol. Survey Bull. 905:1–80.

Cope, E. D.
1882a. Contributions to the history of the Vertebrata of the lower Eocene of Wyoming and New Mexico, made during 1881. Proc. Am. Philos. Soc. 20:139–197.
1882b. A second genus of Eocene Plagiaulacidae. Am.Naturalist 16:416–417.
1882c. Mammalia in the Laramie Formation. Am. Naturalist 16:830–831.
1882d. The Periptychidae. Am. Naturalist 16:832–833.
1884a. Second addition to the knowledge of the fauna of the Puerco Epoch. Paleont. Bull. No. 37, Proc. Am. Philos. Soc. 21:309–324.
1884b. (1885). The Vertebrata of the Tertiary formations of the West. Book I. Report U.S. Geol. Survey Terr., Washington 3:1–1009.
1888. Synopsis of the vertebrate fauna of the Puerco Series. Trans. Am. Philos. Soc., N.S. 16:298–361.
1892. On a new genus of Mammalia from the Laramie formation. Am. Naturalist 26:758–762.

Estes, R.
1964. Fossil vertebrates from the Late Cretaceous Lance Formation, eastern Wyoming. Univ. Calif. Publ. Geol. Sci. 49:1–180.

Estes, R., and P. Berberian.
1970. Paleoecology of a Late Cretaceous vertebrate community from Montana. Breviora 343:1–35.

Estes, R., P. Berberian, and C. A. M. Mesozoely.
1969. Lower vertebrates from the Late Cretaceous Hell Creek Formation, McCone County, Montana. Breviora 337:1–33.

Fassett, J. E.
1979. Comment on "Biostratigraphy and magnetostratigraphy of Paleocene terrestrial deposits, San Juan Basin, New Mexico." Geology 7:69–70.

Feldman, R. M.
1972. Stratigraphy and paleoecology of the Fox Hills Formation (Upper Cretaceous) of North Dakota. North Dakota Geol. Survey Bull. 61:1–65.

Folinsbee, R. E., H. Baadsgaard, and G. L. Cumming.
1970. Geochronology of the Cretaceous-Tertiary boundary of the western plains of North America. Ecologae Geol. Helvetiae 63:91.

Folinsbee, R. E., H. Baadsgaard, G. L. Cumming, J. Nascimbene, and M. Shafiqullah.
1965. Late Cretaceous radiometric dates from the Cypress Hills of western Canada. Alberta Soc. Petroleum Geol., Ann. Field Conf. Guidebk. 15:162–174.

Fox, R.C.
1971. Early Campanian multituberculates (Mammalia: Allotheria) from the Upper Milk River Formation, Alberta. Canadian Jour. Earth Sci. 8:916–938.
1976a. Additions to the mammalian local fauna from the Upper Milk River Formation (Upper Cretaceous), Alberta. Canadian Jour. Earth Sci. 13:1105–1118.
1976b. Cretaceous mammals (*Meniscoessus intermedius*, new species, and *Alphadon*, sp.) from the lowermost Oldman Formation, Alberta. Canadian Jour. Earth Sci. 13:1216–1222.
1977. Notes on the dentition and relationships of the Late Cretaceous insectivore *Gypsonictops* Simpson. Canadian Jour. Earth Sci. 14:1823–1831.

Fox, S. K., and R. K. Olsson.
1969. Danian planktonic Foraminifera from the Cannonball Formation in North Dakota. Jour. Paleont. 43:1397–1404.

Gaffney, E. S.
1975. A phylogeny and classification of the higher categories of turtles. Am. Mus. Nat. Hist., Bull. 155:387–436.

Gibson, D. W.
1977. Upper Cretaceous and Tertiary coal-bearing strata in the Drumheller-Ardley region, Red Deer River Valley, Alberta. Geol. Survey Canada Pap. 76-35:1–41.

Giebel, C. G. A.
1855. Die Säugethiere in zoologischer, anatomischer, und palaeontologischer Beziehung um fassend dargestellt. Leipzig: Ambrosius Abel. 1108 pp.

Gill, J. R., and W. A. Cobban.
1973. Stratigraphy and geologic history of the Montana Group and equivalent rocks, Montana, Wyoming, and North and South Dakota. U.S. Geol. Survey Prof. Pap. 776:1–37.

Gill, T.
1872. Arrangement of the families of mammals and synoptical table of the characters of the subdivisions of mammals. Smithsonian Misc. Colln. 11(1):1–98.

Gingerich, P. D.
1977. *Aletodon gunnelli*, a new Clarkforkian hyopsodontid (Mammalia, Condylarthra) from the early Eocene of Wyoming. Univ. Michigan Contr. Mus. Paleont. 25:1–9.

Granger, W., and G. G. Simpson.
1929. A revision of the Tertiary Multituberculata. Am Mus. Nat. Hist., Bull. 56:601–676.

Gray, J. E.
1821. On the natural arrangement of vertebrose animals. London Medical Repository 15:296–310.

Gregory, W. K., and G. G. Simpson.
1926. Cretaceous mammal skulls from Mongolia. Am. Mus. Novitates 225:1–20.

Hahn, G.
1969. Beiträge zur Fauna der Grube Guimarota Nr. 3. Die Multituberculata. Palaeontographica, Abt. A 133:1–100.
1971. The dentition of the Paulchoffitiidae (Multituberculata Upper Jurassic). Portugal Serv. Geol. Mem. 17:7–39.

Hall, J. W., and N. J. Norton.
1967. Palynological evidence of floristic change across the Cretaceous-Tertiary boundary in eastern Montana. Palaeogeog., Palaeoclimatol., Palaeoecol. 3:121–131.

Hennig, W.
1966. Phylogenetic Systematics. Urbana: Univ. of Illinois Press. 263 pp.
1975. "Cladistic analysis or cladistic classification?": A reply to Ernst Mayr. Syst. Zool. 24:244–256.

Holtzman, R. C., and D. L. Wolberg.
1977. Microcosmodontinae and *Microcosmodon woodi*, new multituberculate taxa (Mammalia) from the late Paleocene of North America. Sci. Publ. Mus. Minnesota 4:1–13.

Irish, E. J. W.
1970. The Edmonton Group of south-central Alberta. Bull. Canadian Petroleum Geol. 18:125–155.

Jacob, A. F.
1973. Elongate concretions as paleochannel indicators, Tongue River Formation (Paleocene), North Dakota. Geol. Soc. Am., Bull. 84:2127–2132.

Jeletzky, J. A.
1962. The allegedly Danian dinosaur-bearing rocks of the globe and the problem of the Mesozoic-Cenozoic boundary. Jour. Paleont. 36:1005–1018.
1978. Causes of Cretaceous oscillations of sea level in western and arctic Canada and some general geotectonic implications. Geol. Survey Canada Pap. 77-18:1–44.

Jeletzky, J. A., and W. A. Clemens.
1965. Comments on Cretaceous Eutheria, Lance *Scaphites*, and *Inoceramus*? ex. gr. *tegulatus*. Jour. Paleont. 39:952–959.

Jensen, F. S., and H. D. Varnes.
1964. Geology of the Fort Peck area, Garfield, McCone, and Valley counties, Montana. U.S. Geol. Survey Prof. Pap. 414–F:1–49.

Jepsen, G. L.
1930a. New vertebrate fossils from the lower Eocene of the Bighorn Basin, Wyoming. Proc. Am. Philos. Soc. 69:117–131.
1930b. Stratigraphy and paleontology of the Paleocene of northeastern Park County, Wyoming. Proc. Am. Philos. Soc. 69:463–528.
1940. Paleocene faunas of the Polecat Bench Formation, Park County, Wyoming. Part I. Proc. Am. Philos. Soc. 83:217–341.

Johnson, J. G.
1971. Timing and coordination of orogenic, epeirogenic and eustatic events. Geol. Soc. Am., Bull. 82:3263–3298.

Kielan-Jaworowska, Z.
1969. Preliminary data on the Upper Cretaceous eutherian mammals from Bayn Dzak, Gobi Desert. *in* Results of the Polish-Mongolian Palaeontological Expeditions, pt. 1, Z. Kielan-Jaworowska, ed. Palaeontologia Polonica 19:171–191.
1970. New Upper Cretaceous multituberculate genera from Bayn Dzak, Gobi Desert. *in* Results of the Polish-Mongolian Palaeontological Expeditions., pt. II, Z. Kielan-Jaworowska, ed. Palaeontologia Polonica 21:35–49.
1971. Skull structure and affinities of the Multituberculata. *in* Results of the Polish-Mongolian Palaeontological Expeditions, pt. III, Z. Kielan-Jaworowska, ed. Palaeontologia Polonica 25:5–41.
1974a. Multituberculate succession in the Late Cretaceous of the Gobi Desert (Mongolia). *in* Results of the Polish-Mongolian Palaeontological Expeditions, pt. V, Z. Kielan-Jaworowska, ed. Palaeontologia Polonica 30:23–44.
1974b. Migrations of the multituberculates and the Late Cretaceous connections between Asia and North America. Ann. South African Mus. 64:231–243.
1975a. Preliminary description of two new eutherian genera from the Late Cretaceous of Mongolia. *in* Results of the Polish-Mongolian Palaeontological Expeditions, pt. VI, Z. Kielan-Jaworowska, ed. Palaeontologia Polonica 33:5–16.
1975b. Evolution and migrations of the Late Cretaceous Asian mammals. Pp. 573–584 *in* Problemes Actuels de Paleontologie-Evolution des Vertebres, J. P. Lehman, ed. Paris: Colloques Internat., Centre Natl. Recherche Sci. No. 218.

Kielan-Jaworowska, Z., T. M. Bown, and J. A. Lillegraven.
1979. Eutheria. Pp. 221–258 *in* Mesozoic Mammals; The First Two-thirds of Mammalian History, J. A. Lillegraven, Z. Kielan-Jaworowska, W. A. Clemens, eds. Berkeley: Univ. of Calif. Press.

Kielan-Jaworowska, Z., and R. E. Sloan.
1979. *Catopsalis* (Multituberculata) from Asia and North America and the problem of taeniolabidid dispersal in the Late Cretaceous. Acta Palaeont. Polonica 24:187–197.

Krause, D. W.
1977. Paleocene multituberculates (Mammalia) of the Roche Percee local fauna, Ravenscrag Formation, Saskatchewan, Canada. Palaeontographica, Abt. A, 159:1–36.

LaBrecque, J. L., D. V. Kent, and S. C. Cande.
1977. Revised magnetic polarity time scale for Late Cretaceous and Cenozoic time. Geology 5:330–335.

Lemoine, V.
1880. Communication sur les Ossements Fossiles des Terrains Tertiares Inférieurs des Environs de Reims. Rheims, 1880 (reprint with additions from Assoc. Fr. pour l'Avance. d. Sciences, Congrès de Montpellier, 1879).

Lerbekmo, J. F., M. E. Evans, and H. Baadsgaard.
1979. Magnetostratigraphy, biostratigraphy and geochronology of Cretaceous-Tertiary boundary sediments, Red Deer Valley. Nature 279:26–30.

Lillegraven, J. A.
1969. Latest Cretaceous mammals of upper part of Edmonton Formation of Alberta, Canada, and review of marsupial-placental dichotomy of mammalian evolution. Univ. of Kansas, Paleont. Contri., Art. 50 (Vert. 12):1–122.

Lillegraven, J. A., Z. Kielan-Jaworowska, and W. A. Clemens, eds.
1979. Mesozoic Mammals: The First Two-thirds of Mammalian History. Berkeley: Univ. of Calif. Press. 311 pp.

Lindsay, E. H., L. L. Jacobs, and R. F. Butler.
1978. Biostratigraphy and magnetostratigraphy of Paleocene terrestrial deposits, San Juan Basin, New Mexico. Geology 6:425–429.

Lucas, S. G., and J. K. Rigby, Jr.
1979. Comment on "Biostratigraphy and magnetostratigraphy of Paleocene terrestrial deposits, San Juan Basin, New Mexico." Geology 7:323–325.

Lull, R. S.
1915. The mammals and horned dinosaurs of the Lance Formation, Niobrara County, Wyoming. Am. Jour. Sci., Ser. 4, 40:319–348.

Lydekker, R.
1889. Palaeozoology—Vertebrata. Pp. 880–1474 *in* A Manual of Palaeontology, 3rd ed., vol. 2, part 3, H. A. Nicholson and R. Lydekker, eds. Edinburgh and London: William Blackwood and Sons.

McKenna, M. C.
1965. Collecting microvertebrate fossils by washing and screening. Pp. 193–203 *in* Handbook of Paleontological Techniques, B. Kummer and D. Raup, eds. San Francisco: W. H. Freeman and Co.
1975. Towards a phylogenetic classification of the Mammalia. Pp. 21–46 *in* Phylogeny of the Primates: A Multidisciplinary Approach, W. P. Luckett and F. S. Szalay, eds. New York and London: Plenum Press.

McNab, B. K.
1971. On the ecological significance of Bergmann's rule. Ecology 52:845–854.

Marsh, O. C.
1889a. Discovery of Cretaceous Mammalia. Am. Jour. Sci., Ser. 3, 38:81–92.
1889b. Discovery of Cretaceous Mammalia, part 2. Am. Jour. Sci., Ser. 3, 38:177–180.
1892. Discovery of Cretaceous Mammalia, part 3. Am. Jour. Sci., Ser. 3, 48:249–262.

Matthew, W. D.
1897. A revision of the Puerco fauna. Am. Mus. Nat. Hist., Bull. 9:259–323.
1918. A revision of the lower Eocene Wasatch and Wind River faunas. Part 5, Insectivora (continued), Glires, Edentata. Am. Mus. Nat. Hist., Bull. 38:565–657.
1921. Fossil vertebrates and the Cretaceous-Tertiary problem. Am. Jour. Sci. (5)II:215–227.
1937 Paleocene faunas of the San Juan Basin, New Mexico. Trans. Am. Philos. Soc. 30:1–510.

Matthew, W. D., and W. Granger.
1921. New genera of Paleocene mammals. Am Mus. Novitates 13:1–17.

Murray, A.
1866. The Geographic Distribution of Mammals. London: Day and Son. 420 pp.

Norton, N. J., and J. W. Hall.
1969. Palynology of the Upper Cretaceous and lower Tertiary in the type locality of the Hell Creek Formation, Montana. Palaeontographica, Abt. B, 125:1–64.

Novacek, M. J.
1977. A review of Paleocene and Eocene Leptictidae (Eutheria: Mammalia) from North America. Paleo-Bios 24:1–42.

Novacek, M. J., and W. A. Clemens.
1977. Aspects of intrageneric variation and evolution of *Mesodma* (Multituberculata, Mammalia). Jour. Paleont. 51:701–717.
Obradovich, J. D., and W. A. Cobban.
1975. A time-scale for the Cretaceous of the Western Interior of North America. *in* The Cretaceous System in the Western Interior of North America, W. G. E. Caldwell, ed. Assoc. Canada Spec. Pap. 13:31–54.
Oltz, D. F.
1969. Numerical analyses of palynological data from Cretaceous and early Tertiary sediments in east central Montana. Palaeontographica, Abt. B, 128:39–42.
1971. Cluster analyses of Late Cretaceous-early Tertiary pollen and spore data. Micropaleont. 17:221–232.
Osborn, H. F.
1891. A review of the Cretaceous Mammalia. Proc. Acad. Nat. Sci. Philadelphia, pp. 124–135.
1893. Fossil mammals of the Upper Cretaceous beds. I: The multituberculates. II: The trituberculates. III: Faunal relations of the Laramie mammals. Am Mus. Nat. Hist., Bull. 5:311–330.
1898. Evolution of the Amblypoda. Part I. Taligrada and Pantodonta. Am. Mus. Nat. Hist., Bull. 10:169–218.
1910. The Age of Mammals in Europe, Asia, and North America. New York: Macmillan. 635 pp.
Osborn, H. F., and C. Earle.
1895. Fossil mammals of the Puerco beds. Collection of 1892. Am. Mus. Nat. Hist., Bull. 7:1–70.
Parrington, F. R.
1947. On a collection of Rhaetic mammalian teeth. Zool. Soc. London, Proc. 116:707–728.
Peyer, B.
1956. Über Zähne von Haramiyden [sic], von Triconodonten and von wahrscheinlich synapsiden Reptilien aus dem Rhat von Hallau, Kt. Schaffhausen Schweiz. Schweizerische Paläontol. Abh. 72:1–72.
Pond, C. M.
1977. The significance of lactation in the evolution of mammals. Evolution 31:177–199.
Rogers, G. S., and W. Lee.
1923. Geology of the Tullock Creek coal field, Rosebud and Big Horn counties, Montana. U.S. Geol. Survey Bull. 749:1–181.
Rohlf, F. J., and R. R. Sokal.
1969. Statistical Tables. San Francisco: W. H. Freeman and Co. 253 pp.
Russell, D. A.
1975. Reptilian diversity and the Cretaceous-Tertiary transition in North America. *in* The Cretaceous System in the Western Interior of North America, W. G. E. Caldwell, ed. Geol. Assoc. Canada, Spec. Pap. 13:119–136.
Russell, D. A., and C. Singh.
1978. The Cretaceous-Tertiary boundary in south-central Alberta—a reappraisal based on dinosaurian and microfloral extinctions. Canad. Jour. Earth Sci. 15:284–292.
Russell, D. A., and W. Tucker.
1971. Supernovae and the extinction of the dinosaurs. Nature 229:553–554.
Russell, L. S.
1937. New and interesting mammalian fossils from western Canada. Trans. Royal Soc. Canada, Ser. 3, Sect. 4, 30:75–80.
1964. Cretaceous non-marine faunas of north-western America. Contri. Royal Ont. Mus., Life Sci. 61:1–24.

1975. Mammalian faunal succession in the Cretaceous System of western North America. *in* The Cretaceous System in the Western Interior of North America W. G. E. Caldwell, ed. Geol. Assoc. Canada, Spec. Pap. 13:137–161.

Sahni, A.
1972. The vertebrate fauna of the Judith River Formation, Montana. Am. Mus. Nat. Hist., Bull. 147:325–412.

Savage, D. E.
1977. Aspects of vertebrate paleontological stratigraphy and geochronology. Pp. 427–442 *in* Concepts and Methods of Biostratigraphy, E. G. Kauffman and J. E. Hazel, eds. Stroudsburg, Pennsylvania: Dowden, Hutchinson, and Ross.

Schiebout, J. A.
1974. Vertebrate paleontology and paleoecology of Paleocene Black Peaks Formation, Big Bend National Park, Texas. Texas Mem. Mus. Bull. 24:1–88.

Scott, W. B.
1892. A revision of the North American Creodonta with notes on some genera which have been referred to that group. Proc. Acad. Nat. Sci. Philadelphia, pp. 291–323.

Shafiqullah, M., R. E. Folinsbee, H. Baadsgaard, G. L. Cumming, and J. F. Lerbekmo.
1968. Geochronology of Cretaceous-Tertiary boundary, Alberta, Canada. Rept. 22, Internatl. Geol. Cong. (1964), 3:1–20.

Shotwell, J.
1955. An approach to the paleoecology of mammals. Ecology 36:327–337.

Simpson, G. G.
1925. A Mesozoic mammal skull from Mongolia. Am Mus. Novitates 201:1–11.
1927a. Mammalian fauna of the Hell Creek Formation of Montana. Am. Mus. Novitates 267:1–7.
1927b. Mesozoic Mammalia VIII: genera of Lance mammals other than multituberculates. Am Jour. Sci., Ser. 5, 14:121–130.
1928. A Catalogue of the Mesozoic Mammalia in the Geological Department of the British Museum. London: Oxford Univ. Press. 215 pp.
1929a. American Mesozoic Mammalia. Mem. Peabody Mus. 3:1–171.
1929b. Some Cretaceous mammals from the Lance Formation. Ann. Carnegie Mus. 19:107–113.
1931. A new classification of mammals. Am. Mus. Nat. Hist., Bull. 59:259–293.
1935. New Paleocene mammals from the Fort Union of Montana. Proc. U.S. Nat. Mus. 83:221–244.
1937a. The Fort Union of the Crazy Mountain Field, Montana, and its mammalian faunas. Bull. U.S. Nat. Mus. 169:1–287.
1937b. Skull structure of the Multituberculata. Am. Mus. Nat. Hist., Bull. 73:727–763.
1945. The principles of classification and a classification of mammals. Am. Mus. Nat. Hist., Bull. 85:1–350.
1947. Holarctic mammalian faunas and continental relationships during the Cenozoic. Geol. Soc. Am. Bull. 58:613–688.
1965. The Geography of Evolution. New York: Capricorn Books. 249 pp.

Simpson, G. G., A. Roe, and P. C. Lewontin.
1960. Quantitative Zoology. New York: Harcourt, Brace and World. 440 pp.

Slaughter, B. H.
1965. A therian from the Lower Cretaceous (Albian) of Texas. Yale Peabody Museum Postilla 93:1–18.
1978. Occurrences of didelphine marsupials from the Eocene and Miocene of the Texas Gulf Coastal Plain. Jour. Paleont. 52:744–746.

Sloan, R. E.
1970. Cretaceous and Paleocene terrestrial communities of western North America. Pp.

427–453 *in* Proc. of the North American Paleont. Convention (1969), E. L. Yochelson, ed. Lawrence, Kansas: Allen Press.

1976. The ecology of dinosaur extinction. Pp. 134–154 *in* Athlon. Essays on Palaeontology in Honour of Loris Shano Russell (C. S. Chrucher, ed.). Misc. Pubs. Royal Ontario Mus., Life Sci.

Sloan, R. E., and L. S. Russell.
1974. Mammals from the St. Mary River Formation (Cretaceous) of southwestern Alberta. Life Sci. Contrib., Royal Ontario Mus., 95:1–20.

Sloan, R. E., and L. Van Valen.
1965. Cretaceous mammals from Montana. Science 148:220–227.

Sokal, R. R., and F. J. Rohlf.
1969. Biometry: The Principles and Practice of Statistics in Biological Research. San Francisco: W. H. Freeman and Co. 776 pp.

Stille, H.
1936. The present tectonic state of the earth. Am. Assoc. Petroleum Geologists Bull. 20:849–880.

Stone, R. W., and W. R. Calvert.
1910. Stratigraphic relations of the Livingstone Formation of Montana. Econ. Geol. 5:551–557, 652–669, 741–764.

Tedford, R. H.
1970. Principles and practices of mammalian geochronology in North America. Pp. 666–703 *in* Proc. of the North American Paleont. Convention (1969), E. L. Yochelson, ed. Lawrence, Kansas: Allen Press.

Thom, W. T. and C. E. Dobbin.
1924. Stratigraphy of Cretaceous-Eocene transition beds in eastern Montana and the Dakotas. Geol. Soc. Am., Bull. 35:481–505.

Trouessart, E. L.
1879. Catalogue des mammifères vivants et fossiles. Insectivores. Rev. et Mag. Zool. Pure et Appliqué, Paris, Sér. 3, 7:219–285.

Tschudy, R. H.
1970. Palynology of the Cretaceous-Tertiary boundary in the Northern Rocky Mountain and Mississippi Embayment Regions. Geol. Soc. Am., Spec. Pap. 127:65–111.
1973. The Gasbuggy core—a palynological appraisal. Four Corners Geol. Soc. Mem. 131–143.

Van Valen, L.
1966. Deltatheridia, a new order of mammals. Am. Mus. Nat. Hist., Bull. 132:1–126.
1967. New Paleocene insectivores and insectivore classification. Am. Mus. Nat. Hist., Bull. 135:217–284.
1969. The multiple origins of the placental carnivores. Evolution 23:118–130.
1976. Note on the origin of multituberculates (Mammalia). Jour. Paleont. 50:198–199.
1978. The beginning of the Age of Mammals. Evolutionary Theory 4:45–80.

Van Valen, L., and R. E. Sloan.
1965. The earliest Primates. Science 150:743–745.
1966. The extinction of the multituberculates. Syst. Zool. 15:261–278.
1977a. Ecology and the extinction of the dinosaurs. Evolutionary Theory 2:37–64.
1977b. Contemporaneity of Late Cretaceous extinctions. Nature 270:193.

Waage, K. M.
1968. The type Fox Hills Formation, Cretaceous (Maestrichtian), South Dakota. Part 1. Stratigraphy and paleoenvironments. Peabody Mus. Nat. Hist. Yale Univ. Bull. 27:1–175.

Webb, S. D.
1969. Extinction-origination equilibria in late Cenozoic land mammals of North America. Evolution 23:688–702.

Wilmarth, M. G.
1938. Lexicon of geologic names of the United States (including Alaska). U.S. Geol. Survey Bull. 895:1–2396.

Wilson, R. W.
1965. Type localities of Cope's Cretaceous mammals. Proc. South Dakota Acad. Sci., Univ. South Dakota Bull. 44:88–90.

Winge, H.
1917. Udsigt over Insektaedernes indbyrdes Slaegtskab. Vidensk. Meddel. Dansk. Naturh. Foren. 68:83–203.

Yanshin, A. L.
1973. O tak nazvaemykh nirovykh transgressiyakh i regressiyakh. Bull. Moskovskogo Ob-va Ispytatelei Prirody, Otdel Geologii 48(2):9–45.